Lecture Notes in Computer Science 4375

Commenced Publication in 1973
Founding and Former Series Editors:
Gerhard Goos, Juris Hartmanis, and Jan van Leeuwen

T0223237

Wolfgang Lehner   Norbert Meyer
Achim Streit   Craig Stewart (Eds.)

Euro-Par 2006 Workshops: Parallel Processing

CoreGRID 2006, UNICORE Summit 2006,
Petascale Computational Biology and Bioinformatics
Dresden, Germany, August 29-September 1, 2006
Revised Selected Papers

 Springer

Volume Editors

Wolfgang Lehner
TU Dresden
Database Technology Group
Nöthnitzer Str. 46, 01187 Dresden, Germany
E-mail: wolfgang.lehner@tu-dresden.de

Norbert Meyer
Poznań Supercomputing and Networking Center
ul. Z.Noskowskiego 10, 61-704 Poznań, Poland
E-mail: meyer@man.poznan.pl

Achim Streit
Forschungszentrum Jülich (FZJ)
Zentralinstitut für Angewandte Mathematik (ZAM)
52425 Jülich, Germany
E-mail: a.streit@fz-juelich.de

Craig Stewart
Indiana University
University Information Technology Services
2711 East Tenth Street, Bloomington, Indiana 47408-2671, USA
E-mail: stewart@iu.edu

Library of Congress Control Number: 2007926755

CR Subject Classification (1998): C.1-4, D.1-4, F.1-3, G.1-2, H.2

LNCS Sublibrary: SL 1 – Theoretical Computer Science and General Issues

ISSN   0302-9743
ISBN-10   3-540-72226-2 Springer Berlin Heidelberg New York
ISBN-13   978-3-540-72226-7 Springer Berlin Heidelberg New York

Springer is a part of Springer Science+Business Media

springer.com

© Springer-Verlag Berlin Heidelberg 2007

Typesetting: Camera-ready by author, data conversion by Scientific Publishing Services, Chennai, India
Printed on acid-free paper   SPIN: 12056460   06/3180   5 4 3 2 1 0

Preface

Parallel and Distributed Computing, although within the focus of computer science research for a long time, is gaining more and more importance in a wide spectrum of applications. These proceedings aim to demonstrate the use of parallel and distributed computing concepts in different application scenarios and attempt to spark interest in novel research directions to advance the embracing model of high-performance computing research in general. The objective of the workshop is to specifically address researchers coming from universities, industrial and governmental research organizations, and application-oriented companies in order to close the gap between purely scientific research and the applicability of the theoretical ideas to real-life problems. The individual research contributions published in the current proceedings are the result of three different workshops that were collocated at the Euro-Par Conference 2006 in Dresden, Germany:

The CoreGRID Workshop on Grid middleware

The general focus of the CoreGRID workshop is first to assess the current state of the art; second, new approaches in the areas of Knowledge and Data Management on Grids, Grid Resource Management and Scheduling, and Grid Information, Resource and Workflow Monitoring Services are discussed; and finally, it serves as a forum for the exchange of ideas between the user community and Grid middleware developers. The CoreGRID workshop on Grid middleware was organized by the CoreGRID Network of Excellence and took place August 28–29, 2006, in Dresden. CoreGRID is the European Research Network for strengthening and advancing scientific and technological excellence in the area of Grid and Peer-to-Peer technologies. In particular, the workshop was organized by Domenico Talia (University of Calabria, Italy), Ramin Yahyapour (University of Dortmund, Germany), and Norbert Meyer (Poznań Supercomputing and Networking Center, Poland).

The UNICORE Summit 2006 Workshop

The UNICORE Grid technology provides seamless, secure, and intuitive access to distributed Grid resources. UNICORE is a mature and well-tested Grid middleware system, which is nowadays used in daily production processes worldwide. Beyond this production usage, the UNICORE technology serves as a solid basis for many European and international projects. The UNICORE Summit is a unique opportunity for Grid users, developers, administrators, researchers, and service providers to meet, get an inside view of UNICORE, share experiences, and discuss future developments. The two-day UNICORE Summit 2006

Workshop was held August 30–31, 2006, in Dresden and was mainly organized by Achim Streit (Forschungszentrum Jülich, Germany) and Wolfgang Ziegler (Fraunhofer Gesellschaft SCAI, Germany).

The Petascale Computational Biology and Bioinformatics Workshop

The general topic of this workshop is to address the issues that bioinformatics or computational biology applications can or should accomplish within petascale computing environments and the kind of obstacles that must be overcome in order to implement and use effective solutions for important problems in the life sciences (biology, biochemistry, environmental sciences, etc.). The workshop was held on August 28, 2006, and it was organized by Craig Stewart (Indiana University, USA), Michael Schroeder (Technische Universität Dresden, Germany) and Matthias Müller (Technische Universität Dresden, Germany).

All workshops were tightly integrated into the 12th Euro-Par Conference, held in Dresden, Germany, from August 29 to Septemper 1, 2006. This specific Euro-Par Conference was part of an annual series of international conferences focusing on the advancement of parallel computing by covering topics from hardware architectures over algorithms and software design to specific application needs. The Euro-Par Conference is therefore the meeting point of the international community in this particular research field and attracts scholars as well as technology-oriented decision makers from all over the world.

Acknowledgments

Many organizations and individuals were involved in the preparation and actual realization of the workshops collocated at the Euro-Par Conference 2006 in Dresden. At the Technische Universität Dresden, the Center of Information Services and High Performance Computing (ZIH) contributed staff and general infrastructure. The ZIH was supported by the Institute of System Architecture (Database Technology Group) and the Institute of Scientific Computing in the organization of the workshops. Specifically, the workshop organizers would like to mention Claudia Schmidt and Guido Juckeland for their support, ranging from tackling all administrative challenges to carefully preparing the current volume of the workshop proceedings. Additionally, we would like to express our sincere 'Thank You' to all other institutions and industrial organizations that contributed in various ways to make the workshops a real success story. Finally, we are also grateful to the huge number of individuals ranging from organizers to authors and reviewers who helped set up a high-quality scientific program. They all did a great job in preparing and actually carrying out the individual workshops. Within this context, we appreciate the support of Springer for agreeing to publish these workshop proceedings; this will definitely help future Euro-Par conferences and collocated workshops to continuously attract top-notch researchers presenting outstanding research results.

It is our hope that these proceedings will help advance research in parallel computing in general and act as a catalyst in promoting the idea of parallel computing to a wide spectrum of applications. It was our pleasure hosting the workshops in Dresden and we hope that the discussions and the current proceedings are beneficial for sustainable growth and awareness of parallel computing concepts in future applications.

September 2006

Wolfgang Lehner
Norbert Meyer
Achim Streit
Craig Stewart

Organization

Euro-Par Steering Committee

Chair
Christian Lengauer   University of Passau, Germany

Vice-Chair
Luc Bougé   ENS Cachan, France

European Representatives

José Cunha	New University of Lisbon, Portugal
Marco Danelutto	University of Pisa, Italy
Rainer Feldmann	University of Paderborn, Germany
Christos Kaklamanis	Computer Technology Institute, Greece
Paul Kelly	Imperial College, UK
Harald Kosch	University of Passau, Germany
Thomas Ludwig	University of Heidelberg, Germany
Emilio Luque	Universitat Autònoma de Barcelona, Spain
Luc Moreau	University of Southampton, UK
Rizos Sakellariou	University of Manchester, UK

Non-European Representatives

Jack Dongarra	University of Tennessee at Knoxville, USA
Shinji Tomita	Kyoto University, Japan

Honorary Members

Ron Perrott	Queen's University Belfast, UK
Karl Dieter Reinartz	University of Erlangen-Nuremberg, Germany

Observers

Wolfgang E. Nagel	Technische Universität Dresden, Germany
Anne-Marie Kermarrec	IRISA Rennes, France

Euro-Par 2006 Local Organization

Euro-Par 2006 was organized by the Center for Information Services and High
Perfomance Computing (ZIH) of the Technische Universität Dresden.

Conference Chairs

Wolfgang E. Nagel	Center for Information Services and High Performance Computing (ZIH) and Dept. of Computer Science, Inst. for Computer Engineering, Technische Universität Dresden
Wolfgang V. Walter	Dept. of Mathematics, Inst. for Scientific Computing, Technische Universität Dresden
Wolfgang Lehner	Dept. of Computer Science, Inst. for System Architecture, Technische Universität Dresden

General Organization

Claudia Schmidt

Technical Support

Kirsten Kern, Ronny Zschitzschmann

Grid Village, Exhibits

Dietmar Augustin, Stefan Pflüger

Proceedings

Guido Juckeland

Secretariat

Jenny Baumann

Euro-Par 2006 Workshop Program Committees

CoreGrid Workshop on Grid Middleware

Program Chairs

Norbert Meyer   Poznan Supercomputing and Networking
                         Center, Poland
Domenico Talia   University of Calabria, Italy
Ramin Yahyapour   University of Dortmund, Germany

Program Committee

Artur Andrzejak   Pierre Guisset
Alvaro Arenas   Domenico Laforenza
Angelos Bilas   Philippe Massonet
Maciej Brzezniak   Salvatore Orlando
Marco Danelutto   Thierry Priol
Bruno Le Dantec   Yves Robert
Marios Dikaiakos   Paolo Trunfio
Ewa Deelman   Rizos Sakellariou
Vladimir Getov   Frederic Vivien
Dick Epema   Paul Watson
Antonia Ghiselli   Roman Wyrzykowski
Sergei Gorlatch   Wolfgang Ziegler

UNICORE Summit

Program Chairs

Achim Streit   Forschungszentrum Jülich, Germany
Wolfgang Ziegler   Fraunhofer Gesellschaft SCAI, Germany

Program Committee

Agnes Ansari   Ralf Ratering
Rosa Badia   Johannes Reetz
Piotr Bala   Mathilde Romberg
John Brooke   Bernd Schuller
Anton Frank   David Snelling
Edgar Gabriel   Stefan Wesner
Alfred Geiger   Ramin Yahyapour
Odej Kao
Paolo Malfetti

Additional Reviewer

Graham Fagg
Björn Hagemeier

Petascale Computational Biology and Bioinformatics

Program Chairs

Craig Stewart	Indiana University, USA
Matthias Müller	TU Dresden, ZIH, Germany
Michael Schroeder	TU Dresden, Biotec, Germany

Table of Contents

CoreGRID Workshop on GRID Middleware

UNICORE Summit 2006

Petascale Computational Biology and Bioinformatics

CoreGRID Workshop on GRID Middleware

Introduction

Norbert Meyer, Domenico Talia, and Ramin Yahyapour

Workshop Chairs

Grids today still need investigation in several areas to provide models and tools for developing high-level scalable applications. Grid middleware is an active research area that is bringing several results to Grid developers and users. The goal of this workshop is to provide a bridge between the application community and the developers of middleware services, especially in terms of parallel and distributed computing, with the following major issues:

- To gather current state of the art and new approaches in the areas of knowledge and data management, resource management and information systems
- To include work-in-progress contributions
- To provide a forum for exchanging the ideas between the users' community and grid middleware developers
- To disseminate existing results and provide input to the CoreGRID Network of Excellence

The CoreGRID Workshop on Grid Middleware was held in conjunction with Euro-Par 2006 in Dresden, August 28-29, 2006 and was organized by the three CoreGRID Institutes: the Institute on Knowledge and Data Management, the Institute on Grid Information, Resource and Workflow Monitoring Services and the Institute on Resource Management and Scheduling.

In the call for papers, the list of the suggested topics included the following specific subjects:

- Distributed data storage on Grids
- Information and knowledge management in Grids
- Distributed data mining and knowledge
- Monitoring and information systems
- Fault tolerance, reliability and sustainability
- Application and kernel level checkpointing
- Accounting and user account management in and across Virtual Organizations
- Workflow frameworks
- Grid scheduling architectures
- Multi-level scheduling strategies
- Evaluation and benchmarking of Grid scheduling systems
- Scheduling of high-performance parallel applications
- Performance prediction

The workshop was chaired by the leaders of the three CoreGRID Institutes: Domenico Talia, Norbert Meyer and Ramin Yahyapour. The Program Committee involved many researchers both from CoreGRID and external research teams.

W. Lehner et al. (Eds.): Euro-Par 2006 Workshops, LNCS 4375, pp. 3–4, 2007.

A total of 24 papers were submitted to the workshop. After the review process, 16 papers were selected by the Program Committee and were presented at the meeting in Dresden in five sessions focused on the following major topics:

- Knowledge and Data Management on Grids
- Grid Resource Management and Scheduling
- Grid Information, Resource and Workflow Monitoring Services

An additional invited talk was given by Graeme Kerr from Oracle (member of the CoreGRID IAB) on distributed data management in Grids. The Grid Middleware workshop was attended by about 40 participants coming from several countries. The atmosphere was very active and several interesting discussions arose during the paper presentations.

Architecture of a Network Monitoring Element

Augusto Ciuffoletti[1] and Michalis Polychronakis[2]

[1] CNAF-INFN, Bologna, Italy
[2] FORTH-ICS, Heraklio, Greece

Abstract. A Network Monitoring system is a vital component of a Grid; however, its scalability is a challenge. We propose a network monitoring approach that combines passive monitoring, a domain oriented overlay network, and an attitude for demand driven monitoring sessions. In order to keep into account the demand for extreme scalability, we introduce a solution for two problems that are inherent to the proposed approach: security and group membership maintenance.

Keywords: network monitoring, passive network monitoring, on demand network monitoring, network monitoring element, scalability issues, security issues.

1   Introduction

Monitoring the network infrastructure of a Grid has a vital role in the management and the utilization of the Grid itself. The Global Grid Forum (GGF) schema [7], splits this activity into three distinct phases: *production, publication,* and *utilization* of measurements. Here we focus on the production and publication, with a special concern for scalability: for measurement production we address the usage of passive network monitoring techniques, while for the publication activity we adopt a domain-oriented overlay network which reduces the complexity of the task.

The challenge comes from the fact that a Grid oriented network monitoring should address network routes, not single links, since this is the kind of information needed to optimize distributed applications. Since each pair of Grid Services should be individually monitored, this makes an $O(n^2)$ complexity for many aspects of network monitoring: from the size of the database containing the results, to the number of pings that probe the system.

We combined a number of ideas in order to limit the complexity of our solution: *i)* monitoring shouldn't address single Grid resources, but pools with similar connectivity; *ii)* monitoring tools shouldn't inject traffic, but observe existing traffic; *iii)* monitoring activity should be tailored on application needs. Only the integration of above ideas can effectively control the problem size, and, in some sense, the first two *open the way* for the application of the third one.

We observe that a Grid *topology* is made of pools of resources reachable through dedicated ingress points: the accessibility of such pools depends on ingress points connectivity, and local administration avoids internal bottlenecks.

W. Lehner et al. (Eds.): Euro-Par 2006 Workshops, LNCS 4375, pp. 5–14, 2007.

Therefore the monitoring topology can be simplified by monitoring *Network Elements* between ingress points of distinct pools.

One Network Monitoring architecture, called GlueDomains [3], has been recently designed and prototyped according to a two levels hierarchical overlay; the purpose of such experiment was mainly the assessment of a number of design principles. A Grid-wide deployment of GlueDomains was carried out during the summer of 2006, as part of the Italian branch of the Large Hadron Collider Computing Grid Project (LCG). Apart from the statistics collected (usual packet loss and roundtrip time, together with an experimental one way jitter measurement tool, published through the GridICE Grid Information Service [2]), the most relevant results from the GlueDomains experiment concern the ease of deployment, as well as the resilience, and stability of the architecture, which were assessed during a one month trial. GlueDomains is included in the current release of the Italian branch of LCG.

GlueDomains architecture centers around a number of specialized units hosting the agents in charge of monitoring the network. Such agents are able to autonomously (re)configure their activity based on a dynamic description of the network monitoring topology, available from a relational database. The monitoring activity was based on a domain partitioning of Grid resources: the target of such monitoring is the *Network Element*, which abstracts the network infrastructure in charge of interconnecting two domains.

One relevant lesson learned from GlueDomains experience is the identification of the role played by the agent that concentrates the network monitoring activity for a domain. This role corresponds to a new resource in the Grid architecture, which is mainly dedicated to network monitoring. In the architecture proposed in this paper we call such agent a *Network Monitoring (NM) Element*: its activity is organized into *Network Monitoring Sessions*.

Another cornerstone concept in our architecture is *passive monitoring*, which is non-intrusive by nature. The internal architecture of NM Elements adopts specific hardware and software solutions to address passive network monitoring.

A third concept that cuts down network monitoring complexity is an *application driven* configuration: this is feasible in a Grid, where applications negotiate computing resources with resource brokers, which can configure *Network Monitoring Sessions* on the fly, providing adequate credentials to NM Elements. The relevant conclusion is that, if network monitoring activity is bound to applications, it will increase linearly with system throughput, not with the square of system size.

A relevant aspect of an *application driven* approach is the interface that a NM Element should offer to the outside. Currently brokers find resource characteristics in the *Grid Information System* (GIS), automatically collected by *preconfigured* network monitoring sessions. This attitude is inappropriate in an *application driven* scenario for its limited scalability, and it would be preferable to connect NM Elements to brokers through a publish-subscribe system. Given that this aspect is still a research topic, we indicate a composite interface, which decouples the input, consisting of monitoring requests, from the output, consisting of observation records.

The scheme described above is based on some knowledge shared by all the NM Elements, which can be assimilated to a *group membership*. Such common knowledge consists of the certificates needed to enforce security, complemented by the composition of Domains. Such data should be readily accessible by any Grid component, although the throughput for access operations can be quite asymmetric: frequent read queries should be performed promptly, while infrequent updates can be treated lazily. We address this problem by replicating this directory on each NM Element, and using an epidemic algorithm in order to maintain consistency of distinct network views.

2   Inside View of a *Network Monitoring Element*

The internal structure of a *Network Monitoring Element* is layered according to the scheme in Figure 1. The upper layer is in charge of implementing the interfaces to the outside, offering a *Network Monitoring Service.*

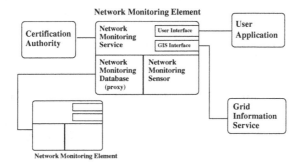

Fig. 1. Interfaces between the NM Element and other Grid components

The NM Service offers three distinct interfaces: one for user applications (resource brokers included), another for the *Grid Information Service* (GIS), and one that interacts with the Certification Authority. The *User Application Interface* allows the submission of a request for a specific monitoring session: the NM Service checks broker credentials and verifies local resources availability before accepting a request. In response, the broker receives an acknowledgement. The User Application Interface also provides users with access to the Network Monitoring database. The *GIS interface* allows the publication of observations coming from network monitoring sessions. We do not explore the architecture of the GIS in this paper, but we note that it should enforce certain access limits: for instance, the results of an *on demand* network monitoring activity should be visible, as a general rule, only to trusted users. The GIS should be informed of such limited access by the NM service which received the request.

The lower layer is composed of two distinct modules that do not interact with each other. The *Network Monitoring Sensor* supports monitoring sessions: the implementation of sessions is delegated to specialized modules that take their

configuration from the upper layer. The results of the monitoring activity are
delivered to the NM Service via dedicated one-way streams from the specific
session to the upper layer. We distinguish between *preconfigured* and *on demand*
sessions: the former are configured directly by the NM Service module using
the Grid topology described by the Network Monitoring Database, while the
latter are configured by an outside user application, through the NM Service.
The *Network Monitoring Database* describes the domain partition of the Grid, as
well as its components: for each element in the Grid (NM elements, Computing,
Storage etc.), the database holds a certificate (which contains a reference to a
domain) for the element, together with other relevant attributes.

2.1   The Passive Network Monitoring Component

The NM Sensor receives measurement requests from the NM Service: its interface
is summarized in Table 1, and Figure 2 describes its internal architecture.

The NM Service creates a new measurement session by sending a `create` request, which specifies the type of the measurement and any measurement-specific
parameters, and returns a measurement identifier (`mid`). The creation of a new
measurement session does not imply that the measurement will immediately
begin upon the receipt of the `create` request; this allows the NM Service to
activate or deactivate the measurement, also depending on resources availability
and timing requirements.

The measurements are carried out using specialized modules implemented on
top of MAPI [18], an API for building passive network monitoring applications.[1]

The basis of MAPI is the *network flow* abstraction, which is generally defined
as *a sequence of packets that satisfy a given set of conditions*. These conditions
range from simple header-based filters, to sophisticated protocol analysis and
content inspection functions.

The back-end of the NM Sensor consists of the basic components of MAPI,
namely the monitoring daemon `mapid` and the communication agent `commd`.
Packets are captured and processed by `mapid`: a user-level process with exclusive access to the captured packets [18]. The monitoring modules are built as
separate applications on top of MAPI. MAPI internally handles all the communication between the modules and the monitoring daemon, making it completely
transparent.

The computed results of a measurement are pushed back to the NM Service
either on-the-fly, or upon the end of the measurement: the desired behavior is
passed in the `create` request.

In the rest of this section we describe the operation of the modules collecting
some relevant network metrics.

The *Network Traffic Load* module provides traffic throughput metrics of varying levels of granularity by passively measuring the number of bytes transferred
through the monitored link. Besides aggregate throughput, fine-grained per-flow
measurements are available for observing the throughput achieved by specific

---

[1] MAPI is available at `http://mapi.uninett.no`

Table 1. API of the NM Sensor

Function	Parameter	Description
create	Type	Measurement type: traffic load, packet loss, or RTT
	Arguments	Measurement-specific parameters
	Return value	Measurement session identifier (mid) or error type
start	Identifier	The mid of the session to be started
	Return value	Acknowledgement or error type
stop	Identifier	The mid of the session to be stopped
	Return value	Acknowledgement or error type
close	Identifier	The mid of the session to be terminated
	Return value	Acknowledgement or error type

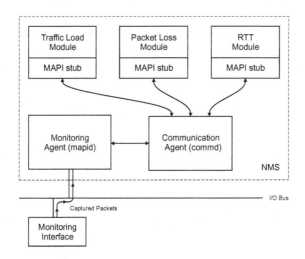

Fig. 2. The architecture of the Network Monitoring Sensor

applications or hosts. MAPI supports generic BPF filters [14], as well as more fine-grained filtering using pattern matching in the packet payloads through string searching or regular expressions.

An estimation of the *Packet Loss Ratio* between two domains is measured by two cooperating observation points, which keep track of packets of specific flows that do not reach the destination within a timeout period. The packet loss module needs traffic information from both ends. This is achieved by creating a network flow in the local sensor, which keeps track of the outgoing packets with a destination IP address that belongs to the remote domain, and a second network flow at the destination domain, specified by the dstdomain parameter. The NM Element of the other domain is instructed to create a second network flow, which keeps track of the incoming packets with a source IP address that belongs to the local domain. The packet loss ratio is then estimated by correlating the data from the two network flows.

The *Round-Trip Time* is estimated using the time difference between the SYN and ACK packets exchanged during the lifetime of existing TCP connections [12]. Each request specifies the destination domain for the end-to-end RTT measurement using the dstdomain parameter. The module then creates a network flow that captures the SYN and ACK packets of existing TCP connections between the two domains, in the unidirectional flow from the local to the remote domain. RTT is estimated from the time interval between the last-SYN and the first-ACK that is sent from the local to the remote domain. The accuracy of the measurement increases with its duration, since a longer duration allows for more TCP connections to be tracked, which gives a better RTT estimation.

2.2   Outline of a Secure Group Membership Scheme

Security issues impose the use of certificates in order to identify the source of configuration inputs to NM elements: this can be assimilated to the secure management of a membership. An efficient and scalable certificates distribution scheme is required [19], conceptually based on the NM Database, where certificates are stored. Access to this database must be secure and scalable, and characterized by a *small* read access latency, a *non bursty* network overhead, and a *predictable* write access latency.

In order to implement such characteristics, the database is replicated: an (almost) complete replica of the whole database is kept at each NM Element. The broadcast of update operations is performed using a number of circulating tokens, each containing a stack of recently issued updates. The number of tokens circulating in the system is tuned automatically, based on a feedback mechanism that enables each NM Database Proxy to inject (or remove) a token when needed.

The peer-to-peer protocol used for token circulation is made secure using the same certificates that are stored in the database itself: upon receiving a token from a neighbor, the NM Database Proxy authenticates it using the public key of the sender retrieved from the local database. In the exceptional case that peer's certificate is not present in the local database, a copy is downloaded from the neighbor.

The protocol is resilient to network and host failures, since it does not follow a preplanned path (or overlay network): tokens wander randomly in the system. Although mostly based on random decisions, the protocol promises an excellent stability and predictability: this conclusion is justified by simulation results reported in [5]. The load is evenly distributed in time and space, while the update latency remains constant.

The *interface* offered by this module to the upper layer consists of the operations outlined in Table 2, and extends the use of the DB to the storage of rather static characteristics of the Network Monitoring topology (like domain partitioning): the select function returns the desired data, while the update returns an acknowledgement. They take as parameters an SQL-like query and the id of the Element for which the NM Service issues the request.

While the select function is clearly synchronous, the update function is not: the acknowledgement reflects the fact that the request has been successfully

Table 2. API of the NM Database Module

Function	Parameter	Description
select	SQL select query	The SQL-like query that returns the desired data
	Submitter	The id of the Element which submitted the query
	Return value	A data structure containing selected data
update	SQL update query	The SQL-like query that modifies the database
	Submitter	The id of the Element which submitted the query
	Return value	A data structure containing the query id
check	Query id	The id of the query
	Return value	A data structure containing the status of the query

queued, not necessarily performed. In order to check the (likely) completion of a requested update, the interface offers the **check** function, which takes as parameter the **update id** returned as an acknowledgement, and returns the current status of the update request, derived from the internal queue. The returned status contains a prediction of the completion time.

3   Related Work

The network monitoring management has been addressed by a number authors: solutions are differentiated in the way they cope with scalability and security.

NWS [22] is the ancestor of network monitoring services, and it shares the same building blocks with the architecture introduced in this paper: sensors and a directory for available monitoring functions. However, NWS did not consider at all scalability and security issues. Therefore, despite its importance as a proof of concept, its applicability is limited to small, protected networks.

TopoMon [9] can be regarded as an evolution of NWS, in a direction which is somewhat complementary to the approach followed in our work. In fact, TopoMon extends NWS with tools and support for managing link level topology, a knowledge we explicitly exclude from our interests. Although we understand that this information is relevant (for instance, in view of a reservation service that cannot ignore the existence of shared links when allocating end to end communication resources), we prefer to explore scalability and security issues, which are not addressed by TopoMon, instead of insisting on tools for exploring communication infrastructure.

The **JAMM [20]** sensor management system has been implemented at LBNL for purposes which are close to ours, and is able to configure sensors upon request from applications. An LDAP based directory service keeps records of available sensors, and data from sensors flow to the user applications through specialized gateways. The authors suggest to use encrypted communication in order to ensure security.

The architecture we present in this paper addresses security and scalability aspects in a different way. In JAMM, gateways are used to decouple producers from users, in order to limit the fan out from the sensors. Our model is characterized by a more composite approach to scalability: a support for domain partition is provided, which limits the size of the problem, directory management is addressed with explicit reference to its complexity, passive monitoring is explicitly supported to contain communication footprint, and finally the solution of the fan out problem is delegated to a GIS, without introducing a new solution to a problem that must be necessarily solved elsewhere.

An interesting approach to the problem of retrieving monitoring data is offered by **Gigascope** [8], a stream oriented database for storing captured network data in a central repository for further analysis using an SQL-like query language.

A large scale project that focuses on a scalable, secure monitoring infrastructure is **NIMI** [1]. The architecture introduced in this paper shares several aspects with such large scale prototype: mainly, the strict separation of concerns regarding making measurements, requesting measurements, analyzing results, configuring probes is reflected in the internal structure of our NME. In our architecture, which is at the design stage, we introduce a decentralized global view of the overall network monitoring system, which serves as a support also to service discovery. Such aspect is not covered by NIMI, which bases the local knowledge of each probe on the information received by Configuration Point of Contacts, without introducing any form of coordination between them.

We employ passive monitoring as a technology that fits our scalability requirements: likewise, this approach is a cornerstone of the **CoMo** project [11]. One purpose of this project is to allow users to query network data gathered from multiple administrative domains in a secure and reliable way, without interfering with resource availability. The white paper which is available does not address the organization of a registry of available sensors, which is needed to address a large, domain structured network.

Sprint's passive monitoring system [10] also collects data from different monitoring points into a central repository for analysis. The authors observe that the amount of data collected becomes rapidly unmanageable: in our design, this drawback is resolved with the introduction of a domain oriented overlay network, and by offering an interface for on demand monitoring.

Arlos et al. [6] propose a distributed passive measurement infrastructure that supports various monitoring equipment within the same administrative domain.

An approach that makes use of passive monitoring is often based on packet analysis libraries, which extract the desired pieces of information from the monitored traffic. The most widely used library for this purpose is `libpcap` [16], which provides a portable API for user-level packet capture. The `libpcap` interface supports a filtering mechanism based on the BSD Packet Filter [15], which allows for selective packet capture based on packet header fields. **CoralReef** provides a set of tools and supports functions for capturing and analyzing network traces [13]. **Nprobe** [17] is a monitoring tool for network protocol analysis. It is based on commodity hardware, and speeds up network monitoring tasks using filters implemented in a programmable network interface.

The passive monitoring components of our system are based on **MAPI** [18], which shares some functionality with the above monitoring systems, but at the same time provides a more expressive programming interface with significantly extended functionality and, in many cases, increased monitoring performance. Additionally, the distributed version of MAPI [21] enables distributed network monitoring applications through a flexible interface that allows the manipulation of many remote monitoring sensors from a single application.

The overall approach described in this paper is derived from the **GlueDomains** [4] prototype, which has been successfully deployed and used in the Grid infrastructure of the Italian National Nuclear Physics Institute (INFN). However, the existence of a centralized repository for configuration data, together with an extended use of active monitoring techniques, limits the scalability of the GlueDomains prototype to approximately 50 domains, which is reasonable only for a small-scale grid.

4 Conclusions

In this paper we outline the internal architecture and the interface of a network monitoring service, specifically addressing security and scalability issues. The basic building block is the *Network Monitoring Element*, a specialized Grid component. A Grid contains several instances of this component, which is responsible for monitoring *Network Elements* between *Network Monitoring Domains*. A Network Monitoring Element carries out its monitoring activity using passive monitoring, virtually without network overhead. Its activity is described by a number of *Network Monitoring Sessions*. We exclude manual intervention for its configuration, which should be carried out automatically, either using pre-configured sessions, or preferably according to requests from resource brokers.

References

1. A. Adams, J. Mahdavi, M. Mathis, and V. Paxson. Creating a scalable architecture for internet measurement. In *Proceedings of INET98*, Geneva, July.
2. C. Aiftimiei, S. Andreozzi, G. Cuscela, N. D. Bortoli, G. Donvito, S. Fantinel, E. Fattibene, G. Misurelli, A. Pierro, G. Rubini, and G. Tortone. GridICE: Requirements, architecture and experience of a monitoring tool for grid systems. In *Proceedings of the International Conference on Computing in High Energy and Nuclear Physics (CHEP2006)*, Mumbai - India, February 2006.
3. S. Andreozzi, A. Ciuffoletti, A. Ghiselli, and C. Vistoli. Monitoring the connectivity of a grid. In *2nd Workshop on Middleware for Grid Computing*, pages 47–51, Toronto, Canada 2004.
4. S. Andreozzi, A. Ciuffoletti, A. Ghiselli, and C. Vistoli. Gluedomains: Organization and accessibility of network monitoring data in a grid. Technical Report TR-05-15, Universit di Pisa, Largo Pontecorvo - Pisa -ITALY, May 2005.
5. S. Andreozzi, D.Antoniades, A.Ciuffoletti, A.Ghiselli, E.P.Markatos, M.Polychronakis, and P.Trimintzios. Issues about the integration of passive and active monitoring for grid networks. In *CoreGRID Integration Workshop 2005*, november 2005.
6. P. Arlos, M. Fiedler, and A. A. Nilsson. A distributed passive measurement infrastructure. In *Proceedings of the 6th International Passive and Active Network Measurement Workshop (PAM'05)*, pages 215–227, 2005.

7. R. Aydt, D. Gunter, W. Smith, M. Swany, V. Taylor, B. Tierney, and R. Wolski. A grid monitoring architecture. Recommendation GWD-I (Rev. 16, jan. 2002), Global Grid Forum, 2000.
8. C. Cranor, T. Johnson, O. Spataschek, and V. Shkapenyuk. Gigascope: a stream database for network applications. In *Proceedings of the ACM SIGMOD international conference on Management of data*, 2003.
9. M. den Burger, T. Kielmann, and H. E. Bal. TOPOMON: A monitoring tool for grid network topology. In *International Conference on Computational Science (2)*, pages 558–567, 2002.
10. C. Fraleigh, C. Diot, B. Lyles, S. Moon, P. Owezarski, D. Papagiannaki, and F. Tobagi. Design and Deployment of a Passive Monitoring Infrastructure. In *Proceedings of the Passive and Active Measurement Workshop*, Apr. 2001.
11. G. Iannaccone, C. Diot, D. McAuley, A. Moore, I. Pratt, and L. Rizzo. The CoMo white paper. Technical Report IRC-TR-04-17, Intel Research, 2004.
12. H. Jiang and C. Dovrolis. Passive estimation of tcp round-trip times. *SIGCOMM Comput. Commun. Rev.*, 32(3):75–88, 2002.
13. K. Keys, D. Moore, R. Koga, E. Lagache, M. Tesch, and K. Claffy. The architecture of CoralReef: an Internet traffic monitoring software suite. In *Proceedings of the 2nd International Passive and Active Network Measurement Workshop*, Apr. 2001.
14. S. McCanne and V. Jacobson. The BSD packet filter: A new architecture for user-level packet capture. In *Proceedings of the USENIX Winter Conference*, January 1993.
15. S. McCanne and V. Jacobson. The BSD Packet Filter: A New Architecture for User-level Packet Capture. In *Proceedings of the Winter 1993 USENIX Conference*, pages 259–270, January 1993.
16. S. McCanne, C. Leres, and V. Jacobson. libpcap. Lawrence Berkeley Laboratory, Berkeley, CA. (software available from http://www.tcpdump.org/).
17. A. Moore, J. Hall, E. Harris, C. Kreibich, and I. Pratt. Architecture of a network monitor. In *Proceedings of the 4th International Passive and Active Network Measurement Workshop*, April 2003.
18. M. Polychronakis, K. G. Anagnostakis, E. P. Markatos, and A. Øslebø. Design of an application programming interface for IP network monitoring. In *Proceedings of the 9th IEEE/IFIP Network Operations and Management Symposium (NOMS)*, pages 483–496, April 2004.
19. A. S. Tanenbaum. *Computer Networks*, chapter 8.5. Prentice Hall, 4th edition, 2003.
20. B. Tierney, B. Crowley, D. Gunter, J. Lee, and M. Thompson. A monitoring sensor management system for grid environments. *Cluster Computing*, 4(1):19–28, Mar. 2001.
21. P. Trimintzios, M. Polychronakis, A. Papadogiannakis, M. Foukarakis, E. P. Markatos, and A. Øslebø. DiMAPI: An application programming interface for distributed network monitoring. In *Proceedings of the 10th IEEE/IFIP Network Operations and Management Symposium (NOMS)*, April 2006.
22. R. Wolski. Dinamically forecasting network performance using the network weather service. Technical Report TR-CS96-494, University of California at San Diego, January 1998.

Support for Automatic Diagnosis and Dynamic Configuration of Scalable Storage Systems

Zsolt Németh[1], Michail D. Flouris[2], Renaud Lachaize[3], and Angelos Bilas[3]

[1] MTA SZTAKI Computer and Automation Research Institute
P.O. Box 63, Budapest, H-1518, Hungary
zsnemeth@sztaki.hu

[2] Department of Computer Science, University of Toronto,
Toronto, Ontario M5S 3G4, Canada

[3] Institute of Computer Science (ICS), Foundation for Research and Technology - Hellas
P.O.Box 1385, Heraklion, GR 71110, Greece
{flouris,rlachaiz,bilas}@ics.forth.gr

Abstract. Distributed storage systems are expected to serve a broad spectrum of applications, satisfying various requirements with respect to capacity, speed, reliability, security at low cost. Virtualization techniques allow flexible configuration of storage systems in order to meet resource constraints and application requirements. Violin provides block level virtualization that enables the extension of storage with new mechanisms and combining them to create modular hierarchies. Creating and maintaining such virtualization hierarchies however, is a complex task where a human system administrator is the most expensive and less efficient element. We introduced Conductor, an automated support system that tries to grasp human expertise with declarative rules that are applied to storage management. So far the initial, static configuration capabilities of Conductor have been elaborated. Static features however, are not sufficient for practical purposes as the storage system evolves, i.e. requirements, workloads, access patterns may change in time. This paper presents work in progress that is aimed at extending Conductor with supporting dynamic features. We introduce the concepts of global and directed reconfigurations and discuss their potential strengths and weaknesses.

Keywords: distributed storage management, virtualization, rule based system.

1   Introduction

As the volume of digital data increases, scalable storage systems provide a means of consolidating all storage in a single system to improve cost-efficiency (Figure 1a). For this reason, storage system architectures are undergoing a transition from directly- to network-attached. This new architecture offers potential for flexible configuration of storage systems to better match application needs and thus improve their performance. This is an important concern because distinct application domains have very diverse storage requirements; Scientific computation, data mining, e-mail serving, e-commerce, search engines, operating system (OS) image serving or data archival impose different

W. Lehner et al. (Eds.): Euro-Par 2006 Workshops, LNCS 4375, pp. 15–21, 2007.

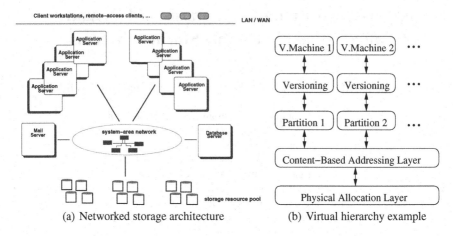

(a) Networked storage architecture   (b) Virtual hierarchy example

Fig. 1. Generic networked storage organization (a) and example virtual hierarchy to consolidate storage for virtual machines

tradeoffs in terms of I/O throughput, latency, reliability, capacity, high-availability, security, data sharing and consistency.

Violin [3] is a kernel-level framework for building and combining virtualization functions at the *block* level. Violin targets commodity storage nodes and replaces the current block-level I/O stack with an improved I/O hierarchy that allows for (i) easy extension of the storage hierarchy with new mechanisms and (ii) flexible combining of these mechanisms to create modular hierarchies with rich semantics. As an example, Figure 1b shows a virtual hierarchy that provides a private virtual disk to each of many virtual machines running on a single system [4]. This system uses partitioning, versioning, and content addressable storage layers to provide the illusion of private disks on top of the same physical storage. More scenarios of advanced virtualization semantics are discussed in [4].

We proposed Conductor [2], a rule-based expert system that is aimed at providing support for configuring and maintaining virtual storage hierarchies in scalable storage systems, such as Violin. Currently, this task relies entirely or mostly on the expertise and intuition of human system administrators. Moreover, most configuration activities are usually complex, not exact, and thus, hard to formalize. Conceptually, Conductor maintains a knowledge base about the storage system as facts, e.g. devices, properties, measured values, structures, and expertise expressed as rules, e.g. how the characteristics of a disk are changed if striped or what are the symptoms of a faulty disk. Based on the facts and rules Conductor will be able to infer new information that may indicate symptoms of problems or may trigger corrective reconfiguration actions.

Storage requirements of applications can be divided in two categories: (a) Statically satisfiable requirements, such as capacity, archival capability, fault tolerance level, encryption, and compression. (b) Dynamically satisfiable requirements that refer mostly to performance characteristics, such as throughput and response time, albeit, some static requirements may also change over time for a given application.

In its current status, as presented in our previous work [2], Conductor is able to deal with static (initial) system configuration. The focus of the work so far has been to suggest optimal configurations that fully satisfy functional (static) requirements, but only approximate performance (dynamic) requirements based on estimated performance values for system components. The performance of a storage element depends both on its physical characteristics as well as the specific application workload. The latter is usually only approximately known at system configuration time. Therefore, initial (static) system configuration usually relies on estimated values for dynamic (performance) characteristics. This is also what happens in Conductor; For instance, when a system needs to provide a virtual volume that offers a certain level of I/O throughput, Conductor relies on estimated values for the throughput of physical devices and heuristics to estimate the throughput of the final virtual device. In essence, Conductor currently addresses two challenges:

- It captures human expertise in the form of rules of a production system. They represent "rules of thumb" that a human administrator would follow, e.g. "to achieve a certain level of throughput stripe a virtual volume over a number of devices". However, to grasp real human expertise and introduce sophisticated rules beyond elementary ones currently implemented, this aspect of Conductor needs to be further investigated.
- It reduces the complexity of searching the configuration space. In this direction, Conductor uses heuristics and tries to reduce search complexity without compromising the quality of the generated solutions significantly.

In this paper, we focus on how to augment Conductor in order to satisfy dynamic requirements, as both workload and system characteristics evolve over time.

Satisfying dynamic requirements requires continuous monitoring of a storage system to detect whether certain goals are being violated. Monitoring is system specific and is usually possible by instrumenting the I/O path at user- or kernel-level. Monitored data are inserted into Conductor's knowledge base in form of facts and subsequently Conductor is able to detect if certain (dynamic) requirements are not met by simply comparing the monitored information to the original specifications; For instance, if throughput of a specific volume drops below a minimum threshold during high traffic conditions, this may imply that the system is not able to satisfy application requests at the agreed rate. Whenever Conductor detects a discrepancy from the original specification it triggers corrective actions that will reconfigure the system. Now, we discuss two main alternative approaches to dynamic reconfiguration: (a) global dynamic reconfiguration and (b) directed dynamic reconfiguration.

2 Global Reconfiguration

One approach to deal with dynamic features is to trigger a full system reconfiguration when problems are detected. This procedure resembles static configuration: a new virtual hierarchy is built from scratch, however, using actual, measured values as opposed to the estimated values used in static system configuration. For instance, if throughput to a specific physical disk is measured to be half of the estimated throughput, then this measured value may lead to more realistic configuration. This scheme can be further refined by establishing certain statistical properties of measured values or relationships between them. As an example it may be projected that encrypted volumes have x percent higher latency where x is established from actual measurements. Hence, not only can measured values replace estimated ones, but also actual experience can refine the way configuration is realized by updating the knowledge base in Conductor. In this sense, measured values also serve as refinements to the existing experience over a longer period; certain trends, relationships between performance and workload, further details of projecting the performance can be established and incorporated into both static and dynamic (re)configuration.

Global reconfiguration is a natural extension of static configuration and re-uses existing mechanisms in Conductor. While initial configuration may not meet performance requirements due to the use of estimated values for performance characteristics, reconfiguration of the system using feedback from the actual system can narrow the gap between required and achieved values. However, this approach has several potential disadvantages.

A main issue is the overhead that a global reconfiguration incurs. Our previous work [2] shows that this is an extensive process. First, configuration itself is an exhaustive search and even though various search strategies have been investigated to reduce complexity by several orders of magnitude, it may still exhibit an exponential behavior. Moreover, reconfiguration affects the entire storage hierarchy independently of the type or focus of the problem, which may not scale in large storage systems. Finally, another potential weakness of global reconfiguration is that although it uses more realistic actual performance values, it omits workload information. Workload behavior may depend on the structure of the hierarchy itself. Thus, rebuilding the full hierarchy from scratch may result in different workload behavior and, as a consequence, reduce the usefulness of the measured values.

3   Directed Reconfiguration

Instead of triggering a global system reconfiguration, an alternative approach is to first detect the type of performance problem as accurately as possible as well as the location where it occurs in a storage hierarchy and then solve it by *directed, local* reconfiguration. This, less intrusive approach requires detecting the origins of discrepancies in dynamic characteristics. Today, this *diagnostic* procedure is the task of experienced human operators that understand both application requirements as well as symptoms of specific performance problems. Automating this procedure is essential for improving the cost-efficiency of large scale storage systems.

Directed reconfiguration relies on the inference mechanism of production systems [5] that is especially appropriate for diagnostic purposes. If the expertise of a human operator is expressed as rules in the production system, then measured facts (monitored values) about system components can be used to infer (diagnose) the location of a problem.

The location where a problem is first detected is not necessarily the origin of the problem. For this reason, diagnosis involves multiple, recursive steps, where inferred information at each step may lead to further decomposing the system to simpler components. For instance, if the bandwidth of a volume is less than expected, diagnosis needs to examine which of the constituent virtual or physical devices of the volume may be responsible. This recursive procedure, essentially follows the structure of virtual volumes, as they are composed out of (possibly multiple layers of) physical devices, storage nodes, and network paths. When a problem is localized with diagnosis, the actions that will be taken by directed reconfiguration represent another form of human expertise and are empirical, inexact, and hard to formalize.

Directed reconfiguration may lead to better decisions compared to global reconfiguration, since diagnosis tries to preserve as much as possible the existing structure. This allows for selecting the most promising reconfiguration actions in a given situation and can lead to better configurations in fewer steps, avoiding a costly (and potentially more disruptive) global reconfiguration that has to try a plethora of possibilities. For example, if there is an indication that 10% more bandwidth is required in a virtual volume, directed reconfiguration may suggest immediately an upgrade of the volume from 2- to 3-way striping, instead of trying all possible configurations using measured performance values.

The main drawback of directed reconfiguration is that in several cases it may not be easy to find the exact scope and cause of a problem. Even human system administrators sometimes can do little more than an intelligent guess – this type of experience can hardly be formalized. It is thus likely that only a subset of the potential problems and their symptoms will be formalized as rules. Nevertheless, we anticipate that significant classes of performance problems can be detected by this method and addressed by efficient system reconfiguration.

4 Related Work

Due to the overall complexity of administering storage systems, the application of abstract control and intelligent methods, have been proposed in recent work. While there are some similarities with our work in certain details, none of them address the issue of configuring and maintaining virtual storage hierarchies.

Polus [7] aims at mapping high level QoS goals to low level storage actions by introducing learning and reasoning capabilities. The system starts with a basic knowledge of a system administrator expressed as "rules of thumb" and it can establish quantitative relationships between actions taken and their observed effects to performance by monitoring and learning.

Ergastulum [1] is aimed at supporting the configuration of storage systems with reducing the search complexity of possible design decisions by utilizing heuristics with randomization and backtracking.

A novel approach presented in [6] tries to predict the effect of certain actions and helps with making decisions at data distribution. It establishes a set of *What... if...* statements where the hypothetical effect (what part) of a certain circumstance (if part) is stored. These relations are obtained by statistical, analytical or simulation methods.

5 Conclusions

Our goal in this work is to examine how we can extend the existing static features of Conductor [2] to automatically configure large scale storage systems so that they satisfy application requirements for dynamic system characteristics. We introduce two potential approaches: The first, global reconfiguration, is a direct extension of Conductor and can be implemented in a straightforward manner. The second one, directed reconfiguration aims at capturing further human expertise when managing large storage systems. One of the main challenges here is introducing appropriate diagnostic rules in the production system.

Global and directed reconfiguration can also be seen as complementary to each other: Global reconfiguration uses measured data but omits structural information. Its effect is global and it is most useful when problematic spots cannot be identified either because they are related to the entire structure with no specific focus or appear too frequently or simply cannot be diagnosed. On the other hand, directed reconfiguration takes into account measured data *and* structural information and tries to locate the problematic spot and the possible causes as precisely as possible. It is more appropriate for "local" problems in the system structure, such as performance hot-spots.

Finally, we are currently implementing the two approaches. This requires addressing the following challenges: (a) Capturing human expertise in the form of rules for diagnosis purposes can happen at various levels of detail. (b) Although detecting a certain problem is easy, it is hard to decide if action must be taken or if the problem is temporary, can be tolerated and should thus, be ignored. (c) Improving the management system by extending the knowledge base gradually with new rules as more experience is acquired with new applications and new system components. (d) Performing experiments with realistic setups that reflect situations encountered in real life.

Overall, we believe that rule-based expert systems, such as Conductor, tuned to the needs of storage applications, can offer significant help in managing large scale storage systems and improving their cost-efficiency.

References

1. E. Anderson, M.Kallahalla, S. Spence, R. Swaminathan, Q.Wang. Ergastulum: Quickly Finding Near-Optimal Storage System Designs. HP Laboratories SSP technical report HPL-SSP-2001-05 (2002)
2. Zs. Németh, A. Bilas, M.D. Flouris, R. Lachaize: Conductor: An Intelligent Configuration Framework for Storage Area Networks. Book on Knowledge and Data Management in Grids, CoreGRID series, Springer Verlag, 3, 2006
3. M.D.Flouris, A. Bilas: Violin: A Framework for extensible Block-level Storage. 22nd IEEE / 13th NASA Goddard Conference on Mass Storage Systems and Technologies (MSST 2005) April 2005, Monterey, CA, USA. IEEE Computer Society 2005.
4. M. Flouris, R. Lachaize, and A. Bilas. Violin: a Framework for Extensible Block-Level Storage. Book on Knowledge and Data Management in Grids, CoreGRID series, Springer Verlag, 3, 2006

5. M.Klein, L.B.Methile: Expert systems: A Decision Support Approach. Addison-Wesley, 1990.
6. E. Thereska, M. Abd-El-Malek, J. J. Wylie, D. Narayan an, G. R. Ganger. Informed data distribution selection in a self-predicting stor age system. Proc. of the International Conference on Autonomic Computing, ICAC-0 6, Dublin, Ireland, June 2006.
7. S. Uttamchandani, K. Voruganti, S. Srinivasan, J. Palmer, D. Pease. Polus: Growing Storage QoS Management Beyond a "Four-year Old Kid". USENIX FAST '04 Conference on File and Storage Technologies, March 2004, San Francisco, CA, USA.

Adding Dynamism to OGSA-DQP: Incorporating the DynaSOAr Framework in Distributed Query Processing

Arijit Mukherjee and Paul Watson

School of Computing Science, Newcastle University,
Claremont Tower, Claremont Road, Newcastle Upon Tyne, United Kingdom
{Arijit.Mukherjee,Paul.Watson}@ncl.ac.uk
http://www.cs.ncl.ac.uk

Abstract. OGSA-DQP is a Distributed Query Processing system for the Grid. It uses the OGSA-DAI framework for querying individual databases and adds on top of it an infrastructure to perform distributed querying on these databases. OGSA-DQP also enables the invocation of analysis services, such as **Blast**, within the query itself, thereby creating a form of declarative workflow system. DynaSOAr is an infrastructure for dynamically deploying web services over a Grid or a set of networked resources. The DynaSOAr view of grid computing revolves around the concept of services, rather than jobs where services are deployed on demand to meet the changing performance requirements. This paper describes the merging of these two frameworks to enable a certain amount of dynamic deployment to take place within distributed query processing.

Keywords: Dynamic deployment, Web Service, Grid, distributed query processing.

1 Introduction

OGSA-DQP[1], [2] is a publicly available service-oriented distributed query processor for the Grid. It provides distributed query functionality on databases spread over the Grid using the commonly used service for data access and integration, OGSA-DAI[3]. OGSA-DQP supports the evaluation of queries expressed in a declarative fashion over one or more services, including data access services and external analysis services. It can be seen as complimentary to other service orchestration mechanisms, such as workflow languages.

Because the services can be potentially located on computational resources distributed across the internet, communication costs can play a major role in the performance of the system. Co-locating the query evaluation service and the analysis service with the data, even with an *on-the-fly* deployment may prove to be potentially beneficial in the long run, especially when frequent, long-running queries are executed. DynaSOAr[5] is a framework, which enables such dynamic deployment of services on available computational nodes. An enhanced

W. Lehner et al. (Eds.): Euro-Par 2006 Workshops, LNCS 4375, pp. 22–33, 2007.

version of OGSA-DQP has been created which incorporates DynaSOAr concepts. This paper briefly describes the DynaSOAr architecture, and its use within the OGSA-DQP context. Experimental results and analyses show how DynaSOAr may benefit service-oriented distributed query processing by moving the analysis and data retrieval code near the actual data.

The paper is organized as follows: Section 2 provides a brief introduction to the OGSA-DQP concept and functionality. Section 3 describes the DynaSOAr architecture in brief, followed by the use of DynaSOAr infrastructure within the OGSA-DQP context in Section 4. The experimental setup, results and analysis are covered in Section 5. Related works are discussed in Section 6, current and future directions in Section 7, with conclusions in Section 8.

2   Brief Description of OGSA-DQP

OGSA-DQP is composed of two major services   (i) Grid Distributed Query Service (GDQS) and (ii) Query Evaluation Service (QES). The GDQS is implemented as an extension to the standard OGSA-DAI service, and is deployed as an OGSA-DAI data service with an exposed data service resource[1]. The DQP data service resource thus exposed, supports querying over a set of OGSA-DAI data services, each wrapping a database on some computational node. It also supports the invocation of analysis services over the query results. An example of a typical query, in OQL, supported by OGSA-DQP is as follows:

```
%print select p.ORF, g.id, calculateEntropy(p.sequence)
from p in protein_sequences, g in goterms, t in protein_goterms
where g.id=t.GOTermIdentifier and p.ORF=t.ORF and
p.ORF like "YBL06%" and g.id like "GO:0000%";
```

In this example, the query spans over three databases (protein_sequence, goterm and protein_goterm) which can be distributed over a large geographical area, and an analysis service exposed as a Web Service is also invoked on each sequence element. Based on the schema and WSDL imported from the data and the analysis services and the resources available to it, a query compiler/optimizer component, Polar*[6], generates a parallel query plan, which is partitioned into sub-plans. These sub-plans are distributed to the participating evaluation services each of which is responsible for evaluating the sub-plan assigned to it and conveying the result back either to the root partition or other evaluation services. Finally, the result is collected at the node evaluating the root partition and sent to the GDQS and hence to the consumer.

---

[1] A data service resource implements the core OGSA-DAI functionality. It accepts perform documents from data services, parses and validates them, executes the data-related activities specified within them and constructs response documents. It can also cache data for retrieval by third-parties (if the data service resource is configured to support asynchronous data delivery). Data service resources are accessed via data services.[4]

3 DynaSOAr Architecture

DynaSOAr is a framework, which provides a generic infrastructure for deploying web services as and when required, on available nodes. DynaSOAr achieves this dynamic deployment by processing the incoming consumer request at different levels between two different components, namely a *DynaSOAr Web Service Provider* and a *Host Provider*, with a defined interface between them.

- The *DynaSOAr Web Service Provider* is the entity with which consumers interact. It advertises the services it can provide, receives SOAP messages from consumers requesting a service from a particular endpoint associated with the message, and is responsible for arranging the processing of the request. The *Web Service Provider* achieves this by choosing an appropriate *Host Provider* and forwarding the message to it with any associated Quality of Service (QoS) parameters and an added element in the message header identifying a *software repository* where the service code can be found in case a dynamic deployment is required.
- The *Host Provider* is responsible for controlling the computational resources, such as a cluster or a grid, on which services can be deployed, and requests for those services can be processed. It accepts the SOAP messages forwarded by the *Web Service Provider* on behalf of the services hosted by it, and sends back any response generated after processing the request.

When a message reaches the *Host Provider*, there can be two different interaction patterns depending on whether or not the requested service is already deployed on the node -

1. If the service is already deployed on the computational node where the request is to be processed, then the *Host Provider* routes the SOAP message to the service on that node. In Figure 1a, the consumer makes a request for service S2, which is already deployed on node *N1* and *N2*. Based on the current information about the system load, the *Host Provider* routes the request to the lightly loaded node *N2* where the request is processed and the response is sent back.
2. The second case is where the consumer makes a request for a service, which is not already deployed on any of the available nodes, such as the request for service S8 sent by the consumer in Figure 1b. In this case, a decision is made about the target node where the service is to be deployed and the message is forwarded to that node. The node downloads the service code from the *software repository*, deploys the service dynamically, and processes the request.

It is to be noted that in the scenarios described above, the consumer is not aware of the resources behind the *Web Service Provider* or the fact that the service has been dynamically deployed. They interact with the *Web Service Provider* by sending SOAP messages which is the standard way of interacting with a service.

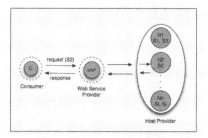

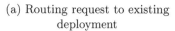

(a) Routing request to existing
deployment

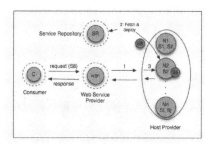

(b) Request service not already
deployed

Fig. 1. Routing requests in DynaSOAr

DynaSOAr has two other components to support dynamic deployment, namely a *Registry Service*, and the *Service Repository*.

- The *DynaSOAr Registry Service* is provided by GRIMOIRES[7], which is a UDDI-based registry, with added support for storing metadata as RDF triplets. Whenever the provider of a service decides to make the service available via DynaSOAr, the service code needs to be uploaded to the *service repository*, as a result of which the registry is updated with the information. The service is added to the registry without any concrete *accessPoints* (in UDDI terms), but a reference to the *Service Repository* web service is added to it. Every time a service is deployed on any of the available nodes, the entry of that service in the registry is updated with the actual endpoint.
- The *Service Repository* manages the *upload* or *download* of the service code. The Host Providers communicate with this service while downloading the service code for a service to be deployed.

The description so far consisted of a single *Host Provider*. However, in reality, several Host Providers may be available to one *Web Service Provider*. It might be advantageous to make a selection between the available *Host Providers* based on certain parameters, such as cost, dependability, QoS, security. To facilitate this, another component, the broker, with the same interface as the *Host Provider*, has been introduced in the architecture to make such decisions. The broker has the knowledge about one or more *Host Providers*, and is able to make the decisions based on the characteristics of the available *Host Providers* and the QoS or security requirements requested by the consumer. Figure 2 describes the generic architecture of DynaSOAr with all its components.

DynaSOAr is a generic framework allowing the structure to grow dynamically to any level or depth. There can be any number of brokers and any number of *Host Providers*, thereby creating the space for a *Web Service Market Place*, where the brokers can choose between all available providers meeting the consumer requirements to process the requests.

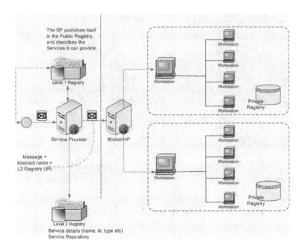

Fig. 2. Generic DynaSOAr Architecture

The generic architecture of DynaSOAr does not restrict the dynamic deployment functionality to web services alone. As described in [5], DynaSOAr enables the dynamic deployment of virtual machines like VMWare[8] and Microsoft Virtual PC[9] and also .NET services and stored procedures over SQLServer.

4   Dynamic OGSA-DQP

The features of OGSA-DQP and the requirements for distributed query processing make it a prime candidate for the use of the DynaSOAr infrastructure. Usage scenarios in OGSA-DQP, which can benefit from the DynaSOAr features, include the following -

1. Frequent and long-running queries can benefit from the *on-the-fly* deployment of an analysis service such that it is co-located with relevant data.
2. A new database wrapped by the OGSA-DAI data service can be deployed to enable the GDQS to serve queries involving the new database.
3. An increased degree of parallelism can be obtained by deploying multiple copies of the analysis service on multiple nodes.
4. Increased performance for a database scan can be enabled by deploying virtual machines containing a copy of the database.
5. Even though the deployment of virtual machines is costly in terms of time required, in the case of frequent and long-running use of the database present in the virtual machine or the service deployed in it, the initial deployment cost can be outweighed by the benefits.
6. Polar* performs some basic optimization based on the information available to it. But this optimization can be enhanced by considering the dynamic deployment scenario, where the scheduler should be able to schedule deployment of new evaluation or analysis services on new computational nodes if it finds the existing deployments to be heavily loaded.

4.1 Implementation

As a proof of concept the publicly available OGSA-DQP has been modified to incorporate the DynaSOAr features into it. In the DynaSOAr-enabled OGSA-DQP, the primary requirement is to create a structure similar to the *Web Service Provider - Host Provider* structure of DynaSOAr. The *Grid Distributed Query Service (GDQS)* corresponds to the *Web Service Provider*, which advertises itself as capable of providing distributed query processing functionality over a set of databases exposed as OGSA-DAI data service resources, and a set of analysis services, either provided by a remote provider, or by the GDQS itself. In the latter case, the analysis service may not be deployed on any node available to the GDQS, but the *Service Repository* stores the service code, and the *registry* contains information about this potentially available service.

As in case of the public OGSA-DQP, a *DQP data service resource* must be created from the deployed GDQS factory data service resource when initializing the service. In the second initialization step, this *data service resource* imports the schema of the databases exposed by OGSA-DAI and the WSDL of the analysis service. During this second phase, the GDQS in the dynamic version of OGSA-DQP tries to co-locate the evaluation services with the OGSA-DAI-wrapped databases by dynamically deploying the *Query Evaluation Service*, which is a standard *web service*, onto the nodes where the data resides. If an analysis service advertised by the GDQS is added to the *DQP data service resource* configuration, the GDQS using the DynaSOAr framework deploys the service on a suitable computational node. Once these new services are deployed, the schema and the WSDL are imported from them in the same way as in case of the standard OGSA-DQP. The complete deployment process is shown in Figure 3.

As the new services are deployed, the *registry* is updated with the corresponding information, so that another *data service resource* for the same GDQS will be able to reap the benefit of the previous deployment by re-using the already deployed services. This is the point where the advantages of DynaSOAr over currently available job-based grid systems become apparent. The services deployed using DynaSOAr will be accessible until they are explicitly removed from the server, or the server becomes unavailable, compared to the jobs, which do not persist beyond a single execution. Thus we can achieve a "deploy once, use many times" philosophy with DynaSOAr, which has a positive effect on the performance of the distributed queries as shall be seen in the analysis of the experimental results.

5 Experiment

5.1 Setup

To analyze the performance of the Dynamic OGSA-DQP system, several experiments have been performed and the results analyzed. The initial experiments primarily concentrated on the dynamic deployment of the analysis services and

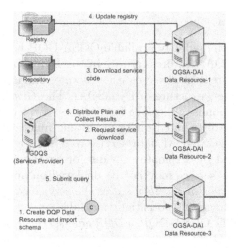

Fig. 3. DynaSOAr enabled OGSA-DQP

the impact made by this on the performance of the distributed query processing activities. The DynaSOAr framework was setup on a set of Linux machines within the Newcastle University GIGA cluster - each of them being a four-processor Intel® Xeon^{TM} CPU 2.80GHz system, with 2GB memory, and Fedora Core 4 installed on them. The GDQS was deployed on another Linux machine - a four-processor Intel® Xeon^{TM} CPU 3.06GHz with 1GB memory and Fedora Core 4 installed on it. The *DynaSOAr Registry* and *Service Repository* service were co-located with the GDQS. The analysis service code, along with the Query Evaluation Service code were uploaded to the *Service Repository*, and a copy of the same analysis service was deployed on a Linux (one-processor Intel® Xeon^{TM} CPU 2.40GHz system with 1GB memory and Red Hat Enterprise 3 Linux) system at the Edinburgh Parallel Computing Centre (EPCC) at Edinburgh University. The network between Newcastle University and EPCC is JANET, which is a high performance gigabit network connecting the universities in the United Kingdom.

Five databases were exposed as OGSA-DAI data resources on five of the Linux systems that were part of the already established DynaSOAr framework. One of the databases used for the test queries was loaded with several tables, each with 100,000 records, and fixed record sizes of 128 bytes, 256 bytes, 512 bytes, 1 Kbytes, 2 Kbytes, 4 Kbytes, 8 Kbytes and 10 Kbytes. The experiments were designed to fetch data out of each table in chunks of 100, 200, 400, 800, 1000, 2000, 5000, 10000, 20000 and 50000 tuples and perform the analysis on each tuple using the analysis service. Results were collected in order to compare the performance of the system with a remote analysis service, to the performance with a local service dynamically deployed using the DynaSOAr framework, i.e to investigate item (1) in Section 4.

5.2   Results and Analysis

In preliminary experiments, the DynaSOAr framework was used to deploy the analysis service on separate hosts. The deployment cost includes the time required to transfer the service code from the repository to the target host and the time taken for the actual deployment within the *web service container*, Apache Tomcat in this case - where the packaged service (packed as a WAR file) is unpacked into a proper directory structure and the various libraries are loaded before the service can be accessed. Figure 4a shows the time taken to deploy the service on different hosts and the average time for deployment.

The average time required for an individual service deployment on a computational node was approximately 32.4 seconds. This is a one-time cost and is incurred only during the DQP initialization phase. Copies of the same service can be deployed in parallel on multiple nodes if required, so that the total deployment cost of all copies becomes equivalent to the cost of a single deployment. Once the service is deployed locally, the performance of the queries executed by this GDQS reaps the full benefit of this one-time deployment, as is evident from the other experiments.

A set of ten queries was executed on a test database, retrieving 100, 200, 400, 800, 1000, 2000, 5000, 10000, 20000 and 50000 tuples from the database. Each query was used to retrieve datasets of different sizes, such as 128 bytes, 256 bytes, 512 bytes, 1 Kbytes, 2 Kbytes, 4 Kbytes, 8 Kbytes and 10 Kbytes. Each query also invoked the analysis service for each retrieved tuple. An example query used in the tests is as follows:

```
%print select p.id, calculateEntropy(p.sequence) from p in
proteinsequence_random_sequence_128s where p.id < 20000;
```

This query retrieves 20,000 tuples from the database and invokes the analysis service (*calculateEntropy* in this case) on the sequence attribute of each tuple. The results of these experiments are shown in Figure 4b and Figure 4c.

Figure 4b compares the invocation cost (in milliseconds) of a local and a remote deployment of the same service for different result cardinalities, ranging from 100 to 20,000. Figure 4c compares the average invocation cost (in milliseconds) of the local and remote service for different sized tuples, from 128 bytes to 10 Kbytes.

It is evident from the plotted results that the invocation cost increases radically for the remote analysis service as the number of tuples increase starting from 100 tuples to 20000 tuples. In Figure 4b, the total cost of invoking the analysis service increases as the number of tuples retrieved from the database increase. But, the rate of increase is far more substantial when the analysis service is remote, than when it is local. In Figure 4c, the average invocation cost per tuple is plotted against the average tuple size, starting from 128 bytes to 10Kbytes. In this case, for both local and remote services, the invocation cost tends to increase as the tuple size increases, but the effect of a remote service is

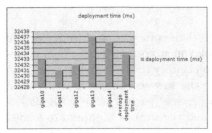

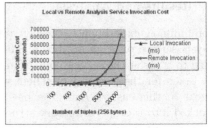

(a) Required deployment time on
different hosts

(b) Comparing average invocation
cost for different tuple sizes

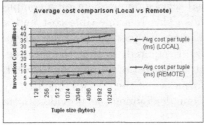

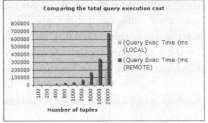

(c) Comparing local and remote
service invocation cost

(d) Comparison of the total evaluation
cost

Fig. 4. Performance Analysis of DynaSOAr-enabled OGSA-DQP

significantly higher than a local service, and it can be inferred that the cost of invoking the remote service will increase further if the data size increases.

Figure 4d shows the total query evaluation cost for two scenarios, (1) when the analysis service was local and (2) when the analysis service was remote. This figure shows that the total query evaluation costs when the analysis service was remote are significantly higher than the total evaluation costs when the service was local. The difference between these two values becomes equal to the average cost of deployment (an average of 32.4 seconds) when the number of tuples is approximately 1000, and starts increasing even more significantly as the number of tuples increase. This data validates the statement made earlier in this paper that the one-time deployment cost can be outweighed by the performance benefits in case of frequent, long-running queries.

These results clearly show that for the queries using analysis services over the data retrieved from the databases, the performance of DynaDQP is much better than the standard OGSA-DQP where the analysis service can be remote from the data. The difference in the performance is quite noticeable considering the fact that a very high-speed Internet backbone exists between the server at Edinburgh Parallel Computing Centre and the Linux cluster at Newcastle University. The performance difference would probably be even more prominent if the analysis service resides much further apart geographically, because a higher communication cost would be incurred in that case.

6   Related Work

Although in the DynaSOAr architecture, the Host Provider sits on top of existing Grid infrastructure as a high level service, it can exploit the results of work producing components on which dynamic deployment frameworks can be built. In particular, the job scheduling fabrics like Condor[20] can be utilized as a means of gathering machine characteristics, and CPU loads. However, deployment of services rather than jobs raises other issues, such as making the decision about whether to deploy a service on a new node or to use an existing but possibly overloaded deployment. The GridSHED project [10],[11] for job scheduling has been investigating this area, and the results are being utilized to design an effective scheduling system for DynaDQP.

There is some work on dynamic deployment as in [18], but this is essentially tied to a particular implementation of Grid middleware and web service container (WS-RF[12] and the Globus Toolkit[13]) without addressing the more widespread deployment scenarios involving commonly used standard toolkits such as Axis and Tomcat. The work described in [21] is built on top of specialized hardware. Moreover, the deployment of different types of components such as virtual machines, stored procedures, .NET services together in one framework is not addressed in any of the current systems.

This paper focuses on the use of the dynamic deployment framework within the context of *Distributed Query Processing*. To our knowledge, there is no current distributed query processing system which is factored out as inter-operable services and allows *on-the-fly* deployment of evaluation and analysis services on available nodes thereby co-locating the data processing and analysis code with the data, as proposed in [19]. The analysis of the results clearly indicate that moving the code to the data, even with an initial deployment cost can potentially be beneficial, especially for frequent execution of long running queries over huge data sets.

7   Current and Future Directions

At present, work is under way to enable the DynaSOAr framework (and hence DynaDQP) to support the usage of virtual machines for dynamic deployment of data access services, evaluation services, analysis services and databases. It has been accepted by OGSA-DAI as well that the availability of a deployment-ready OGSA-DAI service would greatly help the dynamic deployment work, and work is going on in that respect too.

In OGSA-DQP, the compiler/optimizer performs some static scheduling based on a very simple cost model, but that does not consider the inherent dynamism in a Grid system where the dynamics of the environment is liable to change drastically. The effects of changes in resources at runtime have been considered in the investigations into *adaptive distributed query processing* [17], [16]. It will be an effective solution to combine the findings of GridSHED, DynaSOAr and the *adaptive DQP* investigation.

Some work has been done in this area of *fault-tolerant distributed query processing* [14], [15]. The concepts of the dynamic DQP are also relevant to fault-tolerant query processing systems where a failure of an evaluation node can be handled through the deployment of the same service on another node or a virtual machine as a replacement of the failed node, and by replaying certain sections of the query evaluation to regain the state where the processing stopped due to the failure.

Virtual Machines are an important aspect for the Dynamic DQP framework. Some basic work has already been done on deploying Microsoft's Virtual PC systems in DynaSOAr. It is being extended to incorporate VMWare systems, and the deployment of databases in virtual machines. Deploying databases in virtual machines does however raise other key issues such as keeping the copy in sync with the original database.

8   Conclusion

This paper presents an overview of ongoing work on enabling dynamic service deployment in OGSA-DQP using the DynaSOAr framework. We believe that distributed query processing can potentially benefit from the dynamic deployment mechanisms of DynaSOAr by deploying evaluation and analysis services closer to the data, and this claim is supported by the experimental results. It also includes scope of creating *software market places*, where the computational resources can be chosen from a pool of available *Host Providers* based on the cost and (or) the quality of service provided by the host.

The project is continuing to investigate different aspects of the system, such as scheduling new deployments, routing requests between multiple instances of the same service, deploying virtual machines and databases, and the future work involves looking into the utility of the framework for adaptive and fault-tolerant distributed query processing systems and evaluating various transport technologies for transferring the service code.

Acknowledgments. We wish to extend our gratitude to our collaborators in GridSHED: Isi Mitrani and Jennie Palmer. We wish to acknowledge the support from our colleagues in OGSA-DAI, OGSA-DQP and DAIT projects at Manchester University and EPCC. We wish to thank Jim Smith for his helpful discussions. We would also like to thank the UK Engineering and Physical Sciences Research Council, and the DTI for the GridSHED and DAIT projects.

References

1. Alpdemir, M.N., Mukherjee, A., Gounaris, A. et.al.: OGSA-DQP: A Service for Distributed Querying on the Grid. In: Advances in Database Technology - EDBT 2004. Lecture Notes in Computer Science, Vol. 2992. Springer-Verlag, 858–861
2. OGSA-DQP, http://www.ogsadai.org.uk/about/ogsa-dqp/
3. OGSA-DAI, http://www.ogsadai.org.uk/

4. OGSA-DAI Glossary of Terms, http://www.ogsadai.org.uk/documentation/ogsadai-wsrf-2.2/doc/reference/glossary.html

5. Watson, P., Fowler, C., Kubicek, C., Mukherjee, A. et. al.: Dynamically Deploying Web Services on a Grid using Dynasoar. In: Ninth IEEE International Symposium on Object and Component-Oriented Real-Time Distributed Computing, ISORC 2006, IEEE Computer Society 2006

6. Smith, J., Gounaris, A., Watson, P. et. al.: Distributed Query Processing on the Grid. In: Grid Computing 2002. Lecture Notes in Computer Science, Vol. 2536. Springer-Verlag, 279–290

7. GRIMOIRES, http://twiki.grimoires.org/bin/view/Grimoires/

8. VMWare, http://www.vmware.com/

9. Microsoft Virtual PC, http://www.microsoft.com/windows/virtualpc/default.mspx/

10. Palmer, J., Mitrani, I.: Optimal Server Allocation in Reconfigurable Clusters with Multiple Job Types. In: Computational Science and its Applications (ICCSA 2004), Assisi, Italy, 2004.

11. Kubicek, C., Fisher, M., McKee, P., Smith, R.: Dynamic Allocation of Servers to Jobs in a Grid Hosting Environment. BT Technology Journal, Vol. 22, 2004. 251–260

12. Web Services - Resource Framework, http://www.globus.org/wsrf

13. Globus Toolkit, http://www.globus.org/toolkit

14. Smith, J., Watson, P.: Fault-Tolerance in Distributed Query Processing. In: 9th International Database Engineering And Application Symposium. IDEAS 2005, http://ideas.concordia.ca/ideas2005/, IEEE, 329–338

15. Smith, J., Watson, P.: Failure Recovery Alternatives In Grid Based Distributed Query Processing: A Case Study. The University of Newcastle upon Tyne, number CS-TR-957, April 2006

16. Gounaris, A., Paton, N.W., Sakellariou, R., Fernandes, A.A.A. et. al.: Practical Adaptation to Changing Resources in Grid Query Processing. In: The 22nd International Conference on Data Engineering, ICDE 2006

17. Gounaris, A., Paton, N.W., Sakellariou, R., Fernandes, A.A.A. et. al.: Adapting to Changing Resource Performance in Grid Query Processing. In: VLDB Workshop on Data Management in Grids, DMG 2005, http://liris.cnrs.fr/~jpierson/DMG_VLDB05/

18. Qi, L., Jin, H., Foster, I., Gawor, J.: HAND: Highly Available Dynamic Deployment Infrastructure for Globus Toolkit 4, http://www.globus.org/alliance/publications/papers.php#HAND

19. Watson, P., Lee, P.: The NU-Grid Persistent Object Computation Server. In: 1st European Grid Workshop, Poznan, Poland, 2000

20. Tannenbaum, T., Wright, D., Miller, K., and Livny, M.:Condor - A Distributed Job Scheduler. In:Beowulf Cluster Computing with Linux, T. Sterling, Ed.: The MIT Press, 2002.

21. Chrysoulas, C., Haleplidis, E., et. al.:Applying a Web-Services Based Model to Dynamic-Service Deployment. In: International Conference on Intelligent Agents, Web Technology, and Internet Commerce (IAWTIC), Vienna, Austria, November 2005

Review of Security Models Applied to Distributed Data Access

Antonia Ghiselli[2], Federico Stagni[1], and Riccardo Zappi[2]

[1] Istituto Nazionale di Fisica Nucleare sez. di Ferrara,
via Saragat 1 - 44100 Ferrara, Italy
{federico.stagni}@fe.infn.it
http://www.fe.infn.it
[2] Istituto Nazionale di Fisica Nucleare CNAF,
viale Berti Pichat, 6/2 - 40127 Bologna, Italy
{antonia.ghiselli,riccardo.zappi}@cnaf.infn.it
http://www.cnaf.infn.it

Abstract. In this paper, we explore the technologies behind the security models applied to distributed data access in a Grid environment. Our goal is to study a security model allowing data integrity, confidentiality, authentication and authorization for VO users. We split the process for data access in three levels: Grid authentication, Grid authorization, local enforcement. For each level, we introduce at least one possible technological solution. Finally, we show our vision of a SOA oriented security framework.

This work is developed as part of the CoreGRID Network of Excellence, for the Institute on Knowledge and Data Management.

Keywords: Grid, data management, security, authentication, authorization, policy, acl, XACML, SAML.

Introduction

In this report, we will explore the technologies behind the security models applied to distributed data access in a Grid environment. Our goal is to study a security model allowing data integrity, confidentiality, authentication and authorization for VO (Virtual Organizations) users [13]. Although the effort will be to create a generic model, the work will be based on a Grid framework with the following assumptions: Grid users are organized in VOs with existing tools to manage memberships and credentials. In other words, we want to define policies for resource usage on the basis of user credentials, and to enforce them on the basis of Grid status.

The rest of this paper is organized as follows: in section 1 we introduce our approach to security with some definitions. In section 2 we explain some general requirements. In section 3 we describe the technologies to build a security framework. In section 4 we introduce the technologies to build a Grid data access framework.

W. Lehner et al. (Eds.): Euro-Par 2006 Workshops, LNCS 4375, pp. 34–48, 2007.

1 Definitions

Initially, the Grid was referred to as *Computational Grid*, thinking as a way to share computational facilities. However, much of the Grid jobs are data intensive, and to stress this point, today we normally think of Grids in term of *Data Grids*: most large jobs that require Grid services, especially in the scientific domain, involve the generation of large datasets, and their consuption [1]. There is a necessity for the reservation and the scheduling of data repositories, and so we need to express some policies to govern their access. Moreover, thousands of people may want to use storage resources to share him/her data with a limited set of other researchers, or maybe with no one but themself. The future storage systems will contain critical user information for various applications and purposes, like for example life science and financial ones.

Grids need an authorization framework to handle the users privacy necessities, and their limits too. In other words, we want to control the access to the Grid users' data on the basis of some high controlled sharing rules.

1.1 Grid Data Management Systems

A distributed system is a collection of independent computers that appears to its users as a single coherent system. Similarly, a distributed data access system is a distributed data storage, with ubiquitous and transparent data access and migration. A *Grid Data Management System* (GDMS) is a data access system acting in a Grid environment. GDMS offer a common view of storage resources distributed over several administrative domains. Therefore, they must allow the smooth integration or removal of resources, without affecting the integrity of neither the individual independent domains nor the system as a whole. Problems behind the implementations of such a system are:

- *processes communication:* the way distributed processes exchange informations.
- *Naming:* name resolution and localization.
- *Synchronization, consistency and replication:* the way data are synchronized and the definition of policies for the consistency and the replication of data.
- *Security:* the way to gain security for data access.

1.2 Security

We define a security architecture as *a set of features and services that tackles a set of security requirements and can handle a set of cases*[16]. Grid systems in use today do not address security in a systematic way: just to make an example, historically in Globus [17] an authenticated user is a good user. This emphasizes the authentication aspect, but Grids need a strong authorization mechanism. Our aim is to enable new Grid infrastructure developer to create more secure systems, capable to attract new Grid users and applications. Security models should define "who can do what, when and where". A Grid middleware should encompass a security framework, in which we can distinguish two virtual black boxes: the *authentication* box and the *authorization* box:

- *authentication* deals with the verification of the identity of an entity within a network. An implementation should provide an agnostic plug point for multiple authentication mechanisms, and the means for conveying the specific mechanism used in any given authentication operation.
- *Authorization* deals with the verification of an action that an entity can perform after authentication was performed successfully. The goal of an authorization framework is to provide a light-weight, configurable, and easily deployable policy-engine-chaining infrastructure that is agnostic to back-end enforcers and evaluators, as well as the run-time container infrastructure and the state model that hosts them. The framework allows for a combined and flexible decision making process, taking into account information, assertions and policies from a variety of authorities.

We can make a brief comparison between the high-level techniques besides authentication and authorization. The first link in the Grid security chain is authentication. Grid resources authenticate remote users using basically two ways: the first uses a session key, and the second, which is the mostly used too, uses the Public Key Interface (PKI). On the other hand, we need a Privilege Management Infrastructure (PMI): a PMI is to authorization what a PKI is to authentication [2]. Just to make an example, we can express some user's attributes using the X.509 Attribute Certificate (AC), which maintains a strong binding between a user's name and its attributes. Certification Authorities (CAs) digitally sign a public key certificate; in a similar way, the entity that signs an AC is called an Attribute Authority (AA), while the root of trust of the PMI is called the Source of Authority (SOA), which may delegate its power to subordinate AAs. Like Certificate Revocation List (CRL), an AA could issue an Attribute Certificate Revocation List (ACRL) to revoke privileges from an AC. Obviously, ACs is just one of the possible solutions to join users and their attributes.

2   Requirements

Integration, interoperability and trust are the building blocks of the requirements behind a Grid security infrastructure. In this section we give some brief and general guidelines, but we want to point out that more specific requirements will be glean in the proceeding of this paper.

- *Confidentiality* is the property that information doesn't reach unauthorized individuals, entities, or processes. It is achievable by a mechanism for ensuring that only those entitled to see information or data can access them.
- *Integrity* is the assurance that information can only be accessed or modified by those authorized to do so. Data integrity is a nontrivial problem especially when storage hardware and networks are not perfect.

- *Resilience* is an important requirement as the Grid links and nodes are very dynamic in nature and may change over the time. The GDMS security architecture should remain intact and should deliver the promised level of security assurances even if its composition changes over the time. The resilience provides an abstraction layer to hide the architectural changes from the overall security architecture.
- *Data Lifecycle Management (DLM)* is the process of managing data throughout its lifecycle. GDMS should ensure that the data contents will be protected from malevolent entities.
- *Fault Tolerance* is a desirable feature especially when transfers of large data files occur.

2.1   Data Types

In section 1, we've made a really brief history of the evolving of the concept of Grid, from Computing Grids to Data Grids, and we mentioned the data types involved in it. To understand all the security requirements, we have to think to who is using Grid now, who is going to use it soon, and who wish to use it, but can't trust it for some security reason. At the present time, the majority of Grid tools are growing behind some specific needs, mainly HEP[1] experiments. These applications produce and consume a considerably high amount of data with heavy impact on the bandwith, but probably they don't need a high security system, because the main purpose of this activities is to be fast. In the future, much more people is going to use Grids and peer-to-peer systems, not only with the actual purposes. In the next generation file sharing, a user will want to give access to his/her files only to a limited set of people. There's the need for a high control over who is authorized to view them. This means more protection levels, but less performance too. What we want to stress here, is that every data type needs different protection levels, and that a Grid security system must take care of this principle. Different data types can determine the way we achieve data integrity, confidentiality, authentication and authorization.

The next generation storage elements would be able to publish the Quality of Protection (QoP) they can assure to the data they own. In this way, the QoP will be decisive for the entire data storage system. Just to make an example, a user should request the resource provider to not have read access to his/her data: this is a non-trivial challenge, and obviously not all the storage elements will be able to enforce this demand, because this is depending of the locally implemented security system: it determines the QoP and the Service Level Agreement (SLA), which defines how data is protected while in transit over the service. Security negotiations should be used to establish secure sessions between the endpoints. A security infrastructure featuring support for negotiations and establishment of end-to-end and/or hop-to-hop security associations has broader applicability to general networked environments like Grids. Security negotiations require some brokering agent to mediate between the endpoints.

---

[1] High Energy Physics.

3   Data Management Security Technologies

We can roughly divide the process to reach access to a Grid resource in 3 levels: first of all there's a Grid authentication process, then authorization on a Grid-ID base, and finally local enforcement using the resource-specific security framework. In figure 1 you can see the all-round security process.

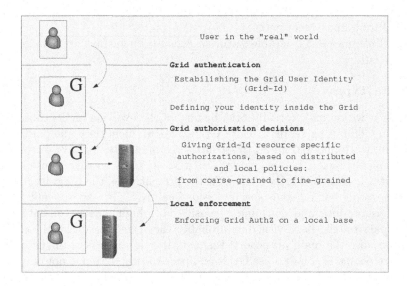

Fig. 1. The security process

3.1   Authentication

Grid computing is, in its essence, about bridging organizational boundaries. In order to do so, we can report here two commonly identified solutions: *virtual organizations* [13] and *federated trust*.[2] We are not going to explain the difference between these two models, because they are quite theoretical and in practice it is often hard to distinguish the boundaries between them. Grid users are traditionally organized in VOs.

In a Grid environment the authentication model is normally based on the concept of *trusted third parties* (TTPs): the first link in the authentication chain is the certification authorities (CAs), which in practice are trust anchors for VOs. This model makes use of the Public Key Infrastructure (PKI) technology: CAs issue X.509 certificates, where essentially a unique identity name and the public key of an entity are bound through the digital signature of that CA. It is possible that some GDMS may require further security controls, but these issues are out of the scope of a Grid authentication service, because they suppose a specific contract between the user and the resource, outside the Grid security infrastructure.

---

[2] For more information, see http://www.projectliberty.org/

An authentication service must define distinctly the Grid identity of any user: this mean that every user inside a Grid is given a background, a description. With description we mean not only user's VOs, but his/her role inside every VO he is member of. In the proceeding of this paper we will refer to this kind of enhanced authentication as of "Grid authentication".

Role Based Access Control. Access Control technologies has evolved from two fundamental types: Discretionary Access Control (DAC), and Mandatory Access Control (MAC). DAC permits the granting and revoking of access control privileges to be left to the discretion of end users, typically the resource owner. MAC is a way of restricting access to objects based on the sensitivity of the information contained in the objects. These policies aren't well suited for VOs authorization requirements, because we need to take access decisions on the basis of the roles that individual users have as part of an (Virtual) Organization. In the Role Based Access Control (RBAC) [5] user access rights are defined by roles in the form of user attributes, letting a separated management access control policy defining what roles are allowed to do what actions on resources. The roles represent typically organizational roles such as secretary, manager, employee, etc. In the authorization policy, they are given a set of permissions, and each user is then assigned to some or more roles. When accessing a target, a user establishes a session and, during it, he can request the activation of some of the roles he is authorized to play. After that, the user will be represented by his/her roles, and so the authorization framework will deal with roles rather users themselves.

We present here two existing estensions. The first is the *hierarchical* RBAC model, which is just a more sophisticated RBAC type, in which the senior roles inherit the privileges of the more junior roles. For example, there might be the following hierarchy:

$$\text{employee} \leq \text{programmer} \leq \text{manager} \leq \text{director}$$

Giving the role "programmer" some permissions means that managers and directors will inherit them. The hierarchical extension to RBAC fits very well the Grid VO requirements, and so we assume that the Grid end systems, like the storage ones, will be able to enforce the capabilities applied to VO roles in a hierarchical fashion.

The second RBAC extension is the *temporal* RBAC model (TRBAC) [6], which supports periodic role enabling and disabling, and temporal dependencies among such actions. Consider for example the case of a part-time staff in a company: what we want to do, is to give him authorization only on working days. With TRBAC, we can assign the part-time staff a role, and enable it only during a temporal interval. The role enabling/disabling depends on some requirements, that can be used to constrain the set of roles that a particular user can activate at a given time. Enabling/disabling actions can be given a priority to help in solving conflicts.

An RBAC system will become a must to manage the future Grid authorizations. Without it, the wide mutable nature of VO-like systems would become a nightmare for all systems administrators, who should take care of granting every single user with his/her capabilities. RBAC simplify the VO's administrator life too, because they have just to assign every user with a somewhat restricted set of roles. In this way the user's identity is managed at VO-level, while the end-systems deal with roles only. We can have the right granularity level with the less possible effort.

Anyway, this isn't perfect yet: assume that a user, like

$$User = John_{Doe}/VO = NeVO/Roles = ExRole, NeRole$$

is trying to do something nasty, for example he is using his *ExRole* capabilities to store a malware in his role shared space. If the system (or the administrator) recognizes it, it should be possible for him to boot that user from his resources without affecting his entire VO/role. There are two possibilities:

- if the end-system doesn't deal with the users authentication names, the only possibility is to do a report to the *NeVO* VO, asking it to reject that user: to do this, there is first the need to recognize the user, and this isn't practical nor fast.
- The second and best option is to let the end-systems to know the effective user names, although the policy end-systems should only use the VO/roles associations to determine the capabilities. In other words, the end-systems authorization frameworks should use the effective user names only for in-depth security reasons.

State of the Art: VOMS. In 3.1 we have stated that, from our point of view, a "Grid authentication" should give a user a complete background and identification. Actually, a framework that can be used to reach this objective is the Virtual Organization Membership Service [8], which is an accepted authentication and authorization framework into existing Grid projects, like for example EGEE[3] and OSG.[4]

VOMS is an Attribute Authority (AA). Users can be organized in a hierarchical structure with groups and subgroups, thus implementing a hierarchical RBAC system. To allow for more flexibility, users are also characterized by two other sets of credentials: roles and capabilities. Roles are used to specify the users' properties as members of some groups. The main difference between groups and roles is that a user can choose which of his roles are to be listed in his credentials, while all his groups are always specified. Capabilities are expressed as free-form strings of characters, and can be used to describe the user's special characteristics. VOMS is traditionally presented as an authorization framework, but in this paper we introduced it as an authentication one. The reason besides this choice is that VOMS is used to define the "Grid user identity" (it can provide a "Grid authentication"), which is not a grant for authorization on any end system: the enforcement of these VO-managed attributes at local level must reflect

---

[3] http://public.eu-egee.org/

[4] http://www.openscienceGrid.org/

the agreements between the VO and the Resource Provider (RP). However it should be possible for an RP to override the permissions granted by a VO, for example banning unwanted users.

3.2   Authorization

Authentication frameworks and Attribute Authorities (AA) can provide a coarse-grained granularity to identify the users' roles and background in a Grid. An authorization system can make use of these information for fine-grained access decisions, using a policy authorization service.

Evaluating the Policies. In this section we will explore the policy interactions and their relations with the Privilege Management Infrastructure (PMI) introduced in section 1.2. First of all, we have to remind some of the requirements for a policy authorization service:

- a future authorization service will be based on a recognized policy expression language and exchange format, and will use a Request/Response protocol to allow intra-site and multiple site scalability. This implies the investigation for the use of "standard" format languages and protocols.
- It will use the principle of ownership in respect to the policy and decision making precedence. This means that the final decision will always reside with the resource owner. It should be able to explicitly accept or reject policies from other domains, and to distribute them.
- A future authorization service will separate authorization infrastructure from the policy itself, providing only secure environment and mechanism for site-authority controlled policy enforcement. The policy evaluation engine will be implemented as a separate service that will be able to call external separate decision points.

We can give a more formal specification for this requirements, using the following definitions:

Policy: The combination of rules and services, where rules define the criteria for resource access and usage.

Policy Decision Point (PDP): The point where decisions about the policies are made. It evaluates *applicable policies* and renders *authorization decisions.* In a loosely coupled distributed environment like Grid, a local to a resource (designated) PDP can call other PDPs requesting for evaluating policy components related to their domain of authority to provide a final decision.

Policy Enforcement Point (PEP): The point where the policy decisions are actually enforced. This is the system entity that performs *access control,* by making *decision requests* and enforcing *authorization decisions.* From a data management perspective, this means that every storage resource should have a local PEP to enforce the policy decisions.

Policy Authority Point (PAP): The point that owns the authority over the PDPs. We should remind that sometimes PAP indicates the *Policy Administration Point*, which is the system entity that creates and administer the policies.

Policy Information Point (PIP): The system entity that act as a source of attribute values.

The PDP-PEP interaction is the key for a good policy distribution. There are two possible basic implementations, the pull model and the push model. The pull model is the more used one, in which a supplicant first ask for the resource PEP to authorize himself, and then the PEP ask to an external PDP for the final decision. We can see a brief example to clarify the way these policy points interact each other. To allow user access on a storage resource, for example a SRM[1] implementation, the storage agent requests via his own PEP an authorization decision from a designated PDP, that evaluates the authorization request against the policy defined for the request, resource and user attributes/roles. During the policy evaluation, the PDP may also request specific user attributes from a Policy Information Point (PIP), or asking an authentication service for user identity confirmation. It should be noted that these controls are a burden for a high percentage of the actual Grid data, but should be a must for some of the future Grid storage uses: data owners and the system administrators should be able to choose how much security controls will be needed. When the PDP identifies the applicable policy instance, it collects the required context information, evaluates the request against the policy, and communicate the decision back to the PEP. After receiving a PDP decision, the PEP conveys the service request to the resource, that may also have a locally determined policy implying additional restrictions on resource usage and/or access. All these communications can be secured using cryptographic technologies like SSL/TLS or MLS.

In essence, when making an authorization decision, we should be able to combine information from a number of different sources. In other words, policies should be defined at different levels, like VO, site, or other stakeholders. It should be noted that every level should have the permission to define different kind of policies, and that sometimes they could overlap each other. For example, a VO-level PDP could force some of his self-managed group/role to not exceed a disk quota of 100 Mbytes, but a resource-level PDP should impose a more restrictive permission. We think that every controversial decision should be resolved in favor of the local decision point, which could be the PEP closest match. This PDP, called "Master PDP", composes the final decision, optionally contacting other PDPs. In figure 2, we show a possible interaction flow between Policy Points.

The solution presented above has known performance problems: requesting a remote PDP decision involves the use of time and resource hungry components, such as building a remote SSL/TLS connection, message parsing, possible remote policy request and PDP/AuthZ service invocation [11]. For this reasons, there's the need for investigation over the PDP-network topology, in order to avoid useless communications. This trade-off can be resolved using distributed policy caching, combining pull and push operation models, using short-validity authorization tickets, or implementing a policy guessing mechanism. In addition,

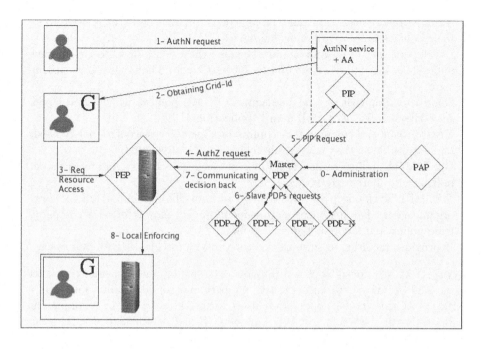

Fig. 2. Interactions between Policy Points in an all-round authorization process. Not all of the shown communication are mandatory.

everyone of the listed services should be a bottleneck for the entire authorization framework; in this situation, consider for example the devastating effects of a Denial Of Service (DOS) attack to anyone of the listed policy points.

Using a standard policy language. Nowadays, the language for writing access control polices that best fit the listed requirements is the eXtensible Access Control Markup Language (XACML) [20], which is an XML based technology developed and standardized by Organization for the Advancement of Structured Information Standards (OASIS).[5] It should be considered a "de facto" standard for expressing policies. XACML includes an access control language, a processing environment and a request-and-response protocol that let developers write policies that determine what users can access on a network or over the Web. XACML can also be used to connect disparate access control policy engines. Every policy is defined for the target triad "Subject-Resource-Action". The processing environment assumes interactions between the policy points described in the previous section.

XACML has many benefits over other access control policy languages:

– One standard access control policy language can replace dozens of application-specific languages.

---

[5] http://www.oasis-open.org/

- Administrators save time and money because they don't need to rewrite their policies in many different languages.
- Developers save time and money because they don't have to invent new policy languages and write code to support them. They can reuse existing code.
- Good tools for writing and managing XACML policies will be developed, since they can be used with many applications.
- XACML is flexible enough to accommodate most access control policy needs and extensible so that new requirements can be supported.
- One XACML policy can cover many resources. This helps avoid inconsistent policies on different resources.
- XACML allows one policy to refer to another. This is important for large organizations. For instance, a site-specific policy may refer to a company-wide policy and a country-specific policy.
- It provides facilities to support the core and hierarchical RBAC approach.

Anyway, XACML doesn't define protocols or transport mechanisms to protect the message security with authenticity, integrity and confidentiality. Full implementation of this model depends on use of other standards, for example the OASIS Security Assertion Markup Language (SAML) [19] [21]. SAML is an XML standard that supports web single sign on, attribute-based authorization and securing web services. There are threes basic SAML components: assertions, protocol, and binding. Assertions can be one of three types: authentication, attribute, and authorization. Authentication assertion validates the identity of the user. The attribute assertion contains specific information about the user, while the authorization assertion identifies what the user is authorized to do. The protocol defines how SAML request and receives assertions. There are several available binding for SAML, that define how message exchanges are mapped to SOAP, HTTP, SMTP and FTP among others.

State of the Art: Gridmap File, CAS, G-PBox. One of the first attempt to provide authorization in Grid was in the form of the Globus Gridmap File. This file was simply a list of the authorized user, identified by a distinguished name, and the equivalent local user account name they are to be mapped into. This solution is infeasible for the next generation Grids, because the resource owner can't set a policy for who is allowed to do what, and maximize the workload of the resource administrator who must keep track of all the authorized users. This system isn't scalable nor flexible [3].

The Globus team developed the Community Authorization Service (CAS) [9]. CAS allows for a separation of concerns between site policies and VO policies. It allows the resource owner to grant access to a portion of his/her resource to a VO. The CAS server acts as a trusted intermediary between VO users and resources: the users first contact the CAS asking for a permission to use a resource, the CAS server consults its policy, and grants or deny the access. CAS does not issue Attribute Certificate's (AC), but whole new proxy certificates, and this isn't a good solutions, because a security system should use standards. Another problem is that CAS completely remove control from site administrators, and that it requires a VO to know everything about the configuration of its farms.

One of the most interesting authorization framework is the Grid Policy Box (G-PBox),[7] which can be used for the representation and management of policies for Grid infrastructures. It's based upon the composition of modular objects, Policy Boxes (PBox), which are policy repositories hierarchically-distributed to independent administrative-based layers, each containing only policies regarding itself. In G-PBox, there are PBoxes at VO, domain, farm and site level, with the possibility to have sub-farm levels. Each and every client that wants to be policy-aware, has a configured PBox that will be contacted whenever a policy decision is required. From a theoretical point of view, we can think at every PBox as being a Policy Authority Point containing a Policy Decision Point, while every resource should have a local Policy Enforcement Point. In G-PBox, the policies are defined using XACML.

3.3   Local Security Enforcement

There are basically two ways to enforce access control over data: the first is to allow Grid access only, the second is to allow local access in parallel to Grid access.

In the *Grid enforced security model*, users can access their files only via Grid tools and services. As stated in [14], "the easiest way of implementing this is to assign all files in a storage element to a service userid, for example *gstorage*, and to add a component, which runs under this identity and interacts with the user": the users should go through the Grid middleware services to gain access to their files. With this type of service, we can easily have a standard authorization service for all the Grid resources, with uniform security semantics, that can take authorization decisions like a centralized authorization service. This model gives support for resources with weak local authorization mechanism.

In the *Site enforced security model*, if we want to allow local access in parallel to Grid access, we have to implement a mapping from Grid identities to local userids. If a Grid service has to act on the user's behalf, then it needs the user's credential to be delegated. With this model, the site storage administrator has full control over his resources, and he will be able to use the local authorization mechanism he prefers.

The choice from a Grid user's point of view should be the Grid security model, because it integrates the site peculiarities into a uniform security model, where every Grid storage site looks the same. Although this last point could be reached also in the site enforced security model using an additional layer for the standardization, we have to remind that an external security service will let site administrators to administer their own local security, in a site technology independent fashion. The real problem besides this is that the Grid security model is not acceptable by some sites due to their local policies, and in addition existing security infrastructure can't be replaced overnight. Each domain typically has its own authentication and authorization infrastructure that is reputed secure and reliable, and site administrators won't replace it in favor of a single new model or mechanism. From the beginning of the Grid computing one of the fundamental requirements was to let every site to use his own security mechanism, and this implies the use of a site enforced security model. In the Globus Toolkit [17], gateways are used to translate between the common GSI infrastructure and local site mechanisms, for example Kerberos Identities or local UNIX users and groups. In LCMAPS [18], we can map Grid users to local ones, and primary and secondary local groups, which are predefined by the resource owner: LCMAPS is used to delegate some global Grid credentials to the local site security system, in this case the UNIX uid/gid match, with the possibility to add ACLs if the file system (and the kernel) can handle them.

Access Control List. Access Control List (ACL) is a means of determining the appropriate access rights to a given object, depending on certain aspects of the process that is making the request, principally the process's user identity. This is a deliberately general definition, because ACLs have been implemented in many ways in different environments.

The POSIX.1e ACLs [12] are an estension of the POSIX.1 permission model, the standard 9-bit access permissions of the UNIX systems. The extended ACLs support more fine-grained and complex permission scenarios, that are difficult or impossible to implement with the minimal model. Unfortunately, the work behind ACLs never became a POSIX formal standard, and at the time of writing there's a wild mix of implementations with subtle differences and incompatibilities. We aren't going to explain how they work, we just say that they can be applied to files and directories, increasing flexibility and security.

For our purposes, the worst problems come when we have to preserve permissions in a distributed system: it's very difficult to implement a system able to preserve as much information as possible. There are a number of complications that make the operation prone to implementation errors, especially when we have different kernels and file systems. The semantics of ACLs differ widely among UNIX systems alone, not to speak of non-UNIX ones. A full ACL support over any kind of distributed system requires a mechanism so that all access decisions are performed in a way that honors ACLs: this means that every remote site should have a system-ACL support. Using only fs-ACLs will will lead to interoperability problems, although the good part is that they are automatically enforced on the end systems. In a Grid environment, there's the need to translate the resource-dependent ACLs in a common format.

In a Grid, we need to map a global security mechanism into a local one, which is independent from the "Grid security infrastructure". This brief discussion on ACL wants to remark the fact that every Grid resource should expose its security capabilities because not all of them are able to enforce the security and privacy requirement of some data types, due to the lack of security potentiality.

A PEP implementation can be used to map a Grid-ID in a local account, using File System ACL to enforce the Grid Authorization response. For our purposes, an example of an SRM implementation that can act as a PEP is StoRM [4].

4   Building a Grid Data Access Framework

Grid middleware should define a Grid security framework, encompassing both authentication and authorization in a standard way, and interfacing with local storage elements. Ensuring integrity, confidentiality and interoperability between heterogeneous systems can be achieved using a Web Service Architecture [22], which is an incarnation of a Service Oriented Architecture (SOA) in the context of the World Wide Web. SOA is the leading architectural style for building the current and future generation Grid technologies. Protocols based on Web Services provide important benefits for Grids, particularly in avoiding the tendency that proprietary binary protocols frequently become closely tied to particular implementations or languages. As stated in [10], "the Grid authorization model should be built on top of upcoming standards in the area of authorization, e.g. XACML, SAML, and WS-Authorization".

We think that a Grid authentication model should include an attribute authority that issues attribute assertions, and that a Grid authorization model should be built over a standard policy language. Different Policy Points should make decisions based on initiator identity and attributes, and so what is needed is a standard attribute language, that allow for interoperability between AA's and and PDPs. In addition, we have to remember that there may be several authorities that assert attributes for users, including other users. In section 3.2 we have already outlined the Policy Points actions, using the pull and push models. We should extend these variants thinking at the interactions between an authorization mechanism and a AA. From a technological point of view, a number of methods for requesting and encoding attributes already exist: for example X.509 Attribute Certificates[15], SAML[19] Attribute Assertions and XACML[20] Attributes. Since the emerging of the use of XACML for policy expression and the capabilities of SAML for attribute encoding, we should be able to combine this upcoming standards building a Grid Data Access Framework.

5   Conclusions

In the past sections we explored the technologies behind security in a Grid environment, focusing on the Grid Data Management Systems security aspects. With this paper, we didn't want to propose a definite solution, instead we defined some of the requirements and boundaries that we'll guide our future works in this field. In the following months, we will outline a Grid-based RBAC model for accessing distributed data, and we'll follow the implementation of a multipolicy authorization framework, based on XACML and SAML specifications. We'll define policies applicable to GDMS, their distribution and consuption, and interactions with monitoring and accounting services. At the same time we'll study methods to increase the performance of the whole authorization system.

References

1. J. Gu A. Shoshani, A. Sim. Storage resource manager: Essential components for the grid. 2003.
2. D. Chadwick. An x.509 role-base privilege management infrastructure. Technical report, 2002.
3. D. Chadwick. Authorization in grid computing. *Information Security Technical Report*, (10):33–40, 2005.
4. E. Corso, S. Cozzini, F. Donno, A. Ghiselli, L. Magnoni, M. Mazzucato, R. Murri, P.P. Ricci, H. Stockinger, A. Terpin, V. Vagnoni, and R. Zappi. Storm, an srm Implementation for lhc Analysis Farms, Computing in High Energy Physics. In *In Proceedings of the International Conference on Computing in High Energy and Nuclear Physics (CHEP2006), Mumbai, India,* Feb 2006.
5. S. Gavrila D.R. Kuhn R. Chandramouli D. Ferraiolo, R. Sandhu. Proposed nist standard for role-based access control. *ACM Transactions on Information and System Security (TISSEC),* (3):224–274, 2001.
6. E. Ferrari E. Bertino, P. A. Bonatti. Trbac: A temporal role-based access control model. *ACM Transactions on Information and System Security (TISSEC),* (4): 191 – 233, 2001.
7. A. Caltroni et al. *G-Pbox: a Policy Framework for Grid Environments.* INFN Grid-it.
8. Alfieri et al. Voms, an authorization system for virtual organizations. In *In proceedings of 1st European Across Grid Conference.*
9. L. Pearlman et al. The community authorization service: Status and future. In *In proceedings at CHEP03, March 24-28 2003, La Jolla, California.*
10. Nagaratman et al. Security architecture for open grid services. memo GWD-I, GGF OGSA Security Workgroup, 2002m revised 2003.
11. Y. Demchenko et al. *Job-centric Security model for Open Collaborative Environment,* pages 69–77. IEEE Computer Society, 2005.
12. A. Grunbacher. Posix access control lists on linux. In *Submitted for publication at the USENIX ATC, San Antonio, Texas, June 2003.*
13. S. Tuecke I. Foster, C. Kesselman. The anatomy of the grid: Enabling scalable virtual organizations. *International J. Supercomputer Applications,* (15(3)), 2001.
14. A. Frohner P. Kunszt. glite data management security model disussion, 2005.
15. R. Housley S. Farrel. Rfc3281: An internet attribute certificate profile for authorization. Technical report, 2002.
16. EGEE security JRA3. Global security architecture. 2004.
17. The Globus security team. Gt 4.0 security. http://www.globus.org/toolkit/docs/ 4.0/security/, 2005.
18. M. Steenbakkers. Guide to lcmaps version 0.0.23. http://www.dutchGrid.nl/ DataGrid/wp4/lcmaps/edg-lcmaps_gcc3_2_2-0.0.23/, 2003.
19. OASIS SAML TC. Oasis security assertion markup language (saml) tc. http://www.oasis-open.org/committees/tc_home.php?wg_abbrev=security, 2005.
20. OASIS XACML TC. Oasis extensible access control markup language (xacml) tc. http://www.oasis-open.org/committees/tc_home.php?wg_abbrev=xacml# XACML20, 2005.
21. OASIS XACML TC. Saml 2.0 profile of xacml v2.0. http://docs.oasis-open.org/ xacml/2.0/access_control-xacml-2.0-saml-profile-spec-os.pdf, 2005.
22. W3C WG. Web services architecture. http://www.w3.org/TR/ws-arch/, 2004.

Security Requirements Analysis for Large-Scale Distributed File Systems*

Syed Naqvi[1], Olivier Poitou[1], Philippe Massonet[1], and Alvaro Arenas[2]

[1] Centre of Excellence in Information and Communication Technologies (CETIC), Belgium
{syed.naqvi,olivier.poitou,philippe.massonet}@cetic.be
[2] CCLRC Rutherford Appleton Laboratory, UK
a.e.arenas@rl.ac.uk

Abstract. This paper presents an analysis of security requirements of large-scale distributed file systems. Our objective is to identify their generic as well as specific security requirements and to propose potential solutions that can be employed to address these requirements. *FileStamp* – a multi-writer distributed file system developed at CETIC is considered as a case study for this analysis. This analysis yields that the existing range of security solutions can be employed to secure large-scale distributed file systems. However, they should be holistically employed to triumph over the security chinks in the *FileStamp*'s armor.

Keywords: security services, requirements analysis, highly scalable systems, distributed data management.

1 Introduction

The exponential growth in the scale of distributed data management systems and corresponding increase in the amount of data being handled by these systems require efficient management of files by maintaining consistency, ensuring security, fault tolerance and good performance in terms of availability and security. Read only systems such as CFS [1] are much easier to design as the time interval between meta-data updates is expected to be relatively high. This allows the extensive use of caching, since cached data is either seldom invalidated or kept until its expiry. Security in a read-only system is also quite simple to implement. Digitally signing a single root block with the administrator's private key and using one-way hash functions allow clients to verify the integrity and authenticity of all file system data. Finally, consistency is hardly a problem as only a single user, the administrator, can modify the file system.

Multi-writer file systems face a number of operational issues not found in the read only systems. These issues include maintaining consistency between replicas, enforcing access control, guaranteeing that update requests are authenticated and correctly processed, and dealing with conflicting updates.

---

* This research work is supported by the European Network of Excellence **CoreGRID** (project reference number 004265). The network aims at strengthening and advancing scientific and technological excellence in the area of Grid and Peer-to-Peer technologies. The CoreGRID webpage is located at www.coregrid.net

W. Lehner et al. (Eds.): Euro-Par 2006 Workshops, LNCS 4375, pp. 49–60, 2007.

This paper is organized in the following manner: an overview of *FileStamp* distributed file system is presented in section 2. Its generic and specific security requirements are elaborated in section 3. Section 4 presents a detailed account of technologies that can be employed to address the security requirements of the *FileStamp*. Finally some conclusions are drawn in section 5.

2 *FileStamp* Architecture

FileStamp is a distributed file system developed at CETIC with the aim of finding a solution to the problems encountered in multi-writer file systems. It is a highly scalable, completely decentralized multi-writer peer-to-peer file system. The current version of the *FileStamp* is based on Pastis [2] architecture. It aims at making use of the aggregate storage capacity of hundreds of thousands of PCs connected to the Internet by means of a completely decentralized network. Replication allows persistent storage in spite of a highly transient node population, while cryptographic techniques ensure the authenticity and integrity of file system data.

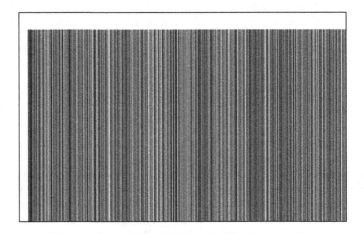

Fig. 1. *FileStamp* Layered Architecture

The layered architecture of *FileStamp* is shown in figure 1. Routing and data storage are handled by the Pastry [3] routing protocol and the PAST [4] distributed hash table (DHT). The good locality properties of Pastry/PAST allow Pastis to minimize network access latencies, thus achieving a good level of performance when using a relaxed consistency model. In Pastis, for a file system update to be valid, the user must provide a certificate signed by the file owner which proves that he has write access to that file.

The format of the Pastis certificate is shown in figure 2. This certificate is issued by the file owner and it grants the write access to a given user. The expiration date allows access revocation.

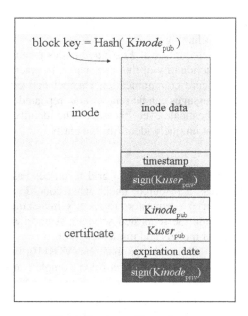

Fig. 2. Pastis certificate format

Authentication of the certificate is performed by the DHT nodes and FS clients. They verify both signatures when storing and/or retrieving a UCB (User Certificate Block).

This certificate has two crucial problems. First, it always gives *write* permission to its users whereas in a real life application, a user may only be given *read* permission while accessing the file. Second, its format is not standardized. It does not correspond with the format of the X.509 certificate and hence it renders compatibility problem with the existing standard credentials. This issue is discussed in detail in section 4.1

3 Security Requirements of *FileStamp*

The security requirements of *FileStamp* are driven by the roadmap of Open Grid Services Architecture (OGSA) [5]. OGSA security model casts security functions as OGSA services. This strategy allows well-defined protocols and interfaces to be defined for these services and permits an application to outsource security functionality by using a security service with a particular implementation to fit its current need.

3.1 Generic Requirements

This is the set of security services that constitutes the fundamental requirements for any data management system.

3.1.1 Authentication

Authentication provides plug points for multiple authentication mechanisms and the means for conveying the specific mechanism used in any given authentication operation. The authentication mechanism may be a custom authentication mechanism or an industry-standard technology. The authentication plug point must be agnostic to any specific authentication technology.

Authentication between two entities of *FileStamp* nodes means that each party establishes a level of trust in the identity of the other party. In practical use an authentication protocol sets up a secure communication channel between the authenticated parties, so that subsequent messages can be sent without repeated authentication steps, although it is possible to authenticate every message. The identity of an entity is typically some token or name that uniquely identifies the entity.

3.1.2 Authorization

Authorization allows for controlling access to grid resources based on authorization policies (i.e., who can access a resource, under what conditions) attached to each service. It also allows for service requestors to specify invocation policies (i.e. who does the client trust to provide the requested service). Authorization should accommodate various access control models and implementation.

In the grid environments, the virtual organisations (VOs) [6] introduce challenging management and policy issues, resulting from often complex relationships between local site policies and the goals of the VO with respect to access control, resource allocation, and so forth. In particular, authorization solutions are needed that can empower *FileStamp* to set policies concerning how resources assigned to the *community* are used without, however, compromising site policy requirements [7].

3.1.3 Availability

Availability of a requested data item is an important performance parameter. A well-known technique for improving availability in distributed systems is replication. If multiple copies of data exist on independent nodes, then the chances of at least one copy being accessible are increased. Aggregate data access performance will also tend to increase, and total network load will tend to decrease, if replicas and requests are reasonably distributed.

3.1.4 Confidentiality

Confidentiality is the property that information does not reach unauthorized individuals, entities, or processes. It is achievable by a mechanism for ensuring that only those entitled to see information or data have access to that information. The confidentiality requirement includes point-to-point transport as well as store-and-forward mechanisms.

3.1.5 Integrity

Integrity is the assurance that information can only be accessed or modified by those authorized to do so. Data integrity is a nontrivial problem especially when storage hardware and networks are not perfect. Data loss and corruption must be timely caught and swiftly fixed. As systems grow in size and complexity, problems may pass unnoticed until recovery becomes difficult and expensive.

3.2 Specific Requirements

This is the set of security services that are specifically needed for *FileStamp*. These services complement the generic set of security services and are needed to enhance the quality of security of the data management system.

3.2.1 Resilience

Resilience is an important requirement as the grid links and nodes are very dynamic in nature and may change over the time. *FileStamp* security architecture should remain intact and should deliver the promised level of security assurances even if its composition changes over the time. The resilience provides an abstraction layer to hide the architectural changes from the overall security architecture.

3.2.2 Data Lifecycle Management (DLM)

Data Lifecycle Management (DLM) is the process of managing data throughout its lifecycle from conception until disposal across different storage media, within the constraints of the entire process. The lifecycle is the time from the moment data is created until it is deleted or stored indefinitely. Security assurances require spanning the entire lifecycle of data. *FileStamp* should ensure that the data contents will be protected from the malevolent entities throughout its lifecycle.

3.2.3 Fault Tolerance

Fault tolerance is a desirable feature especially when transfers of large data files occur. Protocols such as GridFTP [8] allow for *resuming* transfers from the last byte acknowledged. Overlay networks provide *caching* of transfers via store-and-forward protocols. However, caching reduces performance of the overall data transfer and the amount of data that can be cached is dependent on the storage policies at the intermediate network points.

4 Solutions for the *FileStamp* Security Requirements

In this section, solutions to the security requirements of *FileStamp* are provided. The premier objective of this section is to identify the range of existing technologies that can be employed in *FileStamp*. However, solutions to all the security requirements do not already exist. In situations where existing solutions are either inadequate or non-existent, we have discussed the potential solutions and have given reference to our ongoing work in that direction.

The aim of this approach is to workout new solutions which are needed for the security architecture of the grid data management systems without reinventing the wheel.

4.1 Authentication

Most of the current grid tools are built on Grid security Infrastructure (GSI) [9] or Secure Hyper Text Transfer Protocol (HTTPS) [10], both of which use X.509 certificates [11] for securely establishing a grid identity [12].

Other schemes include PGP keys [13], SSH keys [14], and SPKI [15] keys and protocols. SPKI focuses on authorization certificates more than identity certificates. SSH is primarily a private/public key mapping with no real attempt to provide global names. The X.509 scheme has a small set of trusted third parties called Certification Authorities (CAs). These CAs are used to sign identity certificates that contain subscriber's public key. This improves the scaling properties of public key distribution in that only the CA's public key needs to be distributed in an out-of-band secure manner. In systems without a trusted third party, such as PGP, each key holder must find some secure way of establishing the association of his identity with his public key, to each party with which he wishes to establish authenticated communication. In the X.509 infrastructure, the individual subscriber's public key can be transmitted in a public key certificate as part of a TLS connection handshake and can be accepted as valid if the certificate is signed by a trusted CA. Another feature of the X.509 infrastructure is that it supports multiple independent CAs. In a Grid each site may chose which CAs it will accept for binding domain names and public keys.

We recommend the use of X.509 infrastructure for *FileStamp*. It will not only standardize its authentication mechanism (unlike owner's issued certificates) but also facilitate its interactions with the grid world. *FileStamp* with X.509 infrastructure will be easily integrated with any grid platform. Initially, a local CA can be created that will deliver the standard X.509 certificates to the bona fide users of *FileStamp*. Later the certificates of other CAs (such as Belgian Grid CA [16]) can be used for authentication purposes.

4.2 Authorization

FileStamp may simply employ local mapping of the users (like UNIX authorization matrix). This mapping also serves as an access control check – access to the resource is denied if the user is not listed in the local mapping configuration. In this scheme, once the user is mapped to a local identity, local policy management and enforcement mechanisms constrain the user's actions to those allowed by local policy. This approach allows the local operating system to act as a sandbox. Thus, administrators can use normal policy administration tools to configure policy.

This simple approach has the advantage of being easy for site administrators to understand and configure because it uses existing local policy management and enforcement mechanisms with which the administrator is presumably already familiar. However, in the context of the grid environment, this approach has several shortcomings (such as scalability, lack of expressiveness, consistency of policies, etc.).

These problems are addressed in the Community Authorization Service (CAS) [17]. The idea behind the evolution of CAS is inspired from the Role Based Access Control (RBAC) [18]. CAS allows for a separation of concerns between site policies and VO policies. Specifically, sites can delegate management of a subset of their policy space to the VO. CAS provides a fine-grained mechanism for a VO to manage these delegated policy spaces, allowing it to express and enforce expressive, consistent policies across resources spanning multiple independent policy domains. CAS implementations are built on the Globus [19], thus allowing for easy integration of CAS with existing Grid deployments.

Other solutions include VOMS [20], Akenti [21], and PERMIS [22]. VOMS (Virtual Organization Management Service) and CAS are similar architecturally in that both issue policy assertions to a user that the user then presents to a resource for the purpose of obtaining VO issued rights. The primary difference between the two systems is the level of granularity at which they operate. The policy about what memberships a user has is centralized in the VOMS server, but the policy regarding exactly what rights those memberships grant is distributed among the sites. CAS assertions provide the rights directly and do not need interpretation by the resource. This complete centralization of policy can achieve better consistency especially in situations where policies are changing dynamically.

Akenti and PERMIS, while having differences in implementation and features, are architecturally similar in that they provide a resource with an authorization decision in regards to a request. While the CAS implementations provide simple authorization decision functionality, they are limited to supporting CAS policy assertions and do not have as rich a feature set as either Akenti or PERMIS. It is possible that either of these systems, with some modifications, could be used to provide resource-side functionality for CAS (i.e., parse the CAS assertion and use it to authorize the user's request.)

We recommend the use of CAS with the implementation of a local authorization server for *FileStamp*. Local authorization server would accept authorization queries from request servers, apply all applicable local and community policies, and return a yes or no answer. This authorization server would need to be highly trusted by the resource server and highly available. This service could potentially take CAS credentials, forwarded by the resource, and use their credentials in making its decision, or it could contact the CAS server itself. Such a server could be implemented by using Akenti or PERMIS.

4.3 Availability, Confidentiality, and Integrity

Grid technologies enable transparent access to a wider resource pool, across organizations as well as within organizations; they can be used as a building block to realize stable, highly reliable execution environments. In such a complex environment, policy-based autonomous control and dynamic mobility are keys to realizing systems that are highly flexible and recoverable. Availability is often not considered in literature, when it comes to a model design. Nevertheless, in a production environment we cannot expect user not having assurances regarding the availability of what they pay for. GSI provides mechanisms to grant availability of data owned by a user on a remote resource. These are achieved by means of secure communication protocols, such as HTTPS. As far as services availability is concerned, Globus relies on a dedicated module that manages a limited set of grid events.

Use of some adequate encryption technologies is indispensable to guarantee the secure communications across the grid nodes which assure the confidentiality and also integrity. Encryption indirectly assures the availability too; however, the protection against the denial of service attack is addressed in the security policy. There exist a range of encryption technologies from HTTPS (where a layer of security is added on the top of HTTP) to Secure Hash Algorithm (SHA) [23] (where it is computationally difficult or impossible to hack and the integrity check – checksum – is also performed).

Figure 1 shows graphic representations of these two encryption schemes. In figure 1a, the layered architecture of HTTPS is shown. Figure 1b depicts how the SHA works. The quick comparison of these two techniques show that SHA seems quite powerful as it require considerable computing power to break the algorithm; however, in the specific context of the grid applications notably *FileStamp*, we need to consider the overhead incurred due to the encryption operations. Large datasets will consume enormous

Fig. 1a. HTTPS Architecture

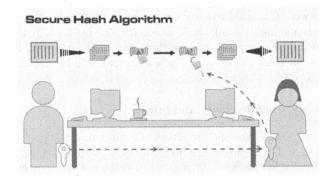

Fig. 1b. SHA Architecture

computing cycles for the SHA processing and HTTPS may not be considered as dependable solution especially when network connections are not reliable.

We recommend the use of encryption technology for *FileStamp* as the data movements across the grid nodes will be subject to potential attacks if there will be plain text data exchange between the nodes. However, the selection of some specific encryption technology is a tricky issue that depends on the nature of data (required security level of the data movement) and the affordability of the total cost of the encryption algorithms. A simple technique such as HTTPS can be employed for generic situations and some more powerful techniques can be used for providing higher level of security assurances. SHA consumes enormous amount of computing power but in return it provides highest security assurances.

4.4 Resilience and Fault Tolerance

General trend for the attainment of resilience and fault tolerance in the distributed systems is to maintain ample number of replicas of the dataset. When some node fails then the load/job is transferred to some other node. The quality of service depends on how efficiently the system recognizes the faulty nodes and how transparently the jobs are migrated from the faulty nodes to working nodes without interrupting operations.

In order to assure resilience and fault tolerance features, *FileStamp* should be able to negotiate the terms of security parameters with the nodes so that new replicas be created if the set of nodes expands resulting in the need of more replicas; or failure of some existing nodes bearing replica sets need to be compensated by generating new replicas.

We recommend the phased approach (as mentioned in [24]) to deal with the resilience and fault tolerance issue. According to this approach:

1. In Phase I, the service providers that need to interact are identified. It is generally assumed that this is undertaken through a manager entity – which is forming the VO in order to undertake a particular activity.
2. In Phase II, the identified providers are asked to join the VO. This phase may involve negotiation between the manager entity and the providers (or directly between the providers) to ensure that a Service Level Agreement (SLA) is established between the entity and each provider (or directly between the providers).
3. In Phase III, the providers interact to perform the particular activity desired by the manager entity.

A set of protocols is needed to perform these negotiations. Negotiation protocols are the set of rules that govern the interaction. They are required to realize SLA-aware resource management system.

We recommend the use of *Service Negotiation and Acquisition Protocol (SNAP)* [25] as negotiations protocol. SNAP is structured around the negotiation of SLAs to solve the negotiation problems at run-time. When SNAP is used to submit a file transfer job to a community scheduler, the scheduler understands that a transfer requires substantial storage space on the destination resource, and substantial network and endpoint I/O bandwidth during the transfer. The distributed applications (common in Grid environments) exacerbate the coordination problems of community schedulers. Not only do SLAs coordinate use of resources by mutually distrustful schedulers, they also coordinate the use of distrustful resources for a single application goal. The file transfer emphasizes such distributed goals by requiring real-time coordination of significant endpoint and network capability.

4.5 Data Lifecycle Management (DLM)

Data lifecycle management (DLM) is a policy-based approach to managing the flow of an information system's data throughout its life cycle – i.e. from creation and initial storage to the time when it becomes obsolete and is deleted. Security assurances require spanning the entire lifecycle of data. Existing Grids are already managing huge quantities of data [26]. Since Grids maximize the utilization of computing resources, their potential to generate new data and consume storage is very high, making storage capacity and DLM critical issues. By targeting data to appropriate storage media (primary disk storage, secondary serial advanced technology attachment (ATA) storage, tape, etc.) DLM solutions can influence on the overall protection of the data besides significantly reducing the cost of Grid storage infrastructures. *FileStamp* should ensure that the data contents will be protected from the malevolent entities throughout its lifecycle.

We recommend a two-tier approach to handle the DLM issue in the *FileStamp* system:

First, the security policy should explicitly mention the desired lifecycle of the data being managed by the *FileStamp* system. The dynamic nature of the grid environments does not permit some rigid definition of any parameter including security; however, the security policy of a VO is generally fixed for that VO and hence the VOs using the *FileStamp* should include a formal description of the stage where the data generated by the VO operations be destroyed from the storage devices.

Second, FileStamp should also employ some secure storage management technique such as HSM (Hierarchical Storage Management) [27]. HSM is policy-based management of file backup and archiving in a way that uses storage devices economically and without the user needing to be aware of when files are being retrieved from backup storage media. The hierarchy represents different types of storage media, such as redundant array of independent disks systems, optical storage, or tape, each type representing a different level of cost and speed of retrieval when access is needed.

5 Conclusions

Global connectivity of computing and storage resources opens up the possibility of misusing information to a degree never seen before. The objective to facilitate use of these resources by protecting them against any misuse must, however, be realistic given the current technical infrastructure. It is important that the security technologies be integrated in these systems from the inception stage rather than considering them as add-on optional features. Security issues should not be overlooked while designing these systems as they are critical to the success of these scalable distributed systems.

In this paper, the security requirements of large-scale distributed file systems are addressed. The *FileStamp* multi-writer distributed file system is considered as a case study for this analysis. Various security requirements are identified and the potential solutions corresponding to these requirements are proposed. However, it is important to remember that the analysis of security requirements is a process, the risk and threat pictures are always changing, and their analysis needs to be continuously updated. In other words, overall infrastructure of large-scale distributed file systems should be subject to constant review and upgrade, so that any security loophole can be plugged as soon as it is discovered.

References

1. Dabek F., Kaashoek M., Karger D., Morris R., and Stoica I., *Wide-Area Cooperative Storage with CFS*, In the proceedings of 18th ACM Symposium on Operating Systems Principles (SOSP'01), chateau Lake Louise, Banff, Canada, October 2001
2. INRIA Project PASTIS http://regal.lip6.fr/projects/pastis/pastis_fr.html
3. Rowstron A. and Druschel P., *Pastry: Scalable, Distributed Object Location and Routing for Large-Scale Peer-to-Peer Systems*, Proceedings of the IFIP/ACM International Conference on Distributed Systems Platforms (Middleware), 2001, pp 329-350

4. Druschel P., and Rowstron A., *Past: Persistent and Anonymous Storage in a Peer-to-Peer Networking Environment*, Proceedings of the 8th IEEE Workshop on Hot Topics in Operating Systems (HotOS-VIII)? 2001), pp. 65-70

5. Welch V., Siebenlist F., Foster I., Bresnahan J., Czajkowski K., Gawor J., Kesselman C., Meder S., Pearlman L., Tuecke S., *Security for Grid Services*, Proceedings of the 12th IEEE International Symposium on High Performance Distributed Computing (HPDC'03), 2003

6. Foster I., Kesselman C., and Tuecke S., *The Anatomy of the Grid: Enabling Scalable Virtual Organizations.* International Journal of High Performance Computing Application, 15 (3), pp. 200-222, 2001

7. Foster I., Kesselman C., Pearlman L., Tuecke S., and Welch V., *The Community Authorization Service: Status and Future*, In Proceedings of Computing in High Energy Physics 03 (CHEP '03), La Jolla, California, USA, March 24-28, 2003

8. Allcock W. et al., *GridFTP: Protocol extensions to FTP for the Grid*, GGF Document Series GFD.20, April 2003

9. Foster I., Kesselman C., Tsudik G., Tuecke S., *A Security Architecture for Computational Grids*, ACM Conference Proceedings 1998, ISBN 1-58113-007-4, pp 83-92

10. Rescorla E., *Hyper Text Transfer Protocol (HTTP) over Transport Layer Security (TLS)*, Internet Engineering Task Force (IETF) draft RFC # 2818, May 2000

11. Chokhani S., *Internet X.509 Public Key Infrastructure Certificate Policy and Certification Practices Framework*, Internet Engineering Task Force (IETF) draft RFC # 2527, March 1999

12. Thompson M., Olson D., Cowles R., Mullen S., Helm M., *CA-based Trust Issues for Grid Authentication and Identity Delegation*, Global Grid Forum (GGF) Certification Authority Operations Working Group Community Practices Document, Oct 2002

13. Garfinkel S., *PGP: Pretty Good Privacy*, O'Reilly & Associates, 1994

14. Barret D. and Silverman R., *SSH: The Secure Shell*, O'Reilly & Associates, 2001

15. Ellison C., *SPKI Requirements*, IETF RFC 2692 1999, http://www.ietf.org/rfc/rfc2692.txt

16. The Certification Authority of Belgian Grid Initiative – www.begrid.be/certification.htm

17. Pearlman L., Welch V., Foster I., Kesselman C., Tuecke S., *A Community Authorization Service for Group Collaboration.*, Proceedings of the IEEE 3rd International Workshop on Policies for Distributed Systems and Networks, 2002

18. Ferraiolo D., Cugini J., and Kuhn D., *Role Based Access Control (RBAC): Features and Motivations*, Proceedings of the 11th Computer Security Applications Conference, pp 241-248, New Orleans, LA, USA, 11-15 December 1995

19. Foster I. and Kesselman C., *Globus: A Metacomputing Infrastructure Toolkit*, International Journal of Supercomputer Applications, 11 (2). 115-129. 1998

20. VOMS Architecture v1.1, http://gridauth.infn.it/docs/VOMS-v1_1.pdf, May 2002.

21. Thompson M., Johnston W., Mudumbai S., Hoo G., Jackson K., and Essiari A., *Certificate-based Access Control for Widely Distributed Resources*, 8th Usenix Security Symposium, 1999

22. Chadwick D. and Otenko A., *The PERMIS X.509 Role Based Privilege Management Infrastructure*, 7th ACM Symposium on Access Control Models and Technologies, 2002

23. National Institute of Standards and Technology, *Secure Hash Standard*, Federal Information Processing Standards Publication 180-1, April 17, 1995

24. Olmedilla D., Rana O., Matthews B., and Nejdl W., *Security and Trust Issues in Semantic Grids*, Proceedings of Schloss Dagstuhl Seminar no. 05271: Semantic Grid: The Convergence of Technologies, Dagstuhl, Germany, July 03-08, 2005

25. Czajkowski K., Foster I., Kesselman C., Sander V., Tuecke S., *SNAP: A Protocol for Negotiating Service Level Agreements and Coordinating Resource Management in Distributed Systems*, Lecture Notes In Computer Science; Vol. 2537, Revised Papers from the 8th International Workshop on Job Scheduling Strategies for Parallel Processing, pp 153-183, ISBN:3-540-00172-7, 2002

26. Silicon Graphics Incorporate (SGI), *SGI and Intel on the Grid – Unique Capabilities for Grid Computing*, Whitepaper, 2005

27. Watson, R., *High Performance Storage System Scalability: Architecture, Implementation and Experience*, Proceedings of 22nd IEEE / 13th NASA Goddard Conference on Mass Storage Systems and Technologies 2005, pp145-159, 11-14 April 2005

Coupling Contracts for Deployment on Alien Grids

Javier Bustos-Jiménez[3], Denis Caromel[1], Mario Leyton[1], and José Piquer[2]

[1] INRIA Sophia-Antipolis, CNRS-I3S, UNSA. 2004, Route des Lucioles, BP 93,
F-06902 Sophia-Antipolis Cedex, France
`First.Last@sophia.inria.fr`
[2] Departamento de Ciencias de la Computación, Universidad de Chile. Blanco
Encalada 2120, Santiago, Chile
`{jbustos,jpiquer}@dcc.uchile.cl`
[3] Escuela de Ingeniería Informática. Universidad Diego Portales Av. Ejercito 441,
Santiago, Chile
`javier.bustos@udp.cl`

Abstract. We propose coupling based on contracts as a mechanism to address the problem of exchanging information between parties that require information to work together. Specifically, we show how our approach can be used to couple the deployment of an application with a Grid infrastructure deployment descriptor using ProActive[11,2].

To achieve this, we identify the properties related with information exchange between parties, and we group the properties of interest into typed clauses. We then propose that interfaces can be built using shared typed clauses. If the interfaces between parties are compatible, the coupling of the interfaces can yield a coupling contract. The clauses belonging to the contract represent *what* information can be shared between the parties, and the type of the clause specifies *how* this information will be shared.

Finally, we show how the deployment of applications on the Grid can benefit from the proposed approach. Unfamiliar applications can couple with deployment descriptors to deploy on alien Grids, without modifying or inspecting neither of them.

1 Introduction

Originally, distributed resources were managed using a centralized approach. This has been shown to be unpractical in the Grid. The resources can be numerous, heterogeneous, with distributed ownership, and having different policies [8,14].

The problem of scheduling an application on distributed resources was addressed using different strategies. This generated a diversity of mechanism for resource acquisition protocols (LSF [16], PBS[10], SGE[9], Globus-gram[8], etc.). At that point in time, application developers were forced to choose and bind an application to a specific resource acquisition protocol. Migrating from one resource acquisition protocol to another required modifying the application.

W. Lehner et al. (Eds.): Euro-Par 2006 Workshops, LNCS 4375, pp. 61–73, 2007.
© Springer-Verlag Berlin Heidelberg 2007

Later, new levels of abstractions were introduced which allowed the application developers to abstract the application, not only from the resource acquisition protocol used, but also from other Grid infrastructure details such as communication protocols, software location, etc.[3].

In the current scenario, we can now imagine having repositories of applications and repositories of Grid infrastructures. The problem of finding a suitable Grid infrastructure for an application can be seen as a problem of classified advertisements and matchmaking [12,13] or a problem of database search like UDDI web services [6].

We set sail from this point. Let us imagine two candidate parties (ex: application and Grid infrastructure) that have already been matched. To work together, each party requires and provides information from the other. We propose coupling based on contracts as a mechanism to address the problem of exchanging information in a generic way between unfamiliar parties. Specifically, we show how our approach can be used to couple the deployment of an unfamiliar application with an unfamiliar Grid infrastructure descriptor using ProActive[11]. Therefore, our objective is the deployment of an application on a Grid infrastructure without modifying or inspecting either.

This paper is organized as follows. In section 2 we review the related work. Then in section 3 we explain our coupling proposal, and in section 4 we show how this proposal is applied for deployment on the Grid using ProActive. Finally we conclude and present our future work in section 5.

2   Related Work

The problem of finding suitable resources for a given application have already been addressed by techniques such as matchmaking in Condor [12,13], collections in Legion [4], or using resource management architectures like Globus[5].

In the case of Condor, the resource acquisition is viewed as a three stage process composed of advertisement, matchmaking, and claiming. The requirements are advertised by the involved parties (jobs and resources), suitable matches are found, and finally the claiming of the resources takes place. To achieve the claiming, the advertised information from each party is exchanged.

While this approach has been acknowledged as suitable for finding matches, *how* the advertised information sharing is done has been overlooked. Up to now, techniques like the ones proposed by Condor allow finding suitable matches by specifying *what* information is exchanged, but no mechanism is provided for defining *how* the information exchange should take place.

For example, if an application is looking for n nodes, and a Grid infrastructure can provide m nodes $(n < m)$, then these two parties will be matched. If no *how* semantics are provided for the claiming face, the following scenario could happen: the application could decide to take advantage of the m nodes provided by the infrastructure, while the infrastructure can decide to provide only the n

nodes advertised by the application. The result would be the application trying to use m nodes, while the infrastructure is only providing n nodes. Therefore, a mechanism is required to specify *how* the information exchange takes place.

To address this issue, we propose the addition of a new stage called *coupling*, thus rendering four stages: advertisement, matchmaking, coupling, and claiming. Once the matchmaking has taken place, the semantics of *how* the information will be shared between the parties will be addressed in the coupling face, before the resources are successfully claimed.

Another related approach corresponds to the Web Services Agreement (WS-Agreement) Specification[1], which is about to become a draft recommendation of the Global Grid Forum[7]. The WS-Agreement is a two layer model: Agreement Layer and Service Layer. Many of the concepts introduced in this paper find their reflection in the Agreement Layer. According to the specification "an *agreement* defines a dynamically-established and dynamically-managed relationship between parties", much like the proposed coupling contracts. Also, the proposed coupling interfaces can be seen as *agreement templates* in WS-Agreement, since they are both used to perform advertisement. Additionally, in the same way that interfaces and contracts are composed of clauses, in WS-Agreement templates and agreements are composed of *terms*. Finally, the concept of constraints is present in both approaches.

The similarity of the proposed approach and WS-Agreement Specification is encouraging when we consider that both were conceived independently. On the other hand, the main difference in the approaches is that the definition of a protocol for negotiating agreements is outside of the WS-Agreement Specification scope. Therefore, we believe that WS-Agreement could benefit from the proposed automated coupling approach, built using typed clauses. From the WS-Agreement perspective, typed clauses can be seen as an automated negotiation approach because they provide an automated mechanism for accepting or rejecting an agreement.

3   Coupling Matches with Contracts

In this section we describe our approach for coupling parties (ex: application and descriptor) that require exchanging information to work together.

Figure 1i shows the problematic. Unfamiliar parties cannot exchange information with each other in a generic way. Our approach proposes to capture the properties of *how* the information exchange occurs into types (Figure 1ii). A group of typed clauses will then form an interface that will specify *what* information is required and provided by each party (Figure 1iii). The coupling of the interfaces will yield a contract, that will allow the parts to couple and work together on a common goal (Figure 1iv).

In the rest of this section we provide the details on how the parties can couple using the proposed approach. Later in section 4 we will show how this approach

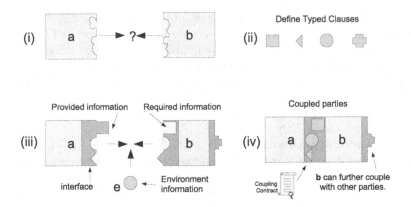

Fig. 1. Coupling Matches with Contracts

can be used to couple distributed application with Grid deployment descriptor using the ProActive[11] Grid middleware.

3.1   Clause Types

Let a and b be matched parties that require information from each other, or from an external source e like the environment to work together. We have identified that the information requirements can be exposed and fulfilled using typed clauses. The type of the clause represents a specific configuration of the following properties:

1. **Ability to set a value.** This defines which party has the ability to set a value for the clause. Possibilities are any permutation of a, b, e: $\{abe, ab, be, ae, a, b, e\}$.
2. **Ability to set empty values.** This defines which party can set this clause as empty. The possibilities are any permutation of a, b: $\{a, b, ab, -\}$.
3. **Ability to set constraints to the values**, thus narrowing the space of possible values. This can be done by providing an explicit list of alternatives, or using comparison operators ($<$, $>$, $=$, ...). The alternatives are permutations of: a, b: $\{ab, a, b, -\}$.
4. **Priority.** If more than one party can set a value, an empty value, or the constraints, this identifies which has the priority. The alternatives are combinations of a, b, e: $\{abe, aeb, bae, bea, ...\}$. The order in which they are expressed defines the priority.

For example, we have identified the types depicted in Table 1. Conceptually the types can be interpreted as:

A The value can only be set by a. Since b can set the value to empty, then b can force a to provide a value.
B Corresponds to the symmetrical of A.

Table 1. Types

Type Name	Set value	Set empty	Set constraints	Priority
A	a	b	-	a
B	b	a	-	b
A-PRI	ab	-	b	ab
B-PRI	ab	-	a	ba
ENV	e	-	-	e

A-PRI The value can be set either by a or b, where b can provide a default value, and a can override the default.

B-PRI Corresponds to the symmetrical of A-PRI.

ENV The value can be set from the environment.

The flexibility of the approach allows defining the types of interest only, and extending the set of types as required. The definition of new typed clauses is possible using these or future imagined properties. For example, we could imagine handling the priorities at a finer grain, thus having to specify three priorities for setting the value, setting the empty value, and setting the constraints. In this work we will focus on the types depicted in Table 1, because these represent the types of interest in section 4.

3.2 Typed Clauses

We will define a typed clause (clause for short) as having the following fields:

1. **Type** Corresponds to one of the allowed clause types. These are: A, B, A-PRI, B-PRI, ENV.
2. **Name** Corresponds to the name of the clause.
3. **Value** The value that will be set, empty or not.
4. **Constraints** The restrictions imposed on the values that can be set, if allowed by the *type*.

We will say that a clause pair named cls_a and cls_b compose a **shared clause** cls if both clauses names match $cls_a = cls_b$. The shared clause cls is **type compatible** if $cls_a.type = cls_b.type$, and incompatible otherwise.

The fields of a type compatible shared clause are defined as:

- Name: $cls = cls_a = cls_b$,
- Type: $cls.type = cls_a.type = cls_b.type$,
- Value: $cls.value = cls.type.priority(cls_a.value, cls_b.value)$
- Constraint: $cls.constraints = cls.type.priority(cls_a.constraints, cls_b.constraints)$

We will say a clause, shared or not, is **valid** if and only if $cls.value \neq empty$ and $cls.value$ satisfies $cls.constraints$ such that: $cls.constraints(cls.value) = true$. Note that two invalid clauses can be separately invalid, but the shared clause composed using both of them can be a valid clause.

3.3   Coupling Interfaces

An coupling interface (interface for short) corresponds to a group of clauses. A party can expose more than one interface, thus allowing coupling with more than one party. An interface is defined by:

1. A name
2. Set of clauses identified by their names

Thus for a party a we can identify an interface by $a.int\_name$. And for identifying a clause belonging to an interface we write: $a.int\_name.cls\_name$.

We will say that two **interfaces can be coupled** ($a.int\_name$ and $b.int\_name$), if there are no type incompatible shared clauses between the interfaces. The result of the interface coupling corresponds to the set of all shared clauses, and will denote it as: $a.int\_name \diamond b.int\_name$.

3.4   Coupling Contracts

A coupling contract (contract for short) corresponds to the interaction between two interfaces of different parties. If there exists two interfaces $a.int$ and $b.int$, such that both interfaces can be coupled, then the contract is defined as a set of clauses:

$$Contract = a.int \diamond b.int \cup (a.int - (a.int \diamond b.int)) \cup (b.int - (a.int \diamond b.int))$$

This means that the clauses contract will contain the shared clauses between the interfaces, the unshared clauses of a, and the unshared clauses of b.

We will say that **two parties a and b can be coupled** if:

1. A contract can be built between them: two interfaces belonging to a and b can be coupled, and
2. the contract is valid: every clause in the contract is valid.

3.5   Matching Parties: Descriptors and Applications Example

Typed clauses can also be used to perform advertisement and matchmaking in the Condor style. Both parties can expose their interface (advertisement) to a matchmaker or broker. To determine if the two parties are a suitable match, the coupling contract can be generated and validated.

The clauses belonging to the interfaces will specify *what* information is shared (provided or required) for the matchmaking. And the type of the clauses will specify *how* the information is shared for the coupling.

4   Coupling Distributed Applications with Deployment on the Grid

In this section we show how the concepts defined in section 3 can be applied. Specifically, we aim at coupling a distributed application with Grid resources using the Grid middleware ProActive. ProActive already provides a mechanism based on deployment descriptors for deploying on the Grid. We will show how this mechanism can benefit from the use of coupling contracts to couple applications with deployment descriptors.

This section is organized as follows. We will first provide some background on ProActive. Then, we will show how coupling contracts have been incorporated into ProActive.

4.1   Background on ProActive Deployment Descriptors

Within the *ProActive Descriptor Deployment Model* [3], it is possible to deploy applications on sites that use heterogeneous protocols, without changing the application source code. All information related with the deployment of the application is described in an XML Deployment Descriptor. Thus, eliminating references inside the application code to: machine names, submission protocols (local, rsh, ssh, lsf, globus-gram, unicore, pbs, lsf, nordugrid-arc, etc.) and communication protocols (rmi, jini, http, etc.).

The Descriptor Deployment Model is shown in Figure 2.

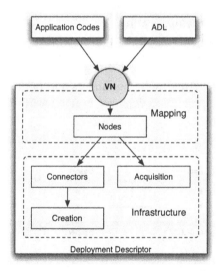

Fig. 2. Descriptor Deployment Model

The infrastructure section contains the information necessary for booking remote resources. Once booked, ProActive Nodes can be created (or acquired) on the resources. To link the Nodes with the application code, a Virtual Node (VN) abstractions is provided, which corresponds to the actual references in the application code. Virtual Nodes have a unique identifier which is hardcoded inside the application and the descriptor.

A deployer can change the mapping of the application → Virtual Node to deploy on a different Grid, without modifying a single line of code in the application.

4.2   The Problematic of Applications and Descriptors

In the traditional approach, the application developer and the descriptor developers need to have a previous agreement on the name of the Virtual Node. This means that the name of the Virtual Node is hardcoded inside the application and the descriptor. If the application wants to use a new descriptor, then either the descriptor or the application has to be modified to agree on the new Virtual Node name.

A possible solution to this problem is passing the Virtual Node name as a parameter to the application. Nevertheless, the problem of figuring out the proper Virtual Node name from the descriptor remains. To find out the name of the Virtual Node, inspection of the descriptor has to be performed, which can be a problem for someone alien with respect to the Grid infrastructure's descriptor.

Furthermore, the Virtual Node name is not the only information sharing problem that the application and descriptor have. For example, a descriptor might be configured to deploy on k nodes, but the application only requires j nodes $(j < k)$. Without shared clauses, the descriptor has to be modified to comply with the requirements of the application.

Modifying the application or the descriptor can be a painfull task, specially if we consider that the person deploying the application (deployer) may not be the author of either. To complicate things further, the application source may not even be available for inspecting the requirements and performing modifications. Figure 3 illustrates the issue. The deployer is not aware of the application or descriptor requirements.

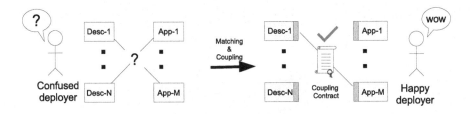

Fig. 3. Matching and Coupling Contracts

Nevertheless, using coupling contracts, the deployment can be further enhanced by enabling automated matchmaking and coupling of applications and descriptors.

4.3   Clause Types

The involved parties are the application (a) and the descriptor (b), and the environment information (e) (given through java properties). To improve the

Table 2. ProActive Deployment Clause Types

Type Name — ProActive Type Name	
A	Application
B	Descriptor
A-PRI	ApplicationPriority
B-PRI	DescriptorPriority
ENV	JavaProperty

clarity of the example, we have renamed the clause types identified in the Table 1 to the names shown in Table 2.

4.4   Clauses in ProActive Descriptors

Clauses can be specified using XML tags as shown in the example of Figure 4 for the descriptor. To define the clauses a new section labeled `clauses` has been added at the beginning of the descriptor to hold the `interfaces`. The clauses shown in the example correspond to:

`PROACTIVE_HOME` & `MAX_NODES` Correspond to descriptor set clauses. The value is set directly in the descriptor, and can be used later on, inside the descriptor or the application.

`VIRTUAL_NODE_NAME` Corresponds to a clause that the descriptor enforces the application to set. If the application doest not set this value, the clause inside the coupling contract will not be valid, and the application will not be allowed to couple with the descriptor. In the example, we force the application to set the name of the Virtual Node.

`LOAD_BALANCING` Corresponds to a clause that the application has set, but the descriptor can override. In the example, we imagine that an application is capable of handling, or not, the load balancing. By default the application will assume that no load balancing is provided by the Grid infrastructure (Figure 5), and thus handle the load balancing at the application level. Nevertheless, the descriptor is aware if load balancing can be done at the Grid infrastructure level and activate it. The application can then access the contract's clauses to learn if the infrastructure is using the load balancing and disable the application load balancing mechanism.

`NUMBER_OF_NODES` Corresponds to a clause that the descriptor has set a value, but the application may override. Additionally, the descriptor has set constraints indicating that the value must be an integer between 1 and `MAX_NODES`.

`USER_NAME` Corresponds to a clause that is set from the environment. In this case, the username can be specified from the environment as a java property.

Figure 4 also shows an example of how the clauses can be used inside descriptors. Note that the value of the clause `VIRTUAL_NODE_NAME` has not been set in the descriptor, since it is of type Application. This means that the value used inside the descriptor will be the one set from the application. Also note, that clauses obtained from the environment can also be used, like the `USER_NAME` clause.

```
<clauses>
  <interface name="descriptor-example-interface">
    <Descriptor name="PROACTIVE_HOME" value="ProActive/"/>
    <Descriptor name="MAX_NODES" value="100/"/>
    <Application name="VIRTUAL_NODE_NAME" value=""/>
    <DescriptorPriority name="LOAD_BALANCING" value="on"/>
    <ApplicationPriority name="NUMBER_OF_NODES" value="1">
      <!--// (NUMBER_OF_NODES>0) && NUMBER_OF_NODES<=MAX_NODES -->
      <integerConstraint>
        <and>
          <biggerThan>0</biggerThan>
          <smallerOrEqualThan>${MAX_NODES}</smallerOrEqualThan>
        </and>
      </integerConstraint>
    </ApplicationPriority>
    <JavaProperty name="USER_NAME" value="user.name"/>
  <interface>
</clauses>
...
  <virtualNodesDefinition>
   <virtualNode name="${VIRTUAL_NODE_NAME}"/>
  </virtualNodesDefinition>
...
<sshProcess class="org.objectweb.proactive.core.process.SSHProcess"
      hostname="example.host" username="${USER_NAME}"/>
```

Fig. 4. Example of clauses in descriptor

4.5   Clauses in ProActive Applications

We have also provided a mechanism for specifying clauses and interfaces from
the application. This can be done through an API, or loading the clauses from an

```
//Create a new interface
ClausesInterface ci= new ClausesInterface("application-example-interface");

//Set the clauses in this interface
  //set(<type>, <clause name>, <value>, [<constraint>])
ci.set(Application, "VIRTUAL_NODE_NAME", "testnode",);
ci.set(ApplicationPriority, "NUMBER_OF_VIRTUAL_NODES", "16");

// LOADBALANCE="on" || LOADBALANCE="off"
OrConstraint oc = new OrConstraint();
oc.add(new EqualsConstraint("on"));
oc.add(new EqualsConstraint("off"));
ci.set(DescriptorPriority, "LOAD_BALANCING", "off", new StringConstraint(oc));

//Parse and load the descriptor using the coupling interface. If the application and
      descriptor can not be coupled an exception will be thrown
ProActiveDescriptor pad = ProActive.getProactiveDescriptor("descriptor.xml", ci);

//Clauses from the coupling contract can be used in the application
CouplingContract cc = pad.getCouplingContract();
String loadBalancing = cc.getValue("LOAD_BALANCING");

//The application can take decisions based on the clauses
if(loadBalancing.equals("on")){...}
else{...}
```

Fig. 5. Example of clauses in application

external XML file. Since the XML approach has already been shown for the descriptor, Figure 5 shows an example using the API. First an *interface* is created, and then the clauses are added to the interface. The interface is then passed as a parameter when parsing the descriptor. The parsing will try to generate a coupling contract using the application's and the descriptor's interfaces.

If the application can be coupled with the descriptor, then the application can retrieve the coupling contract and consult the contract's clauses. For example, using this strategy the application can know if the descriptor activated the infrastructure load balancing, and avoid using the application load balancing.

4.6 Constraints

Constraints are boolean expressions that will be evaluated for each clause when the contract is built. The constraints can be of two types: *integer* or *string*. For each constraint the logical operators: *and*, *or*, *xor* are allowed. Also, boolean operators are provided for each type of constraint. The integer operators are: `biggerThan, biggerOrEqualThan, smallerThan, smallerOrEqualThan, equals`. The string case sensitive operators are: `subString, superString, equals`. Figure 6 shows the constraint grammar specified using XML Schema[15] for the integer type constraints.

```
<xs:element name="integerConstraint">
  <xs:complexType>
    <xs:choice>
      <xs:element name="and" type="intConst"/>
      <xs:element name="or" type="intConst"/>
      <xs:element name="xor" type="intConst"/>
    </xs:choice>
  </xs:complexType>
</xs:element>
</xs:complexType>
  <xs:complexType name="intConst">
  <xs:choice minOccurs="1" maxOccurs="unbounded">
    <xs:element name="and" type="intConst"/>
    <xs:element name="or" type="intConst"/>
    <xs:element name="xor" type="intConst"/>
    <xs:element name="biggerThan" type="xs:string"/>
    <xs:element name="biggerOrEqualThan" type="xs:string"/>
    <xs:element name="smallerThan" type="xs:string"/>
    <xs:element name="smallerOrEqualThan" type="xs:string"/>
    <xs:element name="equals" type="xs:string"/>
  </xs:choice>
</xs:complexType>
```

Fig. 6. Integer Constraint Schema Grammar

Figure 4 shows an example where the clause `NUMBER_OF_NODES` is constrained to be: $0 <$ `NUMBER_OF_NODES` $<=$ `MAX_NODES`. Note that `MAX_NODES` is defined as a `Descriptor` type clause. Figure 5 shows an example using string constraints. The clause `LOAD_BALANCING` is constrained to be either `on` or `off`.

5   Conclusions and Future Work

We have shown an approach for coupling parties that require exchanging information to work together. To achieve this, we have identified the properties related with information exchange between parties, and we have grouped the properties of interest into typed clauses. We have then proposed that interfaces can be built using shared typed clauses.

If two interfaces between parties are compatible, the coupling of the interfaces can yield a coupling contract. The clauses belonging to the contract represent *what* information can be shared between the parties, and the type of the clauses specify *how* this information will be shared.

Using the proposed coupling approach, we have shown how coupling contracts can be applied for automated deployment of unfamiliar applications on alien Grids. For this, we have provided a mechanisms to specify clauses in the application and the deployment descriptor using the Grid middleware ProActive. As a result, the approach can now be used to couple applications with descriptors, without having to modify or inspect either.

Nevertheless, it can be argued that the proposed approach requires each party to know beforehand the names of the clauses used in the coupling. In reality, only a subset of the clauses belonging to the coupling contract have to be known: the ones that must be provided with a value to make the contract valid. Furthere more, if two different interfaces couple with a third generating two valid coupling contracts, the clauses contained in these contracts can be different. While this seems strange, it is a direct result of the proposed approach being boolean: either the contract is valid or not. In the future, we would like to extend this concept by introducing Conformance Levels in coupling contracts. Thus, a minimum conformance level (i.e. minimum set of known clauses) could be provided for basic applications, and higher conformance levels (i.e. a superset of the lower conformance levels) could be used for more advanced features that require more specific clauses.

From the Grid infrastructure side, in the future we would like to identify standard interfaces for coupling applications with different types of Grids. The idea is to be able to release applications packaged with interfaces that certify the deployment of an application with a Grid interface. On the other hand, from the application point of view, we would like to identify interfaces for common structured parallel programming patterns. For example, if an application uses the master-slave pattern, then it can benefit by coupling with a Grid interface optimized by deploying the master on a more powerfull or better connected resource than the regular slaves. Thus, a Grid could provide an optimized interface for applications exploiting different patterns such as: farm, pipe, divide and conquer, etc.

References

1. Alain Andrieux, Karl Czajkowski, Asit Dan, Kate Keahey, Heiko Ludwig, Toshiyuki Nakata, Jim Pruyne, John Rofrano, Steve Tuecke, and Ming Xu. Web services agreement specification (ws-agreement). Draft Version 2005/09. http://forge.gridforum.org/projects/graap-wg.

2. L. Baduel, F. Baude, D. Caromel, A. Contes, F. Huet, M. Morel, and R. Quilici. *Grid Computing: Software Environments and Tools*, chapter Programming, Composing, Deploying, for the Grid. Springer Verlag, 2005.
3. F. Baude, D. Caromel, L. Mestre, F. Huet, and J. Vayssière. Interactive and descriptor-based deployment of object-oriented grid applications. In *Proceedings of the 11th IEEE International Symposium on High Performance Distributed Computing*, pages 93–102, Edinburgh, Scotland, July 2002. IEEE Computer Society.
4. Steve Chapin, Dimitrios Katramatos, John Karpovich, and Andrew Grimshaw. Resource management in legion. Legion Winter Workshop, 1997.
5. Karl Czajkowski, Ian T. Foster, Nicholas T. Karonis, Carl Kesselman, Stuart Martin, Warren Smith, and Steven Tuecke. A resource management architecture for metacomputing systems. In *IPPS/SPDP '98: Proceedings of the Workshop on Job Scheduling Strategies for Parallel Processing*, volume 1459 of *Lecture Notes in Computer Science*, pages 62–82, London, UK, 1998. Springer-Verlag.
6. D. Fensel and C. Bussler. The web service modeling framework WSMF. *Electronic Commerce Research and Applications*, 1(2):113–137, Summer 2002.
7. Global Grid Forum. `http://www.gridforum.org/`.
8. I. Foster and C. Kesselman. Globus: A metacomputing infrastructure toolkit, 1996.
9. Wolfgang Gentzsch. Sun grid engine: Towards creating a compute power grid. In *CCGRID*, pages 35–39. IEEE Computer Society, 2001.
10. R. Henderson and D. Tweten. Portable batch system: External reference specification. Technical report, NASA, Ames Research Center, 1996.
11. ProActive. `http://proactive.objectweb.org`.
12. R. Raman, M. Livny, and M. Solomon. Matchmaking: Distributed resource management for high throughput computing. In *In Proceedings of the Seventh IEEE International Symposium on High Performance Distributed Computing*, 1998.
13. R. Raman, M. Livny, and M. Solomon. Policy driven heterogeneous resource coallocation with gangmatching. In *Proc. of the 12th IEEE Int'l Symp. on High Performance Distributed Computing (HPDC-12)*, 2003.
14. INRIA OASIS Team and ETSI. 2nd grid plugtests report. `http://www-sop.inria.fr/oasis/plugtest2005/2ndGridPlugtestsReport.pdf`.
15. W3C. Xml schema: Formal description. `http://www.w3.org/TR/xmlschema-formal/`.
16. S Zhou. Load sharing in large-scale heterogenous distributed systems. In *Proceedings of the Workshop on Cluster Computing, 1992*.

A Transparent Framework for Hierarchical Master-Slave Grid Computing

Nadia Ranaldo[1] and Eugenio Zimeo[2]

[1] Department of Engineering, University of Sannio,
82100 Benevento, Italy
ranaldo@unisannio.it
[2] Research Centre on Software Technology (RCOST), University of Sannio,
82100 Benevento, Italy
zimeo@unisannio.it

Abstract. The use of grid computing to easily and efficiently execute data and compute-intensive applications strongly depends on new software development approaches able to separate application-domain aspects from non-functional ones, such as task mapping and deployment. In this paper, we present an object-oriented framework that is able to transparently transform non-distributed programs into hierarchical master-slave ones, and to map and schedule them onto a grid computing system. Moreover, the framework is able to leverage services delivered by the underlying middleware platform, such as resource management and communication, to satisfy user requirements. The paper presents the framework architecture, a reflection-based implementation and its evaluation atop of a hierarchical grid middleware.

1 Introduction

Thanks to the increasing amount of resources available across the Internet and to improvements of wide-area network performance, in recent years grid computing is emerging as a viable computing paradigm to execute data and compute-intensive applications.

At the state of the art, two of the main difficulties to wide diffusion of grid technologies are *usability* and *efficiency*: if the computing environment provided by the grid system is seamless, user-friendly and efficient, users will potentially exploit wide-area distributed resources to obtain high performance with a little effort related to the management of the distributed system and the deployment of applications on it. Existing distributed programming approaches based on message-passing (such as MPICH-G2 [1]) adopted for not or limited distributed systems (such as parallel machines or clusters of workstations), or "standard" approaches based on object-oriented technologies (such as Java RMI and CORBA) are hardly applicable to write and execute applications in highly dynamic and geographically distributed computing environments. These approaches, in fact, require to directly deal with problems not encountered for sequential programming, such as non-determinism, synchronization, data partitioning and distribution, load-balancing, fault-tolerance, security, etc.

W. Lehner et al. (Eds.): Euro-Par 2006 Workshops, LNCS 4375, pp. 74–86, 2007.

To overcome the burden of these approaches, new programming models, abstractions, tools and methodologies are required. In this connection, we believe that *object-oriented component frameworks* for high-level distributed programming are strategic to increase the spread of grid computing technologies (even in industrial and enterprise environments) and the productivity of grid programmers. This convincement derives from the analysis of similar technologies, such as Enterprise Java Beans and application servers employed in enterprise environments to separate functional and non-functional aspects in distributed software systems.

To improve efficiency, scalability and adaptability of applications, a framework for grid computing has to: (1) permit the programmer to focus only on domain-dependent aspects of an application, rather than on control and coordination aspects of distribution, which depend on the target environment; (2) be able to reuse the same application logic into different computing environments (such as parallel machines, clusters and Grids).

As concerning distributed computing models, in this work we focus on the master-slave pattern [2], which is a widespread architectural pattern adopted to implement coarse-grained parallel and distributed applications either in local- and wide-area networks. We focus on the hierarchical version of such pattern, since it is particularly effective to be used in intrinsically hierarchical grid computing systems, because of well-defined and limited communication patterns among computing nodes. In these systems, computing nodes are often hosted by heterogeneous resources characterized by limited-bandwidth communication in the levels of the hierarchy close to the user, and high communication performance in the other levels, typically not directly accessible through the Internet because they are often clusters accessible only through a front-end. In future we intend to take into account other widespread patterns currently adopted in the distributed computing, such as divide and conquer and pipeline.

This paper presents a framework to simplify the development of parallel and distributed object-oriented applications for grid systems. The framework, called TMS Framework (*Transparent Master-Slave Framework*), is able to transparently implement hierarchical master-slave applications in a hierarchical grid environment, and to satisfy Quality of Service (QoS) requirements by dynamically exploiting services delivered by underlying middleware platforms. The framework was implemented by leveraging reflection mechanisms provided by a meta-object protocol [3]. We considered, moreover, its customisation for a hierarchical grid middleware [4], which delivers an economy-driven resource broker usable by the TMS Framework to automatically map and schedule distributed tasks satisfying time and cost constraints specified by the user.

The rest of the paper is organized as follows. Section 2 discusses related work. Section 3 presents the TMS Framework. Section 4 describes a reflection-based framework implementation. Section 5 presents an evaluation of the TMS Framework in writing a distributed application and a preliminary experimental analysis, and finally Section 6 summarizes the paper and presents future work.

2 Related Work

Some frameworks for master-slave applications in dynamic and heterogeneous systems have been proposed in literature. The most significant ones are *AppLeS*

Master-Worker Application Template (AMWAT) [5] and *Condor Master-Worker* (MW) [6]. Also Javelin 3 [7] and Satin [8] are interesting proposals.

AMWAT is a library that provides a software template to implement self-scheduling master-slave applications written in C, C++, and Fortran in distributed memory architectures. The AMWAT programming interface specifies the high-level functionalities that the application developer must minimally supply. Such functionalities are provided in form of portable and reusable modules. In particular, the Application Template module contains fifteen application activity functions, which are provided by developers to implement application-specific functions.

Condor MW is a framework proposed for implementing grid-enabled master-slave applications written in C++. Condor MW provides a "top-level" interface to application software and a "bottom-level" interface, called Infrastructure Programming Interface (IPI). The top-level interface permits to parallelize an application and requires the programmer to re-implement some abstract classes, in particular the MWTask, which is the abstraction of one unit of work, and the MWWorker, which represents a slave process. The IPI interface permits to use existing grid computing toolkits without any changes from the view-point of the application developer.

While the AMWAT approach focuses on application performance in terms of execution time, the Condor MW approach emphasizes the delivery of high throughput computing. It typically deals with many processor faults, since the default Condor behaviour is to vacate a running process on a remote machine when it is no longer in idle status.

Even if the approaches described above permit to simplify writing of master-slave applications by hiding distribution, scheduling and communication aspects, they still require to explicitly write code for the distributed version of the problem, requiring a specific implementation of the application for the master-slave pattern and so limiting the programmer productivity and existing code re-use.

A better separation of functional aspects from non-functional ones can be reached through the new programming approach based on skeletons [9], conceived to design easy-to-use structured parallel programming environments. The idea is to capture recurring patterns in parallel and distributed applications in generic software constructs that can be customized by the programmers to write different applications.

A recent proposal is HOC [10] based on Web services, which requires configuring services through application-specific code, such as, in the master-slave pattern, how to split input parameters among the slaves and how to process them. Such customisation is obtained through the implementation of specific interfaces.

Another proposal that focuses on grid systems is Lithium [11], a library based on Java and RMI, which supports common skeletons, including pipelines, task farms, iterative and data parallel skeletons.

As for the skeleton-oriented approaches, our goal is to simplify writing distributed applications, considering the difficulty in learning new paradigms and programming approaches. For this reason, we propose a framework that permits writing (or re-using) an application in a sequential version, hiding the distributed aspects related to the pattern/s adopted for its deployment. Our idea is to configure the pattern-related aspects through a preliminary phase that requires writing a configuration file and the classes for the framework customization. Moreover we focus on a framework implementation that hides pattern-related aspects in some configurable components of the system, able to leverage existing grid services, for example resource discovery and load balancing.

Most of the distributed computing environments for master-slave applications, which deliver scheduling functionalities, use mapping algorithms that try to optimise only the execution performance [12] [13] [5]. In a future commercialisation of grid technologies, the resource price will represent a distinctive property to regulate the supply-and-demand for resources. To this end, the work in [14] represents one of the first effort to introduce economy-driven mapping algorithms for generic applications with no control or data dependencies; whereas a previous paper of the authors [15] defines a heuristic for mapping tasks to the slaves of a master-slave application based on deadline and budget constraints.

3   TMS Framework

The TMS Framework design is based on the following principles: (1) *separation of concerns*: the framework has to permit a programmer to concentrate only on the domain-dependent aspects, without dealing with low-level aspects of distributed computation such as the definition of the number of resources, the distribution of tasks among the resources, synchronization, etc.; (2) *code re-use*: the framework has to permit the re-use, in a distributed computing environment, of existing code written to solve the same problem in sequential manner, so permitting to use, in a nearly-seamless way, the same code for execution on a single workstation, or on a homogeneous cluster or on a heterogeneous wide-area distributed system; (3) *adaptability*: the framework has to dynamically leverage services delivered by the underlying computation architecture in order to automatically optimise application execution and fulfil user QoS requirements.

The main objective of the proposed framework is to re-use existing code for sequential execution to automatically produce a parallel and distributed version of it through the adoption of the hierarchical master-slave pattern at run-time. The hierarchical master-slave pattern consists of extending the single master of the canonical pattern to a hierarchy of masters at different levels. The master at the top controls the overall computation and distributes it among the masters at lower levels, and so on, until the computation is sent to the slaves, which directly process the request. The collection of computed results is performed in the reverse order. With respect to the master-slave pattern, it permits to increase scalability by removing the centralized control of a single master, which could easily become a bottleneck for a high number of resources and limited-bandwidth networks.

The TMS Framework provides a run-time distributed environment in which masters and slaves run. To achieve separation of concerns, it defines a generic architectural skeleton, which can be customized by the user through application-domain code used for sequential version, and some descriptive information for the deployment. The distribution aspects that depend on the underlying computational infrastructure are captured and managed by the framework, without the necessity of application-domain code modification.

The framework is designed so to automatically manage and trigger well-defined coordination activities of the hierarchical master-slave model, which are: (1) splitting of the workload, (2) call to slaves, (3) waiting and gathering of results performed by the master. The idea is to set up such well-defined activities through a configuration phase, which permits to specify the policies to adopt for each activity. In figure 1, the main components of the framework are shown, considering one level of the hierarchy for simplicity.

The framework is used to dynamically parallelize object-oriented applications whose functional aspects are delivered through a method of a class (called in the following *Task* class), which implements a sequential solution to a given problem. To transparently turn the sequential computation of such method into a parallel one, a Task object is used to customize the main framework component, called *TMS Task*. The TMS Task is loaded into the TMS Framework of each computing node and is configured in order to act as master or slave of the computation. For a master node the TMS Task consists of the replication of the original Task object, and a customisable framework component, called *Master Behaviour*, which performs the master functionalities of workload splitting and result gathering.

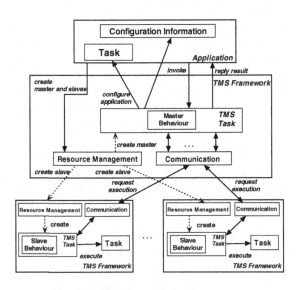

Fig. 1. TMS Framework Architecture

For a slave node the TMS Task consists of the replication of the original Task object, and a framework component, which performs the slave activities, called *Slave Behaviour* component.

The framework contains other two configurable components used to capture and manage the main capabilities required to distribute and manage a master-slave application, that are the *Resource Management* and the *Communication* components. The Resource Management component has the task to schedule and manage the masters and slaves on distributed resources. The communication and synchronization between masters and slaves is managed by the Communication component.

4 Reflection-Based Implementation

The main goal of the TMS Framework is the *implicit* implementation of the hierarchical master-slave pattern, which can be achieved following static or dynamic approaches.

The static approach is based on specialized pre-compilers, which take care of parallelising the application and deploying it on distributed resources. Such approach does not fit in a dynamic distributed environment since it does not permit to perform on-the-fly modifications in order to adapt applications to variations in underlying services and resource availability.

The dynamic approach permits to overcome such limitations and is based on the openness of the system to change, even at run-time, some aspects of its behaviour.

We propose a version of the TMS Framework based on a dynamic approach implemented with reflection mechanisms [2]. A reflection-based framework permits to easily adapt components to changing conditions, and to extend or reconfigure the system to meet new requirements. With this approach, an application is logically divided in two parts: the *meta-level* and the *base-level*. The meta-level is the part of application which provides knowledge of its properties and makes the system self-aware. The system properties available at the meta-level are represented by *Meta Objects*, which encapsulate and represent information about a single system aspect that should be adaptable. The base-level models and implements the application logic and represents the various services the system offers. Its implementation uses information and services provided by the meta-level to remain flexible and independent from those aspects that are likely to be modified.

A reflection-based TMS Framework requires individuating the set of Meta Objects, which capture the incomplete parts of the framework and permit to customize it for the execution of an application. Reflection mechanisms are also used to customize the Resource Management and Communication components so to deliver functionalities exploiting existing basic services of the underlying middleware.

4.1 MOP-Based Implementation

We implemented the dynamic master-slave pattern by exploiting the reflection features provided by Meta-Object Protocol (MOP) implemented in ProActive [3]. It is a proxy-based run-time mechanism, which permits reification of method invocations and constructor calls. It is entirely written in Java and avoids any modification or extension to the JVMs, as opposed to other meta-object protocols.

By using MOP, the TMS Framework permits to employ every existing class to transparently instantiate the set of master and slave active objects (ProActive objects), keeping the application very similar to that used for a sequential computation.

Therefore, the hierarchical master-slave pattern is dynamically implemented and an existing object can be turned in a master able to transparently split the overall task into sub-tasks and in a slave able to perform the assigned part of the overall task.

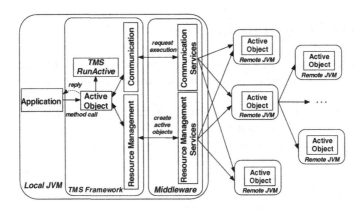

Fig. 2. MOP-based TMS Framework implementation

The behaviour of active objects, with respect to the parallelism exploitation patterns, is specified through the implementation of the RunActive interface of ProActive, delivered by the framework and called TMSRunActive, which specifies the actions executed by the active object when a method execution request is received (see figure 2). In particular, if the active object is a master, the following actions are performed: (1) to collect information on the performance capabilities of each resource available for computation; (2) to perform a partition of input parameters following the policies indicated in a configuration file; (3) to send method calls to the active objects which are masters or slaves of the lower level, using as input parameters: the original input parameter, if it is a non-partitioned parameter, or the corresponding part of the partition, if it is a partitioned parameter; (4) to wait for the collection of each partial results, which are assembled following the policy specified in a configuration file. If the active object is a slave, it directly executes the method.

4.2 Programming Model

The programming model of the TMS Framework permits the parallelisation of one or more tasks, each represented by a method of an existing class. The parallelisation is initialised through the *Configuration Phase*, performed delivering a configuration file and invoking the static method of the TMSFramework class:

```
Object configureDistributedTask(Object original, String configFile);
```

that returns a reified object. The input parameters are original, which is an instance of the class used to perform the distributed task and configFile, which is the name of the configuration file used to configure the deployment of active objects. It is an

XML-based file, called *Job Description Format* (JDF) in which a part depends on the underlying middleware adopted for active objects deployment, while another is common and is used for the reflection mechanisms. The common part contains the following information: (1) the methods whose invocations have to be distributed over the active objects; (2) for each method, the input parameters that have to be partitioned and the policy to partition each of them; (3) for each method, the assembling policy of the output parameter.

A partition policy is specified by the implementation of the following method of the `SplitHelper` interface:

```
public Object[] split (Object[] data, double[] caps);
```

in which, `data` represents the information used to obtain a partition on an input parameter and `caps` the performance information on each active object, used to eventually obtain a load balanced partition.

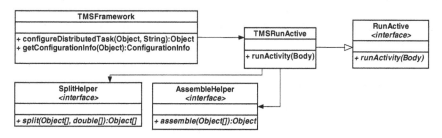

Fig. 3. Class Diagram of MOP-based TMS Framework

An assembling policy is specified by the implementation of the following method of the `AssembleHelper` interface:

```
public Object assemble (Object[] data);
```

in which `data` represents the partial results to assemble into a single object representing the overall result of the distributed computation.

The Configuration Phase is followed by the *Execution Phase*, in which the user performs method invocations in the same way as for standard objects. The method invocation on ProActive active objects is asynchronous, which permits to increase the concurrency among local and remote activities.

4.3 User QoS Requirements

The default version of ProActive leverages Java RMI and, as a consequence, requires the direct handling of scheduling functionalities of resource discovery, selection and task mapping, limiting the capability to fulfil user QoS requirements.

Through the adoption of the ProActive-HiMM adapter [16], the TMS Framework can be configured to transparently leverage HiMM functionalities. HiMM is a Java-based middleware able to exploit hierarchical collections of computers interconnected by heterogeneous networks. Even if HiMM is not a complete grid middleware (it lacks of sophisticated security mechanisms and efficient data access), it delivers all the basic services of resource discovery, management, scheduling, and efficient communication mechanisms useful to implement master-slave applications into a grid system.

HiMM, in particular, provides an economy-driven broker for master-slave applications which is responsible for automatic resource discovery and task mapping on the basis of availability, performance and cost of resources, and on time and cost parameters specified by the user. It is based on the task mapping heuristic proposed in [15] which permits to minimize the total execution time without exceeding a fixed budget. The HiMM resource broker can be adopted by the TMS Framework to transparently deploy the distributed tasks of resources satisfying time and cost constraints specified by the user in the configuration phase. This is obtained following the programming model described above and using a file (JDF – Job Description Format – file) which contains all the information necessary to exploit broker functionalities of HiMM, which are application information (task dependencies, overall complexity, single task complexity, etc), application code, input data and user requirements. The current version of the HiMM broker does not take into account the mapping problem of master hierarchy because it focuses on a grid system with an intrinsic hierarchical topology, in which the masters are naturally hosted on those machines used as front-end for pools of resources such as clusters.

5 Framework Evaluation

To evaluate the usefulness for programming and to analyse the performance delivered by TMS Framework, a simple application is described. It is the well-known multiplication of square matrices implemented with the master-slave pattern and using the strip partitioning of the left matrix: the master partitions the received left matrix and sends the parts to slaves for processing.

A standard class `Matrix`, eventually already written for sequential applications, delivers a constructor to initialise a bi-dimensional array of `float` values, and the `multiply` method that sequentially executes the multiplication between the current matrix, used as right matrix, and the matrix passed as parameter, used as left matrix.

The following code shows the use of the TMS Framework to turn a standard instance of `Matrix` into a transparent master-slave one:

```
...
Matrix rigMat = new Matrix(...); // initialisation
Matrix leftMat = new Matrix(...);
Matrix result = null;
String configFile = null;
// Configuration Phase: definition of the XML-based JDF file
...
rigMat=(Matrix)TMSFramework.configureDistributedTask(
                                     rigMat,configFile);
// Execution Phase
result = rigMat.multiply(leftMat);
...
```

The parallelisation of the multiplication of two matrices requires to specify, in a JDF file, the class which contains the method to parallelise, that is `Matrix`, and the classes which implement the `SplitHelper` and `AssembleHelper` interfaces used, respectively, to split a matrix in blocks of rows and to assemble blocks of rows into one block to return a single matrix.

We underline that such classes could be already available in a library included in the framework or delivered by a third-part developer. A section of the JDF file for this application is reported below.

```
<APPLICATION-STRUCTURE>
  <DISTR-PROG-MODEL>Master-Slave</DIST-PROG-MODEL>
  <MIDDLEWARE-SPECIFIC-INFORMATION>
   <USER-REQUIREMENTS>
    <DEADLINE>50000</DEADLINE>
    <BUDGET>100</BUDGET>
    <MAPPING-POLICY>TIME_OPTIMIZATION</MAPPING-POLICY>
   </USER-REQUIREMENTS>
   ...
  </MIDDLEWARE-SPECIFIC-INFORMATION>
  <TASKS>
    <TASK>
     <TASK-CLASS-NAME>Matrix</TASK-CLASS-NAME>
     <METHOD-NAME>multiply</METHOD-NAME>
     <METHOD-PARAMETERS>
       <METHOD-PARAMETER>Matrix</METHOD-PARAMETER>
     </METHOD-PARAMETERS>
     <RETURN-TYPE>Matrix</RETURN-TYPE>
     <DATA>
      <DISTRIBUTED-INPUTS>
       <INPUT>
          <INPUT-TYPE>Matrix</INPUT-TYPE>
          <INPUT-INDEX>0</INPUT-INDEX>
          <PARTITION>
           <PARTITION-CLASS-NAME>TMSFramework.util.MatrixSplitHelper
           </PARTITION-CLASS-NAME>
            <PARAMETERS>
             <PARAMETER>
              <INPUT-TYPE>Matrix</INPUT-TYPE>
              <INPUT-INDEX>0</INPUT-INDEX>
             </PARAMETER>
            </PARAMETERS>
          </PARTITION>
       </INPUT>
      </DISTRIBUTED-INPUTS>
      <ASSEMBLING>
          <ASSEMBLING-CLASS-NAME>TMSFramework.util.MatrixAssembleHelper
          </ASSEMBLING-CLASS-NAME>
            <PARAMETERS>
             <PARAMETER>
              <INPUT-TYPE>Matrix</INPUT-TYPE>
              <INPUT-INDEX>-1</INPUT-INDEX>
             </PARAMETER>
            </PARAMETERS>
      </ASSEMBLING>
     </DATA>
    </TASK>
  </TASKS>
</APPLICATION-STRUCTURE>
```

Figure 4 shows the components that must be provided by the developer to configure the framework and Figure 5 shows the deployment of the components on a pool of distributed resources through a broker for resource management. During the configuration phase, the broker is adopted to discover and select a pool of resources able to satisfy user performance and cost requirements specified in the JDF file. Selected resources are adopted to build a hierarchical virtual machine managed through HiMM.

We conducted a preliminary performance analysis on a network of workstations composed of fourteen homogeneous machines, each equipped with Intel Pentium Xeon 2.8 GHz, a RAM of 1GB, running Custer-Linux Rocks ver. 4 operating system, and inter-connected by a Fast Ethernet network. The used software packages are Java 2 SDK 1.4.2, ProActive version 3.0 and HiMM version 1.1. We used the multiplication of two square matrices as benchmark and adopted the time

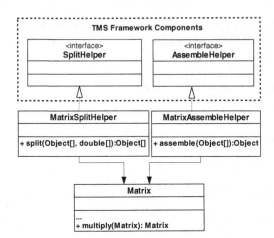

Fig. 4. Configuration of the TMS Framework for the matrix multiplication application

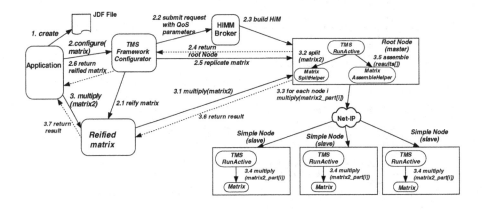

Fig. 5. Dynamics of the TMS Framework components for the execution of matrix multiplication application

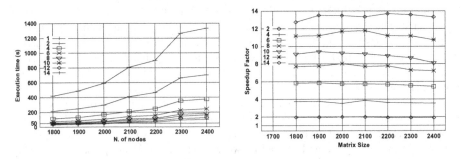

Fig. 6. (a) Execution times (b) Speedup factors

minimization heuristic considering the same performance parameters for each resource so to obtain roughly the same execution time on each of them. We measured the overall execution times and evaluated the speedup factor considering various matrix sizes and various numbers of available resources. The execution times and speedup factors are reported in figure 6 (a) and (b), whose trends show the system efficiency.

6 Conclusion

We defined a component framework able to automatically implement the hierarchical master-slave pattern in a distributed environment leveraging the application code for the sequential solution. We described a reflection-based implementation that exploits reflection to use the services of the underlying grid middleware. The usability of the TMS Framework for writing distributed applications and the results of an experimental analysis to prove the system efficiency were shown. In future, we will test the system scalability of the TMS Framework for a heterogeneous hierarchical environment. Moreover, we intend to customize the TMS Framework so to leverage more efficient communication mechanisms based on IP multicast for master-slave interactions and other middleware services, such as the WSRF-complaint services delivered by Globus [17].

References

1. Karonis, N., Toonen, B., Foster, I.: Mpich-G2: A Grid-enabled Implementation of the Message Passing Interface. Journal of Parallel and Distributed Computing, 63(5) (2003) 551-563
2. Bushmann, F., et al.: Pattern-Oriented Software Architecture: A System of Patterns. J. Wiley and Sons (1996)
3. Caromel, D., Klauser, W., Vayssiere, J.: Towards Seamless Computing and Metacomputing in Java. Concurrency: Practice and Experience, Vol. 10 (11-13) (1998) 1043-1061
4. Di Santo, M., Frattolillo, F., Russo, W., Zimeo, E.: A Component-based Approach to Build a Portable and Flexible Middleware for Metacomputing. Parallel Computing, Elsevier, 28(12) (2002) 1789-1810
5. Berman, F., et al.: Adaptive Computing on the Grid Using AppLeS. IEEE Trans. Parallel and Distributed Systems, 14(4) (2003) 369-382
6. Linderoth, J., Kilkarni, S., Goux, J. P., Yoder, M.: An Enabling Framework for Master-Worker Applications on the Computational Grid. Proceedings of the Ninth IEEE Symposium on High Performance Distributed Computing, Pittsburgh, Pennsylvania, (2000) 43-50
7. Neary, M. O., Cappello, P.: Advanced Eager Scheduling for Java-Based Adaptively Parallel Computing. Proceedings of the joint ACM-ISCOPE Conference on Java Grande, (2002)
8. van Nieuwpoort, R. V., Kelmann, T., Bal, H. E.: Efficient Load Balancing for Wide-Area Divide-and-Conquer Applications. Proceedings of the 8-th ACM SIGPLAN Symposium on Principles and Practices of Parallel Programming, Utah, (2001) 34-43
9. Cole, M. I.: Algorithmic Skeletons: a Structured Approach to the Management of Parallel Computation. MIT Press & Pitman, (1989)

10. Gorlatch, S., Dunnweber, J.: From Grid Middleware to Grid Applications: Bridging the Gap with HOCs. In Future Generation Grids, Springer-Verlag, (2005)

11. Aldinucci, M., Danelutto, M., Teti, P.: An Advanced Environment Supporting Structured Parallel Programming in Java. Future Generation Computer Systems, 19(5) (2003) 611–626

12. Banino, C., Beaumont, O., Carter, L., Ferrante, J., Legrant A., Robert, Y.: Scheduling Strategies for Master-Slave Tasking on Heterogeneous Processor Platforms. IEEE Transaction on Parallel and Distributed Systems, 15(4) (2004) 319-330

13. Martino, V., Mililotti, M.: Scheduling in a Grid Computing Environment using Genetic Algorithms. International Parallel and Distributed Processing Symposium, Florida, USA, (2002)

14. Buyya, R., Murshed, M., Abramson, D.: A Deadline and Budget Constrained Cost-Time Optimization Algorithm for Scheduling Task Farming Applications on Global Grids. In Proceedings of Par. and Distr. Processing Techniques and Applications, USA, (2002)

15. Ranaldo, N., Zimeo, E.: An Economy-driven Mapping Heuristic for Hierarchical Master-Slave Applications in Grid Systems. 15-th Heterogeneous Computing Workshop. In Proceedings of the International Parallel and Distributed Processing Symposium, Greece (2006)

16. Di Santo, M., Frattolillo, F., Ranaldo, N., Russo, W., Zimeo, E.: Programming Metasystems with Active Objects. Proceedings of the International Parallel and Distributed Processing Symposium, France, (2003)

17. WSRF. http://www.globus.org/wsrf

A Multi-level Scheduler for the Grid Computing YML Framework

Sébastien Noël[1], Olivier Delannoy[3], Nahid Emad[3], Pierre Manneback[1], and Serge Petiton[2]

[1] Members of CoreGrid Institute on Resource Management and Scheduling
Faculté Polytechnique de Mons and CETIC, Mons, Belgium
{Pierre.Manneback,Sebastien.Noel}@fpms.ac.be
[2] Member of CoreGrid Institute on System Architecture
INRIA-Futurs, LIFL, USTL, Villeneuve d'Ascq, France
Serge.Petiton@inria.fr
[3] PRiSM - Laboratoire d'informatique - UVSQ, Versailles, France
{Nahid.Emad,Olivier.Delannoy}@prism.uvsq.fr

Abstract. This paper presents the integration of a multi-level scheduler in the YML architecture. It demonstrates the advantages of this architecture based on a component model and why it is well suited to develop parallel applications for Grids. Then, the multi-level scheduler under development for this framework is presented.[1]

Keywords: Grid Computing, YML, Scheduling, Resource Management, Workflow.

1 Introduction

High Performance Computing has emerged as a common need in many current applications. In order to solve such applications, Grid computing infrastructures have been developed to allow a high number of heterogeneous resources from different Virtual Organizations (VO) to be shared across a common network. Each cluster in each VO has its own management system. For example, availability of resources, access policies, Local Resource Manager (LRM), usage cost, etc. are usually different from site to site. Therefore common tools have to be provided to deal with resource heterogeneity and to facilitate the interconnection between them. Moreover, resource states are highly dynamic and volatile and increase the difficulty of managing a Grid infrastructure which is accessed concurrently by multiple users.

The development of Grid applications requires thorough knowledge of internal mechanisms and generally involves a preliminary step of identifying parallelizable parts of the application. This identification step leads to the creation of components, which are unitary tasks computed by one node of the Grid. An

---

[1] This research work is carried out under the FP6 Network Of Excellence CoreGRID funded by the European Commission (Contract IST-2002-004265).

W. Lehner et al. (Eds.): Euro-Par 2006 Workshops, LNCS 4375, pp. 87–100, 2007.

application is divided into components initialized with different input parameters and launched taking into account precedence constraints. Such workflow application is usually represented as a Direct Acyclic Graph and requires a high level of control of the Grid infrastructure.

In this paper, we study the use of a framework called YML for developing HPC applications on Grids and propose a multi-level scheduling architecture for it. The paper is organized as follows: in section 2, we succinctly present the YML framework and the associated workflow language YvetteML. Section 3 is devoted to the description of a general architecture of a multi-level scheduler for YML. Finally, section 4 presents some conclusions and perspectives

2   YML Framework

YML is a framework providing tools for parallelizing applications and has been developped at PRiSM laboratories in collaboration with Inria-Futurs/LIFL [3]. It focuses on two major aspects: the development of parallel applications and their execution in a Grid environment. YML makes this development independent of the Grid middlewares used underneath and hides the differences between them.

In the YML context, an application is divided into different computing sections, each of them containing some tasks sequentially or concurrently executed. A task, called a component, is a piece of work that can be mapped to one node in a parallel environment. It has some input and output parameters and is generally reusable in different parts of the application as well as in different applications. YML provides a special type of component, called *graph component*, which consists in the description of a subgraph. As we will see in 3.2, this kind of component will be exploited for the distribution of an application.

YML divides the development of a parallel application into three major steps:

1. *Definition of new components.* This definition consists of an abstract description and implementation component description, which are both presented in the next section.
2. *Description of the parallel application.* This description is independent of any underlying middleware and makes use of the components as functional units. It specifies the parallel and sequential parts of the application using the YvetteML graph description language and provides notifications to synchronise the execution of dependent components. This description is directly deduced from the graph representation of the application. More information on YvetteML is provided in subsection 2.3.
3. *Compilation of the application.* This step analyses and transforms the application graph into a list of parallel tasks taking into account the precedence constraints.

These three steps are all middleware independent and ensure that no Grid relevant knowledge is necessary to develop parallel applications.

After the compilation of the application, the execution can be started using a Workflow Scheduler which will interact with the underlying middleware. This

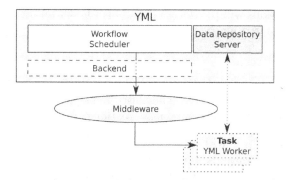

Fig. 1. YML Workflow Scheduler interaction with the middleware

interaction, represented in figure 1, requires the use of a specialized backend dedicated to the corresponding middleware. The execution of the application is directed by the Workflow Scheduler, which will submit tasks to the middleware through the dedicated backend. Each task is launched by a YML worker which will contact the Data Repository Server to obtain component binary and input parameters to start the computation.

2.1 YML Advantages

The comparison of YML with other workflow compatible frameworks like Unicore[5] or DAGMan[1] for Condor[6] points up several advantages.

YML helps the developer in the whole process of parallelizing applications. It starts at the early stage of component creation and goes through to the execution of strongly constrained workflow applications on a Grid. Moreover, YML allows the user to test and validate those applications on his own computer thanks to a special backend, which relies on the multithreading capabilities of the underlying operating system.

As we will see in the next subsection, component creation in YML is relatively simple. Existing code can be reused by importing libraries as new components without any adaptation. Those components are called by the application when computational tasks have to be started. Moreover, the notions of abstract and implementation descriptions of components add three interesting features to the Grid scheduler that could be used in the framework:

– data migration at the start and at the end of the application can easily be quantified from the abstract definition;
– the data used by a component is clearly defined in the abstract and implementation definitions; therefore this can be used in a checkpointing process to move a component from one node to another;
– computation time of a component can be evaluated from the implementation definition.

The use of Data Repository Servers hides the data migrations from the developer and ensures that the necessary data are always available to all application components.

The next subsections will present an example of how to use the YML framework to create a squared-matrix product application.

2.2   Component Creation

A component has to be defined and registered in the YML catalog in order to be used in a parallel application. This will be illustrated by a short example: a matrix multiplication component. The component creation can be done in three steps.

Definition of custom datatypes: New datatypes can be defined in new classes or in existing libraries (the YML compiler allows to include libraries, thus improving reusability of code). Functions for serializing and deserializing data have to be defined: the prototypes and their corresponding definitions (which are not represented here) are required for I/O operations made by the final component. Primitive datatypes such as integer, real and strings are already provided by the YML framework.

```
#ifndef MATRIX_HH
#define MATRIX_HH 1
#include <matrix.h>

typedef math::matrix<int> Matrix;

template <> bool param_import(Matrix& param,
        char* filename);
template <> bool param_export(const Matrix& param,
        char* filename);

#endif
```

This new datatype is called *Matrix* and makes use of the Matrix TCL Lite library [2] which does not require any modification.

Abstract definition: This definition includes a name for the component, a short description and a list of input and output parameters. This list specifies a name and a type for each parameter.

```
<?xml version="1.0" ?>
<yml-query login="userName" password="pass">
  <component name="MatrixProduct" type="abstract"
  description="Product of two matrices">

    <param name="result" type="Matrix"  mode="out"/>
    <param name="mat1"   type="Matrix"  mode="in" />
    <param name="mat2"   type="Matrix"  mode="in" />

  </component>
</yml-query>
```

This abstract definition is included in an XML request providing username and password for authentication purposes. The *Matrix* type is a custom datatype and has been defined at the previous step. The names of the three parameters match the names of the variables in the implementation part.

Implementation: It is based on the abstract definition. The output will be automatically sent to the Data Repository Server and made available for other components. This implementation is currently done using C/C++ but other programming languages can be added into all backends.

```xml
<?xml version="1.0" ?>
   <yml-query login="userName" password="pass">
      <component name="MatrixProduct_Impl" type="impl"
      abstract="MatrixProduct"
      description="Product of two matrices">
         <globals>
            <![CDATA[

              #include <matrix.h>

            ]]>
         </globals>
         <source lang="CXX">
            <![CDATA[

              result = mat1 * mat2;

            ]]>
         </source>
      </component>
   </yml-query>
```

This simple example demonstrates how easily components are created with YML. After the creation of the components, the graph description language YvetteML can be used to describe the application. Application creation with YvetteML is presented in the next subsection.

2.3 Application Creation with YvetteML

YvetteML provides different features for creating applications. These features are described in an illustrative example in figure 2, i.e. a parallel squared-matrix product. This application makes use of:

– *Component calls.* Their role is to submit a new task to the Local Resource Manager (LRM) providing the name of the component defined earlier and the different input parameters (lines 15, 16, 27 and 35 of figure 2).

```
1    <?xml version="1.0"?>
2    <yml-query login="userName" password="pass">
3
4      <application>
5        <source>
6    size     := 4;
7    div      := 2;
8    url1     := "http://www.prism.uvsq.fr/cni/yml/matrix1.csv";
9    url2     := "http://www.prism.uvsq.fr/cni/yml/matrix2.csv";
10
11   par
12     par (i:= 1; div)          # i = index of the row
13         (j:= 1; div)          # j = index of the column
14     do
15        compute MatrixLoad(mat[1][i][j],url1,size,div,i,j);
16        compute MatrixLoad(mat[2][i][j],url2,size,div,i,j);
17        notify(evtMatrixLoaded[1][i][j]);
18        notify(evtMatrixLoaded[2][i][j]);
19     enddo
20   //
21     par (i:= 1; div)
22         (j:= 1; div)
23         (k:= 1; div)
24     do
25        wait(evtMatrixLoaded[1][i][k]);
26        wait(evtMatrixLoaded[2][k][j]);
27        compute MatrixProduct(result[i][j][k],mat[1][i][k],mat[2][k][j]);
28     enddo
29   endpar
30
31   seq (i:= 1; div)
32       (j:= 1; div)
33       (k:= 1; div)
34   do
35     compute MatrixComp(final,i,j,size,div,result[i][j][k]);
36   enddo
37
38      </source>
39      </application>
40    </yml-query>
```

Fig. 2. Squared-Matrix Product Application using YvetteML

- *Parallel sections.* They are used to explicitly define sections which will be executed in parallel (lines 11, 20 and 29 of figure 2) or to execute a parallel loop with iterators (lines 12 and 21 of figure 2).
- *Sequential loops.* They are loops with iterators, which are executed sequentially (line 31 of figure 2).
- *Conditional statements.* They can be used to test the value of iterators.

– *Event notifications.* They are used to synchronize the different parts of the execution when a precedence constraint has to be respected (lines 17, 18, 25 and 26 of figure 2). For instance in line 17, a new event called *evtMatrixLoad* is defined with an index ([1][i][j]) equal to that of the matrix that has just been loaded by the *MatrixLoad* component. After this notification, the corresponding wait call (in line 25) will stop blocking the execution of the iteration in the parallel loop.

The application described in figure 2 is presented for illustrative purpose. It makes use of three components: *MatrixLoad* (which loads part of a file into a *Matrix* datatype), *MatrixProduct* (which computes the product of two matrices) and *MatrixComp* (which composes the result *Matrix* by aggregating all submatrices).

This section has briefly presented the YML framework. More details can be found in [4]. Next section will describe the architecture of a scheduling model that we are considering.

3   A Multi-level Scheduling Model in YML

We describe in this section a multi-level scheduling model based on the YML framework. This model has multiple objectives:

1. to schedule a set of YML components with input data and precedence constraints issued from one or more users;
2. to provide computing resources for these components in a multi-middleware environment;
3. to offer users a guarantee in terms of completion time of the application;
4. to dynamically reorganise the schedule if unexpected events occur.

The following subsections develop different aspects of the model and present a case study.

3.1   An Economic Model

The context we focus on is characterized by the following points:

– the objective of the Grid is High Performance Computing;
– the applications are mostly compute-intensive rather than data-intensive;
– the resources are owned by different providers and part of different VOs;
– the number of resource providers ranges from several dozen to several hundred;
– the architecture is not centralized.

Each cluster:

– is composed of homogeneous resources;
– has a single access point;
– has a previously negotiated access policy to one or more other sites;
– is managed by an LRM (which may be different from one cluster to another).

The model we propose is based on an economic approach of resources and defines different entities which will interact within the Grid infrastructure. An entity can be a resource provider or a consumer, or both. Consumers require resources owned by different providers and available on the Grid. When a provider receives a request from a consumer, he will answer by proposing a set of suitable schedules and associated cost for parts of the application depending on access policy of the consumer and availability of local resources. He can possibly subcontract parts or the whole application to other resource providers without mentioning anything to the consumer.

This model can be used in different scenarios: either cooperation or competition between sites in the Grid infrastructure. Moreover, a hierarchy with different layers of scheduling instances, as presented in [9], can be built.

Technically, the main idea is to provide a YML server for each LRM. This YML server has 3 main purposes:

- to communicate with other YML servers and therefore, connect the different clusters in a common Grid;
- to interact with the underlying LRM using a specialized backend;
- to provide the features missing in the LRM.

The following subsection presents a typical scenario with this economic model.

3.2 Scheduling Scenario

A typical scheduling scenario is as follows:

1. the user/consumer submits his application to the local YML server;
2. the YML server analyses the application and decides whether it can provide the resources or not;
3. the YML server may forward the whole or parts of the request to other resource providers;
4. suitable schedules are sent back in return of each request;
5. the local YML server gathers the information and proposes differents prices to the user/consumer.

Steps 2, 3 and 4 are executed consecutively each time a YML server receives a scheduling request. The different sequences of the above scenario are explained in more detail in the next subsections.

Submission of the application. As described in section 2, the user makes use of YvetteML to describe a parallel application. Within the submission request, the user provides a completion time for the whole application or for some parts of it depending on the requirements. The local YML server handles the user requests in compliance with its local access policy. When the policy forbids access or the user has no authorization, the request fails and the computation is stopped. Otherwise, the scheduling process goes on to the next step.

Analysis of the application graph. Taking into account the amount and types of local resources on the one hand, and the current resource reservations on the other hand, the YML server will attempt to find suitable schedules for the whole or parts of the application. It will try to schedule successively:

1. the whole application;
2. parallel sections;
3. graph components;
4. tasks in the parallel sections.

If local resources are able to compute the whole application and meet the user's constraints, the scheduling process is either stopped or forwarded to other YML servers in the Grid. In the first case, reservation is made on local resources and the computation is started. In the latter case or in case the local infrastructure cannot provide sufficient resources for the whole application, the scheduling continues with step 3.

Forwarding of the request. The local server can decide whether it forwards the whole request or only parts of it (this decision can be made in compliance with the access policy). In the latter case, the request is split into different sub-requests and is sent to other sites. To forward a request in the Grid infrastructure, the local server will interact with other resource providers with whom an access policy has been negotiated.

Return of suitable schedules. Each server will aggregate the suitable local schedules as well as schedules from other resource providers. Then, a reply is sent to the user.

3.3 Access Policy

When an instance wants to join the Grid infrastructure (to provide or to use resources), it has to negotiate access policies with one or more other scheduling instances. We propose an access policy divided into two sections, each containing static or dynamic information. This can be used to get a highly customizable contract between both scheduling instances. The static information will be used first to filter the list of resource providers without any interaction. The resulting list will be filtered again by querying each resource provider.

A non-exhaustive list of possible parameters that can be set in an access policy may include:

- time intervals;
- cost per node per time unit;
- constant/variable cost;
- application size;
- number of nodes;
- nodes description;
- number of providers;

- resource reservation;
- forwarding policy of the requests;
- failure compensation;
- access priority;
- etc.

Some or all of these parameters have to be set so as to define an access policy which can then be used in the resource discovery process. As presented in [7], this process essentially involves two filtering steps: an authorization filtering and a minimal requirement filtering.

The application requirements are defined by the YML Compiler, which analyzes the YvetteML code of the application provided by the user; this information is used for the minimal requirement filtering.

Figure 3 presents the two-step reduction of the set of suitable resources. First, a static filtering is applied to obtain a reduced list B. Then, if the list B is not empty, requests are sent to the resource providers to obtain dynamic information which will be used as a second filter to get a resource list C. This two-step filtering aims to reduce the number of requests exchanged between the scheduling instances.

Each resource provider in list C will be queried for possible schedules of the application. This process is illustrated in the next subsection.

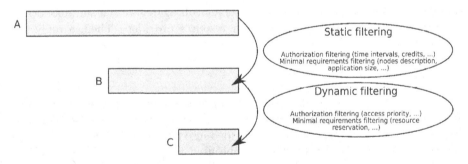

Fig. 3. Resource discovery process: static and dynamic filtering

3.4   Case Study

To describe the scheduling model, we will focus on an example Grid, presented in figure 4: the Grid infrastructure contains 5 clusters from different Virtual Organizations. An arrow from a server to another means that the first has an access policy to contact the latter. For instance, *YML server 1* has two resource providers, namely servers *2* and *4*; the rest of the Grid (servers *3*, *5* and *6*) is not visible to server *1*. When a server receives a request, it can handle the entire request or ask other resource providers. An access policy has previously been

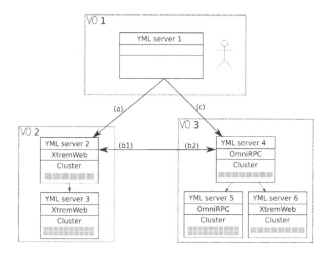

Fig. 4. Example of Grid infrastructure with different Virtual Organizations

negotiated and can be different for each client of a single site. Therefore, a direct request to a server may be less interesting than going through an intermediary. For instance, in figure 4, policy (c) could be more expensive than (a)+(b2); in this case, the client located in VO1 can ask for resources from server *2* which will negotiate resources with server *4* in VO3. The negotiation between *2* and *4* is not visible to the first client. As in VO1, a *YML server* can have no local resources; therefore, it acts only as a client and will contact other sites to get computational resources.

An example application is represented in figure 5 by a graph showing the interdependence between the tasks.

The start of the application is represented at the top of the figure and the end at the bottom. Large dashed squares represent parallel sections of the application, described by the user in the YvetteML code. Each task (which is a component with input parameters) is represented by a plain arrow. Notification arrows (dotted) are used to synchronise tasks and introduce precedence constraints into the application. The example presented in figure 5 has two parallel sections; the first is a preprocessing stage needed to start the computation process of the second one. For instance, the preprocessing tasks can be an initialization of the data. The duration is 3 for a preprocessing task and 10 for a computing task.

The scheduling process or mapping the application in figure 5 on the Grid presented in figure 4 will be simplified to help comprehension.

We suppose that:

- *YML servers 3* and *4* are unavailable for computation;
- 3 nodes of *YML server 2* are unavailable;
- the dialog between *YML servers 4, 5* and *6* is not described;
- single task allocation is not presented but is effective in our model;

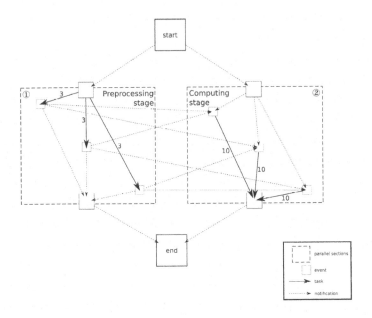

Fig. 5. Example of application graph

– the parameters of access policy (c) are such that no schedules will be proposed.

The response to each request is presented below from a YML server to another. Symbols ① and ② refer to the parallel sections in figure 5.

1. *Response from server 4 to server 1.* Access policy (c) is such that no schedules will be returned to YML server 1.
2. *Response from server 4 to server 2*
   Table 1 presents 3 different sets of schedules. The whole application can be coallocated to the resources of server *4* (or those of subcontractors which is not indicated to server *2*) but this coallocation cannot start before time 45. This means that many reservations have already been made or that the access policy cannot provide enough resources before this time. Other propositions consist in scheduling a parallel section (① or ②), which can be started earlier (on time 3).
   We suppose that the cost per node per time unit equals 2 in access policy (b2) (which is static information). The cost for the different schedules can

Table 1. Set of schedules proposed by server *4* to server *2*

application part	starting time	cost
① and ②	[45,∞]	(3*3+3*10)*2=78
①	[3,17]	3*3*2=18
②	[3,10]	3*10*2=60

Table 2. Set of schedules proposed by server *2* to server *1*

application part	starting time	cost
① and ②	$[45,\infty]$	$(3 * 3 + 3 * 10) * (2 + 1) = 117$
①	$[1,4]$	$3 * 3 * 1 = 9$
①	$[3,17]$	$3 * 3 * (2 + 1) = 27$
②	$[3,10]$	$3 * 10 * (2 + 1) = 90$

be evaluated: 3 nodes x 3 time units x 2 for parallel section ① and 3 nodes x 10 time units x 2 for parallel section ②.

YML server 2 will aggregate those prices with its local suitable schedules.

3. *Response from server 2 to server 1*

We suppose that the cost per node per time unit is not fixed in the access policy and is therefore a dynamic information. This means that server *2* is allowed to ask a different price at each request, depending on local considerations.

The coallocation of the whole application can only be done by server *4* (or subcontractors): this is not indicated to server *1* which will see server *2* as only resource provider. Server *2* will increase the cost of the resources by 1 to take into account bandwidth use to access server *4*.

A new schedule for parallel section ① is proposed by server *2*, which aims to enhance the use of local resources by applying an attractive cost of 1 per node per time unit.

These schedules are received by *YML server 1* which will choose some of them and start resource reservation by requesting server *2*.

3.5   Features for the Scheduling Model

As presented in [8], a Grid scheduling architecture should provide different important features. These features are discussed in this subsection using the economic model described in 3.1.

The resource discovery process is not a major feature in our context. Each LRM is responsible for managing resource status and for providing suitable schedules depending on the resource availability.

In the same way, the status monitoring is not centralized and is only accessible by the local YML server, which will ask the LRM to provide the necessary information. This information can be accessed differently according to the installed LRM.

The reservation of resources is not supported by all LRMs and can therefore be managed by the YML server if necessary. In such Grid, the resource administrator has to ensure that YML is the only way of submitting tasks to the local resources.

The accounting and billing features will be managed at the YML level.

4 Conclusions and Perspectives

In this paper, we have presented the YML Grid computing framework which can be used to develop parallel applications and execute them in a Grid environment. We have also described a multi-level scheduling model which can be used to build cooperative or competitive Grids using a customized access policy between scheduling instances of the Grid. This scheduling model is currently being integrated into the YML framework and will provide multi-middleware capabilities.

We aim to validate this scheduling model by testing it on the YML framework. This testing phase will open up new perspectives and show what will be needed to be adapted in the current model.

References

1. Directed acyclic graph manager website. http://www.cs.wisc.edu/condor/dagman.
2. Techsoft - matrix TCL website. http://www.techsoftpl.com/matrix.
3. YML website. http://www.prism.uvsq.fr/cni/yml.
4. O. Delannoy, N. Emad, and S. Petiton. Workflow global computing with yml. To appear in Proceedings of GRID2006, Barcelona.
5. D. W. Erwin and D. F. Snelling. Unicore: A grid computing environment. *Lecture Notes in Computer Science*, 2150:825–834, 2001.
6. S. Santhanam, P. Elango, A. Arpaci-Dusseau, and M. Livny. Deploying virtual machines as sandboxes for the grid. In *Second Workshop on Real, Large Distributed Systems (WORLDS 2005)*, San Francisco, CA, December 2005.
7. J. M. Schopf. Ten actions when grid scheduling. *Grid Resource Management*, pages 15–23, 2004.
8. U. Schwiegelshohn and R. Yahyapour. Attributes for communication between grid scheduling instances. *Grid Resource Management*, pages 41–52, 2004.
9. N. Tonellotto, P. Wieder, and R. Yahyapour. A proposal for a generic grid scheduling architecture. volume TR-0015 of *CoreGRID Technical Report*, November 2005.

Virtual Environments - Framework for Virtualized Resource Access in the Grid

Michał Jankowski[1], Paweł Wolniewicz[1], Jiří Denemark[2], Norbert Meyer[2],
and Luděk Matyska[23]

[1] Poznań Supercomputing and Networking Center,ul. Noskowskiego 10,
61-704 Poznań, Poland
{jankowsk,meyer,pawelw}@man.poznan.pl
[2] Faculty of Informatics, Masaryk University, Botanická 68a,
602 00 Brno, Czech Republic
jirka@ics.muni.cz
[3] Institute of Computer Science, Masaryk University, Botanická 68a,
602 00 Brno, Czech Republic
ludek@ics.muni.cz

Abstract. To assure secure access to any computer resources one must
provide an adequate level of authentication, authorization job isolation
and possibility of auditing user actions. In the grid environment that
comprises a large number of users and resources in different administrative domains, these features are challenging. Grid economy and accounting related to it are becoming more and more important in an emerging
aspect of grid commercialization. Also, the requirements of the users
and administrators are becoming more and more sophisticated: checkpointing and migration of jobs, detailed software requirements, quality
of service, collaborative work, and load balancing, to name a few. Virtualization techniques, nowadays more and more matured and advanced,
seem to help solve the above-mentioned problems. In the present paper
we discuss some of these techniques as well as existing solutions and then
propose a framework for Virtual Environments. The framework focuses
on resource access control, but the benefits of virtualization are wider.

1 Introduction

Controlled and secure access to grid computational resources requires authentication, authorization, an adequate level of job isolation and possibility of auditing user actions. This should be realized with as little administrative effort
as possible, though providing the administrators and Virtual Organization (VO)
managers with enough control on their resources and users. Grid economy, which
introduces accounting and billing requirements, also becomes more and more
important. From the users point of view the whole Grid should be seen as a
single computer with appropriate software, hiding all the technical details connected with physical locations, middleware, operating systems, etc. The mentioned groups of requirements (described in detail in [13,14]) are closely related
on the conceptual level.

W. Lehner et al. (Eds.): Euro-Par 2006 Workshops, LNCS 4375, pp. 101–111, 2007.
© Springer-Verlag Berlin Heidelberg 2007

We have researched a number of existing solutions and found that there are tools that provide for at least part of the functionality we are interested in, however none of them addresses all the issues. These tools are widely used in numerous projects and some of them have become standard, so it seems to be reasonable to be compatible with them. Moreover, different users, resource owners and VO managers may have different and often conflicting needs. For example, there is no isolation level that would always be suitable; sometimes complete encapsulation of jobs is a must, sometimes jobs need collaboration or strict isolation is too heavy solution. Hence we have several components ("building blocks") that could be used to build a perfect solution for a given situation, but we need a way to put them together into a framework that will combine these tools and their features gaining the synergy effect.

Virtualization technology has a long history in computer science [1]. It allows for partitioning or combining real components of computer infrastructure (hardware, software, networking, etc.) into virtual entities. This technique abstracts from internal details of physical elements, provides isolation and common interface for virtual elements, even if they share physical entities. Examples are virtual memory, virtual machines and virtual networks. This technology seems to be especially promising for grid middleware as it must cover large, loosely coupled and heterogenous distributed systems and should hide its complexity from the user.

In our previous papers we have introduced a concept of the *Virtual Environment*, which we understand as encapsulation of user jobs in order to give a limited set of privileges and be able to identify the user and organization on behalf of which the job acts. Depending on requirements we may virtualize user accounts [10],[11] or virtual machines [7]. The concept of "combining real components" opens a way for dynamic construction of an environment. The virtualization technique simplifies the assignment of jobs to resources either by discovering a statically created environment or by expressing parameters of a dynamically created one, because the user's requirements specified in an abstract language may specify abstract features of the environment. In this work we describe an idea of a framework for the creation and managing of Virtual Environments. The structure of the paper is as follows: in section 2 we discuss advantages and disadvantages of Virtual Accounts (VA) and Virtual Machines (VM); section 3 describes the concept and implementation of Virtual Workspaces (VW), which we found especially useful in constructing our framework; section 4 provides architecture of our framework; accounting issues will be discussed in section 5, and finally section 6 concludes the paper.

2   Comparison of VA and VM

Both methods of virtualization: Virtual Accounts and Virtual Machines allow for running jobs in separated Virtual Workspaces, but they are best suitable for different purposes. Virtual Accounts is just a simple implementation of assigning users to different Unix accounts. Different directories and Unix accounts are used to separate jobs. In case of workflow, tasks are run in the same account. Complete knowledge about mapping from real to the Virtual Account is stored locally and can be used to resolve all Unix accounts from standard accounting procedures. Virtual Machines have far more possibilities. In fact Virtual Machines run several instances of the operating system at the same time and thus provide complete job separation. Virtual Machines are best suitable for resource centers where job requirements differ, e.g. operating system requirements are different or even the grid infrastructure for different users group is incompatible.

Virtual Machines can be used in two ways. One way is to set up static Virtual Workspaces, for example, to run two different grid infrastructures, or to run different grid testbeds for different VOs. Virtual Machines (or rather Virtual Clusters) can also be set up on demand, for the lifetime of the job. This, however, causes some overhead because either the Virtual Machine must be created and started (which is time-consuming) or the Virtual Machine was created before and must now be resumed and reconfigured (which is memory-consuming).

Accounting is very important for the system administrator. The resources used must be calculated and stored. Standard accounting stores all information locally. This causes problems for Virtual Machines, because when the machine is deleted, all detailed information is lost. On the other hand, when the ma chine is migrated, the information may be inaccurate. Estimated accounting is still available from the Virtual Machine Management System (e.g. Xen[22,23]), but they combine all details into one set of numbers. For instance, it is not possible to distinguish between user time and system time, and information about executed command names are not available. The solution is to use an external database and send all information there during shutdown.

A similar limitation is connected with audit. Logging the operations performed locally on the VM, on its virtual resources is usually not interesting for the physical machine administrator, but access to some physical devices (e.g. laboratory equipment) or network connections may be the subject for audit. However, this may be difficult as some relevant logs may be located on VM and lost on deletion. Also if the VM user has root privileges, he may maliciously or accidentally modify or remove the logs.

Using Virtual Machines it is possible to provide a service level agreement (SLA). Resources assigned to the given Virtual Machine can be managed easily. SLA for systems with Virtual Accounts is limited. To some extent it can be achieved by careful configuration of the operating system and the queuing system.

The integration of virtual environments with the grid infrastructure is especially important. The Virtual Accounts can be easily integrated with Globus [15] or gLite [17], because it is just a plugin to the grid middleware. With Virtual Machines things are more complicated. Dynamic creation of Virtual Machines is not compatible with existing grid resource brokers. Resource broker does not know about wirtual environment, therefore it can not create wirtual workspace. The resource broker just contacts the head node and submits the job to the cluster. But in case of the Virtual Cluster the head node is not created yet and it should be created after the resource broker submits the job to the site. Therefore

Table 1. Summary of Virtual Accounts and Virtual Machines

	Virtual Accounts	Virtual Machines
Purpose	small clusters simple needs	Many VOs, many OS-es, Many jobs at a time, SLA
Flexibility	in some extent	very flexible
Job separation	limited	full
Accounting	full	limited
Audit	full trusty	limited may be untrusty
Administration	easy	difficult
SLA	limited	yes
integration with grid systems	easy	difficult
resource consumption	insignificant	small to large

the submission process must have two steps, and an additional module (GRAM Proxy) is needed. First, GRAM Proxy accepts the job from the resource broker and then creates the Virtual Machine and submits jobs there

For Virtual Machines there is also a problem of administration. Virtual Machines are set up from partitions stored somewhere on a hard disk. But Virtual Machines restored from the partition must be up to date, which means that after startup some configuration must be updated, e.g. gridmap files, certificate revocation list, some security patches etc. This causes additional time overhead and delays job starting.

Summary of features for virtual systems is presented in table 1.

In general, Virtual Machines have a big potential, but quite often all site requirement can be fulfilled with Virtual Accounts. For sites with thin nodes (single or dual processors) a typical configuration of job management systems allows for running only one job per node. In this way jobs are completely separated. For large nodes with many jobs running at the same time, dynamic assignment can be a very good solution, especially in the context of a service level agreement.

An intermediate solution with Virtual Machines set up statically to share the same hardware between different grid infrastructures and with Virtual Account used to ensure virtualization inside Virtual Machines is also possible.

3   Virtual Workspaces Approach

An architecture called virtual workspaces [3,4,5,6] has been designed to automate the creation and management of distributed dynamic virtual environments in the Grid. The architecture comprises several services used to create and manage virtual environments. When users want to submit a job to a Grid resource, they contact an appropriate service to create a dynamic virtual environment for them. For existing environment users can use another service to manage the environment, e. g., to change the environment's lifetime, to configure or terminate the environment. During the lifetime of the virtual environment, standard Grid services such as GRAM can be used for submitting jobs.

The Virtual Workspaces architecture does not enforce any virtual environment implementation. Currently, the implementation of Virtual Workspaces based on Dynamic Accounts and Virtual Machines is available. The background technology is not pluggable—it is chosen at install time and then only the selected implementation is available.

Virtual Workspaces can be simple (or atomic), and jobs are submitted directly into it, or it may consist of several (either atomic or complex) workspaces. Complex workspaces are used to create virtual clusters with a set of definitions of Virtual Workspaces for both the head node and worker nodes of a real cluster. According to a specification of a virtual cluster, several Virtual Machines are deployed on physical cluster nodes and set up to form an isolated private IP network. The virtual machine running on a head node is the only part of a virtual cluster with a public IP address. After all virtual machines forming the virtual cluster are up and running, a GRAM service is started on the virtual head node. Clients then use this GRAM service to start their jobs on the virtual cluster.

To support Virtual Workspaces, each node of a physical cluster must run the Xen Virtual Machine Monitor and several services for staging, starting and managing Virtual Machines.

As the Virtual Workspace is just an environment for submitting users' jobs, it must be accessible from everywhere for users to be able to contact its services. In other words, each Virtual Workspace (except for worker nodes of a virtual cluster) has to be provided with a public network address. This may cause problems especially when more than one Virtual Workspace is allowed to be created on a single physical machine.

On the other hand, users are provided with a way of deploying their own environment which perfectly suits their needs. However, if users or VO administrators are allowed to provide a complete image of a Virtual Machine, it must be done in such a way that site administrators are willing to trust the image.

If more detailed information on using specific resources is needed for accurate accounting, coarse runtime data obtained from the Virtual Machine Monitor may not be enough, and special monitoring tools providing data from the inside of a Virtual Machine have to be deployed. Similar tools might be useful for logging user activities.

4 Architecture of the Framework

In our previous papers [13,14] we described a set of different requirements for user management and access to resources. We stated that there are numerous tools that provide at least part of the required functionality, however none of them addresses all the issues. These tools are used in working Grids. We proposed to put them into a pluggable framework that will combine the features gaining the synergy effect.

Section 3 has described Virtual Workspaces effort in detail. Conceptually the Virtual Workspace is the same as the Virtual Environment defined in our previous papers. Moreover, VW implementation seems to fulfill most of our requirements and its architecture is quite similar to our framework. Similarly to our proposition VW employs WS–Stateful Resource [9] for modeling of the workspace and managing its life cycle. Hence we would like our framework implementation to be based on VW. In this section we will discuss how the VW fits our framework, and which elements should be added or modified.

As we discussed in the previous section, both Virtual Environment implementations (Dynamic Accounts aka Virtual Accounts and Virtual Machines) have their pros and cons. The decision on using or not VE and which one is the preferable implementation is up to the resource administrator. This fact should be transparent from the user point of view, but the VW require explicit create and life time management operations and to make things worse, these two implementations provide slightly different interfaces. As a result, the party that requests the job run (either a resource broker or directly the user - let's call them both "client") must take care of workspace management and be conscious of actual interface. These operations may be necessary for advanced global schedulers that support SLA or checkpointing and workspace migration. Explicit workspace management is still redundant from the point of view of the clients in most cases. The user just wants to run a job with specified parameters and creation of the workspace is a technical detail that should be hidden. Also most of existing brokers would require modification in order to support the VW operations. The appreciated scenario is that the client calls the resource manager service (like Globus GRAM) directly and without a previous request for the workspace creation.

We propose to hide the creation and lifetime management inside the resource manager, that will take care of the creation automatically. Any special user requirements concerning hardware (number of nodes, memory, etc.), operating system and software may be expressed in the job description. These information passed to the resource manager may be used for the Virtual Environment creation. Any creation parameters, that are not explicitly specified may obtain default values. The Virtual Environment must live until the job is finished at least, then it may be destroyed. The destruction may be performed periodically or when the resources occupied by the VE are needed by someone else.

The described architecture is shown on figure 1. The newly proposed parts are modified Globus GRAM (may be both WS and pre-WS one) which accepts job management requests, VE database and VE Create & Mapping module that

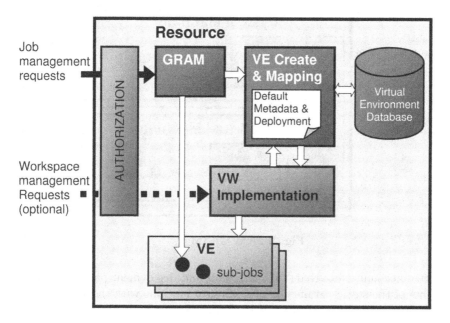

Fig. 1. Architecture of the Framework

interfaces the GRAM with VE database and VW implementation. Webservice interface of VW may be accessible outside optionally and if this is the case, VW operations must be synchronized with the Create & Mapping.

One of the most important features connected with the resource management is fine grained and flexible authorization. The GT 4.0 Authorization Framework [18,19] allows for a variety of authorization schemes, including a gridmap-file, an access control list defined by a service, an SAML-based authorization service and any custom authorization handler. The security descriptors allow for flexible security configuration on different levels: container, service, and even resource. There is a number of existing authorization systems and mechanisms that already are or easily may be plugged into this Globus framework and fulfill our authorization requirements. The administrator may properly configure the Virtual Workspaces and WS GRAM services according to the local needs. The pre–WS authorization is not equally flexible, but it is still possible to implement its own, fine grained authorization using callouts mechanisms [20,21].

Note, that the authorization is closedly related to the workspace creation and mapping user to the workspace. The limitations put on the workspace (e.g. privileges of the virtual account or resources allocated to virtual machine) are simply security enforcement mechanisms. Moreover, the following job run or file transfer requests with the same credencials should be mapped to the same environment.

In case of VA implementation, creation of the environment is virtually equivalent to the mapping operation and may be easily realized by the GRAM mapping module. The environment is simply a record in the VE database that binds user

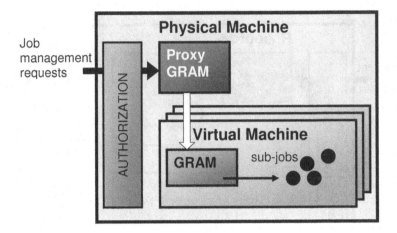

Fig. 2. Proxy GRAM

to a virtual account. Note that VM meta data and deployment parameters are equivalents of the static parameters of a physical machine with the VA system. In case of VA the parameters are only evaluated if they fit to the real resources (e.g. if required software is installed).

In case of VM implementation, virtual cluster matching the user requirements must be actually created by the resource manager. The resource manager, that actually runs the job, is located on the head node of the virtual cluster. In order to make this fact invisible for the client, all client requests are accepted by the GRAM that is located on the physical machine (e.g. on domain0 of Xen). It is called "proxy GAM" in that case. The proxy will access the VE database in order to set/get the current user – VE mapping, create the VM if necessary and forward the job request to the "internal" GRAM see figure 2.

The Virtual Workspaces implementations are missing a database that might be used for storing history of mappings user – Virtual Environment which is crucial for accounting and auditing purposes. The following section describes in more detail what should be stored in this database, how these data might be obtained from the underlying system and how the information would be exposed outside.

5   Accounting and Audit

The auditing or accounting data is normally bound to a local account or Virtual Machine instance. However, in a grid system one is interested in this information in the context of the global user identity and his Virtual Organization. The records of VE operations together with the standard system logs and accounting data provide complete information on user actions and resource usage, but these two sources must be combined. Virtual Environment Information Subsystem enables this feature. It consists of VE Database, VE Information Service and framework for collecting accounting and audit events - see figure 3.

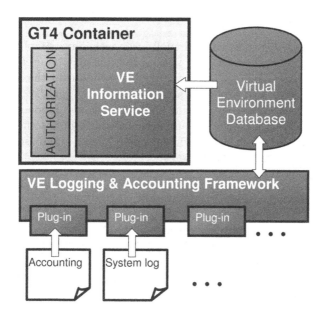

Fig. 3. Architecture of Virtual Environment Information Subsystem

The Virtual Environment Database stores all the relevant information connected with the VE creation and deletion, just on the request of the VE services. The database is also capable of storing any type of accounting data, both standard and nonstandard, and any kinds of events described in string values, all of this unified and connected to the grid user. These data may be collected periodically or on request (e.g. just after the VE is deleted), by analyzing sources like Unix accounting records, system logs etc. The sources may be quite different, depending on VE implementation, operating system, used software etc. so we use a pluggable framework. A plugin must be implemented for each source.

The Virtual Environment Information Service is a frontend for the Virtual Environment Database. Access to the data must be authorized and depends on the users role: all the users have rights to read the accounting data referring to themselves, managers of virtual organizations are able to read data referring to all VO members, owners of resources are allowed to read all the data connected to the resource.

As stated in section 2, the Virtual Machine implementation introduces some problems connected with the accounting and audit data gathering. This may be overcome by careful configuration and running some software on VM that will put the relevant information to the VE database on the physical machine. This workaround, however, will result in lower flexibility of the machine (e.g. the VM is not fully transparent for the migration process).

6 Conclusions

In the paper we have shown how the virtualization techniques simplify access and administration of grid resources, and which solutions may be useful depending on the situation. We have discussed the leading solution in the area: Virtual Workspaces. We have also proposed a framework based on VW which allows for easy integration of numerous existing grid middleware components. VW Our contribution to VW is as follows: automatic (transparent for the client: user or resource broker) creation of the virtual environment, database and service supporting accounting and audit features.

Acknowledgment

This work has been supported by the CESNET Research Intent (MSM6383917201) and by the EU CoreGRID NoE (FP6-004265). Implementation of Virtual Accounts System is included in EU BalticGrid project (RI-026715) and in Clusterix - National Cluster of Linux Systems, project co-funded by the Polish Ministry of Sciences.

References

1. A.Singh: An Introduction to Virtualization.
   http://www.kernelthread.com/publications/virtualization/ (2004)
2. I.Foster, C.Kesselman, S.Tuecke: The Anatomy of the Grid: Enabling Scalable Virtual Organizations. International J. Supercomputer Applications **15(3)** (2001)
3. I.Foster, T. Freeman, K.Keahey, D.Scheftner, B.Sotomayor, X.Zhang: Virtual Clusters for Grid Communities. CCGRID 2006, Singapore (2006).
4. K.Keahey, I. Foster, T. Freeman, X. Zhang: Virtual Workspaces: Achieving Quality of Service and Quality of Life in the Grid. Scientific Programming Journal, Volume 13, **4/2005** (2005)
5. X.Zhang, K.Keahey, I.Foster, T.Freeman: Virtual Cluster Workspaces for Grid Applications. ANL/MCS-P1246-0405 (April 2005)
6. K.Keahey, I.Foster, T.Freeman, X.Zhang, D.Galron: Virtual Workspaces in the Grid. Europar 2005, Lisbon, Portugal (September 2005)
7. K.Keahey, K Doering, I.Foster: From Sandbox to Playground: Dynamic Virtual Environments in the Grid. 5th International Workshop in Grid Computing (Grid 2004), Pittsburgh, PA (November 2004)
8. K.Keahey, M.Ripeanu, K.Doering: Dynamic Creation and Management of Runtime Environments in the Grid. Workshop on Designing and Building Web Services (GGF 9), Chicago, IL (October, 2003)
9. I.Foster, J.Frey, S.Graham, S.Tuecke, K.Czajkowski, D.Ferguson, F.Leymann, M.Nally, I.Sedukhin, D.Snelling, T.Storey, W.Vambenepe, S.Weerawarana: Modeling Stateful Resources with Web Services, version 1.1. http://www-128.ibm.com/developerworks/library/specification/ws-resource/(March 2004)
10. M.Kupczyk, M.Lawenda, N.Meyer, P.Wolniewicz: Using Virtual User Account System for Managing Users Account in Polish National Cluster. HPCN, Amsterdam, (June 2001)

11. M.Jankowski, P.Wolniewicz, N.Meyer: Virtual User System for Globus based grids. Cracow '04 Grid Workshop, (December 2004)
12. J.Denemark, M.Jankowski, A.Krenek, L.Matyska, N.Meyer, M.Ruda, P.Wolniewicz: Best Practices of User Account Management with Virtual Organization Based Access to Grid. 6th International Conference, PPAM 2005, Springer-Verlag LNCS 3911, Poznan, (September 2005)
13. J.Denemark, M.Jankowski, L.Matyska, N.Meyer, M.Ruda, P.Wolniewicz: User Management for Virtual Organizations. CoreGRID Integration Workshop, Pisa (2005)
14. J.Denemark, M.Jankowski, L.Matyska, N.Meyer, M.Ruda, P.Wolniewicz: Core-GRID Technical Report TR-0012: User Management for Virtual Organizations. (2005)
15. http://www.globus.org
16. http://workspace.globus.org
17. http://glite.web.cern.ch/glite/
18. http://www.globus.org/toolkit/docs/4.0/security/authzframe/
19. B.Lang, I.Foster, F.Siebenlist, R.Ananthakrishnan, T.Freeman: A Multipolicy Authorization Framework for Grid Security. Accepted by the IEEE NCA06 Workshop on Adaptive Grid Computing (to appear in Proc. Fifth IEEE Symposium on Network Computing and Application), Cambridge, USA (July 2006)
20. http://www.globus.org/toolkit/security/callouts/
21. GSI Admission Control and Identity Mapping Callout Specification, Draft. (July 1, 2003)
22. P.Barcham at al: Xen 2002. University of Cambridge Computer Labolatory Technical Report UCAM-CL-TR-553, Cambridge, (January 2003).
23. P.Barham, B.Dragovic, K.Fraser, S.Hand, T.Harris, A.Ho, R.Neugebauer, I.Pratt, A.Warfield: Xen and the Art of Virtualization. Symposium on Operating Systems Principles (SOSP '03) (October 2003)

Grid Meta-Broker Architecture: Towards an Interoperable Grid Resource Brokering Service

Attila Kertész[1,2] and Péter Kacsuk[2]

[1] Institute of Informatics, University of Szeged
H-6720 Szeged, Arpad ter 2, Hungary
keratt@inf.u-szeged.hu
[2] MTA SZTAKI Computer and Automation Research Institute
H-1518 Budapest, P. O. Box 63, Hungary
kacsuk@sztaki.hu
CoreGRID Institute on Resource Management and Scheduling

Abstract. Grid computing has gone through some generations and as a result only a few widely used middleware architectures remain. Using the tools of these middlewares different resource brokers have been developed to automate job submission over different grids. As grid resources were grouped to Virtual Organizations, users seem to become isolated by these groups. Enhancing interoperability among these VOs and grids will be the main issue of future generation grids. This paper describes a meta-brokering architecture that shows how to enable the interoperability of various grids through their own resource brokers.

Keywords: Grid Computing, Meta-Broker, Resource Broker, Grid Portal.

1 Introduction

The Grid was originally proposed as a global computational infrastructure to solve grand-challenge, computational intensive problems that cannot be handled within reasonable time even with state of the art supercomputers and computer clusters [1]. Grids can be realized relatively easily by building a uniform middleware layer, on top of the hardware and software resources, the programming concept of such distributed systems is not obvious.

Executing a job in a grid environment requires special skills like how to find out the actual state of the grid, how to reach the resources, etc. As the number of the users is growing and grid services are starting to become commercial, resource brokers are needed to free the users from the cumbersome work of job handling. Though most of the existing grid middlewares give the opportunity to choose the environment for the user's task to run, originally they are lacking such a tool that automates the discovery and selection. Brokers meant to solve this problem. To enhance the manageability of grid resources and users Virtual Organizations were founded. This kind of grouping started an isolation process in grid development,

W. Lehner et al. (Eds.): Euro-Par 2006 Workshops, LNCS 4375, pp. 112–115, 2007.

too. As resource management is a key component of grid middlewares, many solutions have been developed [2]. Interoperability among these "islands" will play an important role in grid research. This paper introduces a meta-brokering approach to reach different grids through a common interface. Grids are typically accessed through portals that serve as both grid application developer and executor environments. This graphical interface helps the users to utilize grids, therefore it is important to provide a portal for user-oriented grid services.

2   Related Work

In the past decade several projects targeted to build an efficient resource broker. A proper solution should follow the standards of grid communities [8], the requirements of user groups and the results of the latest grid middleware research.

Focusing on interoperability, the Grid Interoperability Project [4] has some results on resource brokering between Unicore [6] and Globus [7] Grids. The goal of their work was to create a semantic matching of the resource descriptions. Their ontological mappings specialize only in these two middlewares. The Gridbus Grid Service Broker [5] is designed for computational and data-grid applications and supports all Globus middlewares and Unicore in experimental phase. Both solutions aim at accessing resources from different grids, but their architecture stays on the level of direct resource brokering.

3   Abstract Architecture

Utilizing the existing, widely used and reliable resource brokers and managing interoperability among them could be new point of view in resource management. The following figure (Fig. 1.) introduces an abstract architecture of a Meta-Broker that enables the users to access resources of different grids through their own brokers.

Designing such an interoperable Meta-Broker, the following guidelines are essential: As standards play an important role of today's grid development, the interfaces must provide standard access. The architecture must be "plug-in based" - the components should be easily extended by all means. The properties of the underlying components are also important; we need to be aware of the recent Grid Resource Brokers. The most efficient and widely used ones should be selected in order to make this solution usable.

There are 4 major parts of this architecture. The Translator component is responsible for translating the user requests to the language of the appropriate broker that the Meta-Broker wants to invoke. It should "speak" the languages of the interconnected brokers. The Information Collector stores the properties of the reachable brokers and historical data of the previous submissions. This information shows whether the chosen broker is available, or how reliable it is. This database can be extended with the information of the resources reachable by the utilized brokers. This can also limit or broaden the usability of the appropriate broker. The Matchmaker selects the proper broker for a user request.

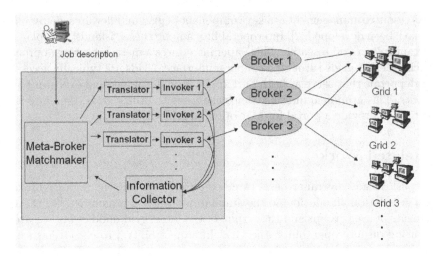

Fig. 1. Grid Meta-Broker Architecture

The job description contains the user request; this should be an exact specification of the user's job, including quality of service requirements and certificate information about the user of the job - this can be used as a filter during matchmaking. The Information Collector provides the broker information needed for the Meta-Broker to decide where to submit the job. The resource broker properties and historical information stored in this component help the Matchmaker to select the proper environment for the actual job defined by its job description. The Invokers are broker-specific components. They communicate with the interconnected brokers, invoke them with job requests and collect the results. Data handling is also an important task of this component. After the user uploaded the job and input files to the Meta-Broker, the Invoker should take care of transferring them to the selected broker's environment. After submission it should stage back the output files, and upgrade the historical data stored in the Information Collector with the appropriate broker's log.

The user's job description is independent from the execution environment, and the Meta-Broker does not need to know how to access resources of different grids. The interconnected brokers' tasks are to perform the actual job submissions; to find the best resource within their scopes, i.e. the VOs they have access to. The Meta-Broker only needs to communicate with them. In this sense meta-brokering stands for brokering over resource brokers instead of resources.

Grid portals give a user friendly access to grid resources and other grid services. Using a Web-based portal, the user can submit a job easily, regardless of location. The P-GRADE Portal [3] is a workflow-oriented, multi-grid portal that provides all the functions needed for job submission. P-GRADE portal is already connected to different grids and brokers. Integrating the Meta-Broker to this portal will be the next step supporting interoperability in grids.

4   Conclusions

The introduced meta-brokering approach opens a new way for interoperability support. The design and the abstract architecture of the Grid Meta-Broker follow the latest results and standards in grid computing. This architecture enables a higher level brokering called meta-brokering by utilizing resource brokers for different middlewares. This service can act as a bridge among the separated "islands" of the current grids, therefore it enables more beneficial resource utilization and collaboration.

Our future work aims at examining and summarizing the prevalent resource brokers and developing the components of the Meta-Broker architecture according to the properties of these brokers.

References

1. I. Foster, C. Kesselman, "Computational Grids", The Grid: Blueprint for a New Computing Infrastructure, Morgan Kaufmann, 1998. pp. 15-52.
2. K. Krauter, R. Buyya, and M. Maheswaran, "A Taxonomy and Survey of Grid Resource Management Systems for Distributed Computing", International Journal of Software: Practice and Experience, Wiley Press, New York, USA, May 2002.
3. Csaba Németh, Gábor Dózsa, Róbert Lovas, Péter Kacsuk, "The P-GRADE Grid Portal", Lecture Notes in Computer Science, Volume 3044, Jan 2004, pp. 10-19.
4. John Brooke, Donal Fellows, Kevin Garwood, Carole Goble, "Semantic Matching of Grid Resource Descriptions", Lecture Notes in Computer Science, Volume 3165, Jan 2004, pp. 240-249.
5. Srikumar Venugopal, Rajkumar Buyya and Lyle Winton, "A Grid Service Broker for Scheduling e-Science Applications on Global Data Grids", Journal of Concurrency and Computation: Practice and Experience, Wiley Press, USA (accepted in Jan. 2005).
6. D. W. Erwin and D. F. Snelling, "UNICORE: A Grid Computing Environment", In Lecture Notes in Computer Science, volume 2150, Springer, 2001, pp. 825-834.
7. I. Foster C. Kesselman, "The Globus project: A status report", in Proc. of the Heterogeneous Computing Workshop, IEEE Computer Society Press, 1998, pp. 4-18.
8. http://www.ggf.org

A Super-Peer Model for Multiple Job Submission on a Grid

Pasquale Cozza[1], Carlo Mastroianni[2], Domenico Talia[1], and Ian Taylor[3]

[1] DEIS University of Calabria, 87036 Rende (CS), Italy
{pcozza,talia}@deis.unical.it
[2] ICAR-CNR, 87036 Rende (CS), Italy
mastroianni@icar.cnr.it
[3] Computer School, Cardiff University, UK
Ian.J.Taylor@cs.cardiff.ac.uk

Abstract. Submission of multiple jobs in a distributed and heterogeneous environment is required by applications that rely on the "public-resource computing" paradigm. We present here a scientific scenario for the analysis of astronomical data, where some nodes are responsible for maintaining and advertising job description files and other so called worker nodes, are dispersed over the Grid to execute the jobs. Job assignment is performed through a mechanism that matches adverts, containing job descriptions, with job queries that are sent by available workers across the Grid exploiting an underlying super-peer topology. With an analogous mechanism, a worker locates the input data file needed to run a job and downloads it from a data center node. This paper presents a super-peer protocol for the submission of a very large number of jobs on a Grid environment. The super-peer architecture enables the replication of data files on multiple data centers, which helps reduce the processing load and speed up the application. A simulation analysis has been performed to evaluate the impact of application and network parameters on performance results.

1 Introduction

Recently, academic and industrial researchers have been promoting the convergence of two paradigms for distributed computing, namely Grid and peer-to-peer (P2P), which in the beginning tended to evolve separately [8]. Super-peer systems have been proposed [7, 9] to achieve a balance between the inherent efficiency of centralized networks, and the autonomy, load balancing and fault-tolerant features offered by P2P networks. A super-peer node can act as a centralized resource for a limited number of regular nodes (peers) of a Grid organization. At the same time, super peers connect among them to form a P2P network at a higher level, thus enabling distributed computing on a very large scale.

This paper reports on a distributed model based on the super-peer paradigm for the support of applications that require the distributed execution of a large number of jobs in a similar fashion to public-resource computing. The term "public resource computing" [1] is used for applications in which jobs are executed by private-owned

W. Lehner et al. (Eds.): Euro-Par 2006 Workshops, LNCS 4375, pp. 116–125, 2007.

computers that use their spare CPU time to support a large scientific computing project. The pioneer project SETI@home [3] has attracted millions of participants wishing to contribute to the digital processing of radio telescope data to search for extra-terrestrial intelligence. A number of similar projects are today supported by the BOINC software system (Berkeley Open Infrastructure for Network Computing [2]), for example: the Einstein@home project [5] aims at detecting certain types of gravitational waves, such as those produced by spinning stars; whereas the Climate@home [4] focuses on long-term climate prediction. The BOINC infrastructure is composed of a scheduling server and a number of clients installed on users' machines. The client software periodically contacts the scheduling server reporting host's hardware and availability, and receives a set of instructions for downloading executable and input files. After that the client runs the assigned job and uploads the resulting output files to the scheduling server.

The BOINC middleware is suited for CPU-intensive applications but it is inappropriate for data-intensive tasks because of its centralized nature. BOINC allows a project to configure a fixed static set of data servers that have to be administered by an entity. Although this scheme enables a number of servers to help load balance the network, the topology is static and is incapable of scaling proportionately as the network grows and more bandwidth is needed for data transfers. In BOINC, an administrator must configure these static machines, which are generally dedicated for a specific project. Such machines are not only costly (to purchase and maintain) but also they are centrally administered and therefore cannot generally be used by other BOINC projects. In the scheme proposed here, the Job initiator is lightweight and sends the data once to the network, which propagates it across the data nodes as and when required. This helps to distribute the data load dynamically in a decentralized fashion, both in topology and administratively, making it far more suitable to the Grid domain. For example, inherent in BOINC-like networks is the need to send a data file needed by several workers several times due to the unreliability of the nodes. This replication represents an evident waste of server-based bandwidth that could be avoided through a caching mechanism that replicates the data across the network when it is first transferred, thereby not only relieving the central bottleneck of the system but also it can place the data in a location closer to where the work is being performed. Further, a number of projects require many nodes to process the same data, with different parameters for example, which can be exploited by such an overlay described here. Our gravitational-wave example described here employs such an algorithm.

The super peer job submission protocol described in this paper enables caching of the input data files in multiple *data centers*, i.e. in super-peers, which have sufficient data storage facilities. Benefits of this replication strategy range from a larger degree of reliability and fault-tolerance to a more efficient use of bandwidth and CPU resources. The job submission protocol requires that job execution is preceded by two *matching* phases, the first one for job assignment and the second one for downloading of input data. Since input data files can be very large, focus is especially on the download phase which is the most bandwidth consuming. A set of simulation runs have been performed to evaluate the impact of the caching and replication mechanism on a set of performance indices, such as overall time to execute all the jobs, throughput, mean time to download a data file, and load experienced by data centers and worker nodes. In particular, we simulated the behavior of the super-peer protocol in a Grid containing 25 super-peers and 250 ordinary peers. The experimental results show that the use of several data center can bring benefits to the Grid applications in terms of lower total execution times, higher throughout and load balancing among worker nodes. The study can also be used to determine the number of data centers that, for a given number of jobs, maximizes the utilization of data center nodes.

In section 2 the super-peer model and the related protocol are presented in more detail, whereas performance is analyzed in section 3. Conclusions and future work are discussed in section 4.

2   Job Assignment and Data Download

A data-intensive Grid application can require the distributed execution of a large number of jobs with the goal to analyze a set of data files. One sample application scenario defined for the GridOneD project [6] shows how one might conduct a massively distributed search for gravitational waveforms produced by binary stars orbiting one around the other. In this scenario, a data file of about 7.2 MB of data is produced every 15 minutes and it must be compared with a large number of templates (between 5,000 and 10,000) by performing fast correlation. Data can be analyzed in parallel by a number of Grid nodes to speed up computation and keep the pace with data production.

The scenario evaluated in this paper assumed the existence of a Grid network in which nodes are organized in a super-peer topology. The *job manager* node (i) receives data from a detector, (ii) produces the job description files (or *job adverts*), and (iii) collects output results. Simple peers, or *workers*, are available for job execution: they issue a job query to get a job description and then a data query to collect the corresponding input data file to be analyzed. Super-peer interconnections are exploited to make job and data queries travel the network rapidly; super peers play the role of *rendezvous nodes*, since they can store job and data adverts (and potentially the data files themselves), and compare these files with queries issued to discover them; thereby acting as a meeting place for both job or data providers and consumers. Since input data files can require a large amount of storage memory, it is assumed that only some of the peers in the network will cache such files. Such peers are referred to as *data centers* (*DC*) nodes and can be located on super peers or worker peers. We envisaged that the same user-driven process is used to configure a peer; that is, each user decides if they want to be a super peer and/or data center, as well as a worker. In the BOINC scenario, the existing dedicated machines would form the obvious data-center backbone and other peers (with high storage and network capacity) would also make themselves available in this mode.

Figure 1 shows a sample topology with 5 super-peers (2 of which are also data centers), and the sequence of messages exchanged among workers, super-peers and data centers to perform the job submission protocol. These messages are related to the execution of a job by a single worker, labeled as W0. Note that here, we do not use normal peers as data centers but we will be comparing this approach in later studies.

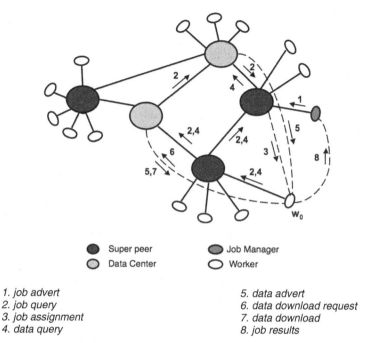

Super peer   Job Manager
Data Center   Worker

1. job advert   5. data advert
2. job query   6. data download request
3. job assignment   7. data download
4. data query   8. job results

Fig. 1. Super-peer job submission protocol: sample network topology and sequence of exchanged messages to execute one job

The proposed protocol requires that job execution is preceded by two matching phases that exploit the features of the super-peer network: the *job-assignment* phase and the *data-download* phase.

In the *job-assignment* phase the *job manager* generates a number of *job adverts*, which are XML documents describing the properties of the jobs to be executed (job parameters, characteristics of the platforms on which they must be executed, information about required input data files, etc.), and sends them to the local rendezvous super-peer, which stores the adverts (step 1 in figure 1). Each worker, when ready to offer a fraction of its CPU time (e.g., worker W_0 in the figure), sends a *job query* that travels the Grid through the super-peer interconnections (step 2). In particular, a query message is sent to the directly connected super-peer, which in turn forwards it to its neighbor super-peers and so on, until the message TTL parameter is decremented to 0 or the job query finds a matching job advert. A job query is expressed by an XML document and can contain the main hardware and software features of the requesting node and CPU time and memory amount that the node offers. A job query matches a job advert when the job query parameters are compatible with the information contained in the job advert, for example concerning the characteristics required for the host that will execute the job. Whenever the job query gets to a rendezvous super-peer that maintains a matching job advert, such a rendezvous assigns the related job to the requesting worker by directly sending it a *job assignment* message (step 3).

In the *data-download* phase, the worker that has been assigned a job inspects the job advert, which contains information about the job and the required input data file (e.g. size and type of data). Then the worker sends a *data query* message to discover the input file (step 4). In a similar fashion to the job assignment phase, the data query travels the super-peer network searching for a matching input data file stored by a data center. Since the same file can be maintained by different data centers, the data center that receives a data query, in order to avoid multiple transmissions of the same file, does not send data directly to the worker. Conversely, the data center sends only a small *data advert* to the super peer connected to the worker and then to the worker itself (step 5). The worker initiates the download operation after receiving the first data advert (steps 6 and 7), and discards the subsequent adverts. After receiving the input data, the worker executes the job, reports the results to the job manager (step 8) and immediately issues a query for another job.

Replication of input data files on multiple data centers allows for a significant saving of time in the querying phase and enables the concurrent retrieving of files from different data centers. In the simulated scenario, it is assumed that all the data centers possess the data files before starting the job submission process. In a more dynamic scenario, the data file is initially maintained by only one data center, and the other data centers could cache the file during the download phase. For example, if in the network depicted in figure 1 the super-peer connected to the worker Wo becomes capable of playing the data center role, it can store the data file downloaded by that worker and provide it for successive requests issued by other workers.

In the job assignment phase the protocol works in a way similar to the BOINC software, except that job queries are not sent directly to the job manager, as in BOINC, but travel the super-peer network hop by hop. Conversely, the data download phase differs from BOINC in that it exploits the presence of multiple data centers in order to replicate input data files across the Grid network.

3 Performance Evaluation

A simulation analysis has been performed by means of an ad hoc event-based simulator, written in C++, to evaluate the performance of the proposed super-peer protocol. The parameters of the astronomical application mentioned in Section 2 were used for the test case (e.g., file size, single job execution time, etc.). It is assumed that all the jobs have similar characteristics and can be executed by any worker.

Simulation parameters, and corresponding values, are reported in Table 1. The Grid network is composed of 25 Grid organizations, each containing one super-peer node and 10 regular nodes on average. The super-peer overlay network is organized so that each super-peer is connected to at most 4 neighbor super-peers. It is assumed that local connections (i.e. between a super-peer and a local simple peer) have a larger bandwidth and a shorter latency than remote connections. To compute download times with a proper accuracy, a data file is split in 100 KB segments, and for each segment the download time is calculated assuming that the downstream bandwidth available at a data center is equally shared among all the download connections that are simultaneous active from the data center to different workers.

In this preliminary study, it is assumed that the data centers download input data files before the workers join the system and issue their job queries. In future work, analysis will focus on a more complex scenario in which data files are replicated, as a whole or in parts, during the execution of jobs.

The last two rows of Table 1 are related to parameters that were given varying values in the simulation runs, specifically the number of jobs N_{job} and the number of data centers N_{dc}, i.e. the number of super-peer nodes able to cache data files.

Table 1. Simulation parameters

Parameters	Values
Grid size: overall number of nodes = super-peers + workers	275= 25 + 250
Maximum number of neighbors for a super-peer	4
Size of input data files	7.2 MB
Latency between two adjacent super-peers (or between two remote peers in a direct connection)	100 ms
Latency between a super-peer and a local simple peer (worker)	10 ms
Bandwidth between two adjacent super-peers (or between two remote peers in a direct connection)	1 Mbps
Bandwidth between a super-peer and a local simple peer	10 Mbps
TTL parameter for job and data queries	4
Mean Job execution time	500 s (±10%)
Number of jobs, N_{job}	from 250 to 10000
Number of data centers, N_{dc}	1, 2, 3, 5, 9, 13

Performance indices are listed in Table 2. The index T_{exec}, the overall time to execute all the jobs, is crucial to determine the rate at which data files can be retrieved from an astronomic telescope while guaranteeing that the workers are able to keep the pace with data. By the throughput index T_{hr} it is possible to evaluate the efficiency of the job submission system. The remaining performance indices help determine the load that is experienced by data centers and by workers in different scenarios.

Table 2. Performance indices

Performance index		Definition
Overall execution time	T_{exec}	Time to execute all the jobs (s)
Throughput	T_{hr}	Average number of jobs completed per time unit (jobs/s)
Percentage of activity time	P_{act}	Average percentage of time in which a data center is active, i.e. has at least one download connection in progress
Mean download time	T_{dl}	Average time that it takes for a worker to download a data file from a data center (s)
Max number of executed jobs	J_{max}	Maximum number of jobs executed by a single worker

Figure 2 shows that the overall execution time decreases as more data centers are made available in the network, for two main reasons: (i) data centers are less heavily loaded and therefore data download time decreases, (ii) workers can exploit a higher parallelism both in the downloading phase and during the execution of jobs. However, depending on the number of jobs to be executed, it is possible to determine a suitable number of data centers, beyond which the insertion of a further data center produces a performance increase which does not justify the related cost. For example, if 10,000 jobs are to be executed, a significant reduction of Texec is perceived as the number of data centers is increased up to a value of 9, whereas if the number of jobs is not greater than 1,000 two or three data centers are sufficient to achieve a good performance level. Analogous comments can be made about the throughput index, reported in Figure 3. A further consideration is that the throughput increases with the number of jobs because download and execution periods are alternated more efficiently if workers execute a larger number of jobs. But this increase tends to be negligible as the number of jobs is so large that the job submission system begins to approach a stable working condition.

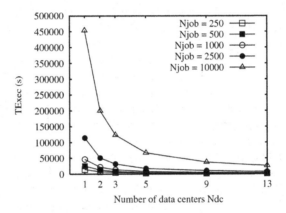

Fig. 2. Performance of the job submission super-peer protocol: overall execution time w.r.t. the number of data centers, for different numbers of jobs

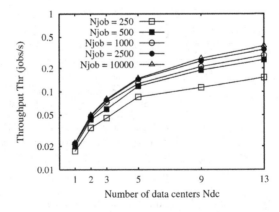

Fig. 3. Performance of the job submission super-peer protocol: throughput w.r.t. the number of data centers, for different numbers of jobs

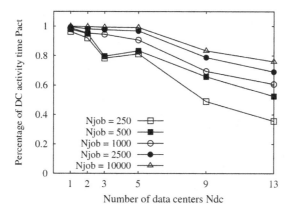

Fig. 4. Percentage of activity time of data centers, for different numbers of jobs

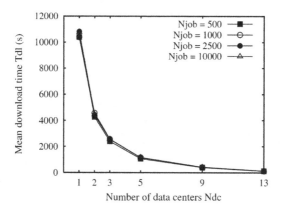

Fig. 5. Mean time to download an input data file w.r.t. the number of data centers, for different numbers of jobs

Figure 4 reports the average percentage of time in which a generic data center supports at least one download connection. Results confirm that the presence of an excessive number of data centers can be inappropriate, especially if the number of jobs is not very large. Indeed when the percentage of activity time decreases below 60%, machine utilization is very low resulting in a poor return of investment (ROI).

Figures 5 and 6 show performance results related to workers. Figure 5 proves that the download time decreases as the number of data centers is increased, resulting in smaller overall execution time. On the other hand, the download time hardly depends on the number of jobs because the simultaneous number of connections that a data center must serve is only related to the number of workers (250), not to the number of jobs. Finally, figure 6 compares number of jobs executed by a worker on average (obtained as $Njob/250$) to the maximum number of jobs executed by a single worker.

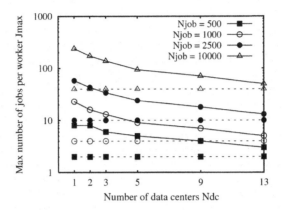

Fig. 6. Maximum number of jobs executed by a single worker w.r.t. the number of data centers, for different numbers of jobs. This index is compared to the average number of jobs executed by a single worker (dotted lines).

It is interesting to note that the two indices approach one another as the number of data centers is increased, leading to a fairer load balancing among workers.

4 Conclusions

This paper reports the first results of a work in progress research on a decentralized architecture for data intensive scientific computing on Grids according to the "public-resource computing" paradigm. We presented a super-peer protocol for the submission of a very large number of jobs in a Grid environment. In the discussed scenario some Grid nodes maintain and advertise job description files, whereas a number of worker nodes, dispersed over the Grid, execute single jobs. Job assignment is performed by matching job descriptions with the job queries that when issued by available workers, travel the Grid by exploiting an underlying super-peer topology.

Simulation analysis has been performed to evaluate the impact of application (the number of jobs) and network parameters (the number of data centers) on performance indices such as the overall time to execute all the jobs, throughput, efficiency of data centers, load experienced by workers. Results show that the use of several data centers can bring benefits to Grid applications in terms of lower total execution times, higher throughput and load balancing among worker nodes. However, since a large number of data centers also causes a smaller utilization of a single data center, the study can also be used to determine the number of data centers that can maximize the return of investment related to the deployment of new data centers.

Future work will move along a number of interesting research avenues, such as: the analysis of (1) redundant computing for applications that require multiple executions of each job; (2) caching of data file fragments on the P2P network, instead of storing entire files, to improve data download performance; (3) performance of the super-peer protocol in the case that input data is progressively fed as a data stream by an external source.

Acknowledgements

This research work is carried out under the FP6 Network of Excellence CoreGRID funded by the European Commission (Contract IST-2002-004265). We would also like to thank Eddie Al-Shakarchi, Tom Goodale, Andrew Harrison, Ian Kelley Matthew Shields and Ian Wang for their help defining the distributed architecture presented here.

References

1. Anderson, D.: Public computing: Reconnecting people to science, Proc. of Conference on Shared Knowledge and the Web, Madrid, Spain, November 2003, pp. 17-19
2. Anderson, D. P.: BOINC: A System for Public-Resource Computing and Storage, 5th IEEE/ACM International Workshop on Grid Computing, November 2004, Pittsburgh, PA, pp. 365-372
3. D. P. Anderson, J. Cobb, E. Korpela, M. Lebofsky, and D.Werthimer: SETI@home: An experiment in public resource computing, Communications of the ACM, November 2002, Vol. 45 No. 11, pp. 56-61.
4. http://climateprediction.net/
5. http://einstein.phys.uwm.edu/
6. http://www.gridoned.org/
7. Mastroianni, C., Talia, D., Verta, O.: A Super-Peer Model for Resource Discovery Services in Large-Scale Grids, Future Generation Computer Systems, Elsevier Science, Vol. 21, No. 8 (2005) 1235-1456.
8. Talia, D., Trunfio, P.: Towards a Synergy between P2P and Grids, IEEE Internet Computing 7(4) (2003) 94-96
9. Yang, B., Garcia-Molina, H.: Designing a Super-Peer Network, 19th Int'l Conf. on Data Engineering, IEEE Computer Society Press, Los Alamitos, CA, USA (2003)

A Scheduling Algorithm for High Performance Peer-to-Peer Platform

Nabil Abdennadher and Régis Boesch

University of Applied Sicences, 4 Rue Prairie, 1202,Geneva, Switzerland
{nabil.abdennadher,regis.boesch}@hesge.ch

Abstract. This paper describes a scheduling algorithm used to execute parallel and distributed applications on a Global Computing (GC) environment, called XtremWeb-CH (*XWCH*). *XWCH* is an improved version of a GC tool called XtremWeb (*XW*). XWCH is an enrichment of *XW* allowing it to match P2P concepts: distributed scheduling, distributed communication and development of symmetrical models. The scheduling algorithm takes into account the heterogeneity and volatility of nodes. This paper illustrates the performance of *XWCH* in a real CPU time consuming application.

Keywords: Peer-To-Peer, High Performance Computing, Scheduling Algorithm.

1 Introduction

High Performance Computing (HPC) landscape has radically changed since the end of the last decade. Based initially on the use of parallel and vectorial computers equipped with specific development environments, computing power consumers are adopting a new approach which takes advantage of the Internet development. The idea consists on deploying High Performance applications on anonymous connected computers by using their available resources. Indeed, the challenge today is to extract, at low cost, a reasonable computing power from a widely distributed platform (by executing interactive applications) rather than extracting the maximum power from a local supercomputer (by executing batch applications). In another words, the majority of the world's computing power is no longer in supercomputer centers and institutional machine rooms. Instead, it is now distributed in a hundred of thousands of personal computers all over the world. This concept is known as Global Computing (GC).

The majority of GC projects adopted a centralized structure based on a Master/Slave Architecture: SETI@home [1], Entropia [2], United Devices [3], Parabon [4], XtremWeb [5], etc. A natural extension of the GC consists on distributing the "decisional degree" of the master in order to avoid any form of centralization. Thus, architectures such as Clients/Servers and Master/Slaves would be withdrawn. This concept, known as Peer-To-Peer (P2P), was successfully used to share and exchange files between computers connected to Internet. The most known projects are Gnutella [6] and Freenet [7]. Indeed, file sharing is well adapted to this model. However, the use of P2P in the field of HPC raises several theoretical and practical problems. Dynamic scheduling algorithms for parallel/distributed applications can not be easily distributed. P2P Computing also goes against the

W. Lehner et al. (Eds.): Euro-Par 2006 Workshops, LNCS 4375, pp. 126–137, 2007.

policies and safety techniques largely used nowadays on Internet: Firewalls, NAT addresses, etc. The objective of these techniques is to protect resources connected to Internet from any voluntary or involuntary abusive use. Internet is then partitioned in several protected zones which are unable to cooperate mutually. Problems related to the development of a true P2P environment for HPC needs remain open.

This document describes a GC environment, called XtremWeb-CH (*XWCH*), which converges towards a P2P system. *XWCH* is an improved version of a GC tool called XtremWeb (*XW*). *XWCH* tries to enrich *XW* in order to match P2P concept: distributed scheduling, distributed communication, development of symmetrical models, etc. In P2P systems, nodes are assumed to be customers and servers at the same time. Although it is utopian, this idea is retained as guide line in the *XWCH* project.

This document is organized as follows: section 2 presents the features that should be satisfied by a GC platform in order to be considered as a real P2P system. Section 3 introduces the *XW* tool in its original version. Section 4 details the new concepts *XWCH* introduces compared to *XW*. It also describes the features of the scheduling algorithm supported by *XWCH*. Section 5 presents the experiments carried out in order to evaluate *XWCH*. Lastly, the section 6 gives some perspectives of this research.

2   What Is a Real Peer-to-Peer System?

A true P2P environment should satisfy three criteria:

- *Platform heterogeneity*: The system should support heterogeneous architectures (hardware) and platforms (software and operating systems). Since these resources are anonymous, the system should take into account all administration policies implemented by local administrators.
- *Natural scalability*: A P2P system should support a huge number of resources. It should be scalable by itself and not by "doping". For that purpose, the performance of the system should be provided by its distributed structure: distributed algorithms, distributed warehouses, distributed scheduling algorithms, etc. This structure should allow open access and search procedures. The search engine should take into account the dynamic nature of the network. The system should be based on a demand-driven computation model: users' queries are only processed when needed and prior results are stored in warehouses, where they can be accessed later on.
- *Symmetric view*: a node belonging to a P2P platform should be server and client at the same time.

File sharing systems like *Gnutella* and *Freenet* satisfy all these criteria. High performance GC environments such as *XtremWeb, Seti@home, Entropia* do not satisfy any of these criteria. They are based on a non symmetric view (Master/Slaves). They are not scalable since the master is overloaded when the number of slaves increases. The only HP oriented tool which seems to satisfy all these constraints is WOS (Web Operating System) [8]. Unfortunately, this tool remained in a purely conceptual state and no prototype was born.

3 XtremWeb

XW is a GC research project carried out at Université d'Orsay (France). Like other Large Scale Distributed Systems (LSDS), *XW* platform uses remote resources (pocket computers, PCs, workstations, servers) connected to Internet to execute a specific application (client). The aim of *XW* is to investigate how a LSDS can be turned into a High Performance Parallel Computer. *XW* belongs to the more general context of Grid research and follows the standardisation effort towards Grid Services [9]. *XW* satisfies the three main constraints imposed by any Large Scale Distributed Environment: volatility, heterogeneity and security.

Security is particularly difficult in the context of LSDS because it's impossible to trust hundreds of thousands resources. Three main security problems, linked to GC and P2P systems, are considered in the context of *XW* project:

- Data integrity/privacy: This problem could be resolved by applying the well known solutions of encryption, public/private keys, etc.
- Protection of participating resources: No aggressive application should be able to corrupt data or system of any participating resource. Sandboxing is the well known technique to resolve this problem. The idea consists on filtering the system calls which appear to be the main security holes of recent operating systems. [10] explains how does *XW* use the sandboxing to resolve the resource protection problem.
- Result certification procedure: This problem is linked to the lack of trust regarding the result provided by the remote resource. Indeed, there is no way to control precisely what happens on a participating resource. Faulty and malicious behaviour must be detected.

A typical *XW* platform is composed of one coordinator and several workers (remote resources). The coordinator is a three-tier layer allowing connection between clients and workers through a coordination service. This layer is designed so as it allows the mobility of clients and the volatility of workers.

3.1 The Coordinator

The coordinator is a three-tier architecture which adds a middle tier between client and workers. There is no task direct submission/result transfer between clients and workers. The coordinator accepts task requests coming from several clients, distributes the tasks to the workers according to a scheduling policy, transfers application code to workers if necessary, supervises task execution on workers, detect worker crash/disconnection, re-launches crashed tasks on any other available worker, collects and store task results to client upon request.

The coordinator is composed of three services: the repository, the scheduler and the result server. The repository is an advertisement services. It publishes services (client applications) to make them available through standard communication ports (Java RMI, XML-RPC). These applications/services are first read from a database and inserted into the task set. The scheduler is the service factory. It instantiates applications and manages their life cycle. It starts them on workers (a task is an instantiation of service or application), stops them as expected and corrects faults (if any) by finding available workers to re-launch them. Finally the result server collects results as they are provided by workers.

3.2 Workers

The worker architecture includes four components: the task pool, the execution thread, the communication manager and the activity monitor. The activity monitor controls whether some computations could take place in the hosting machine regarding parameters such as CPU idle time and mouse/keyboard activity. The tasks pool (worker central point) is managed by a producer/consumer protocol between the communication manager and the execution thread. Each task should be in one of the three states: *ready* to be computed, *running* or *saving*. The first state concerns downloaded tasks, correctly inserted into the pool. The second state is for tasks being computed. The last state corresponds to tasks which need to upload result file to the coordinator. The communication manager ensures communication with the coordinator; it downloads task files (binaries and input data) and upload results, if any. When download completes, the task is inserted into the task pool. The execution thread extracts the first available task from the pool, recreates the task environment as provided by the client (binary code, input data, directories structure, etc.), starts computation and waits for the task to complete. When the task completes, the execution thread finally marks the task state as completed, allowing the communication manager to send results to the coordinator.

In its original version, *XW* applications are standalone modules. The system does not support any interaction between different tasks. However, developers can use asynchronous Remote Process Call called *XWRPC* in order to distribute (parallelize) their applications [11].

4 XtremWeb-CH

XtremWeb-CH (*XWCH*) is an upgraded version of *XW*. The aim of *XWCH* is to build an effective Peer-To-Peer LSDS which satisfies the three criteria detailed in section 2. *XWCH* adds four functionalities to *XW*:

1. Automatic execution of Parallel and Distributed Applications.
2. Automatic detection of the optimal granularity that can be implemented according to the number of available workers and scheduling of tasks.
3. Support of direct communication between workers.
4. *XWCH* provides a set of monitoring tools allowing users to visualize the execution of their applications.

4.1 Automatic Execution of Parallel and Distributed Applications

In *XW*, jobs submitted to the system are standalone. In case of parallel/distributed applications, communicating modules are executed as separate jobs (tasks). It's the user responsibility to link manually output and input data of two communicating tasks. Contrary to this approach, *XWCH* supports the execution of a whole parallel/distributed application represented by is a set of communicating tasks. This

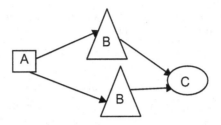

Fig. 1. Data flow graph representing a parallel/distributed application

application is modeled by a data flow graph where nodes are tasks and edges are communications inter-tasks (Fig. 1). Tasks can have the same or different codes. In Fig. 1, tasks having the same shape have the same code.

The data flow graph is represented by an XML file whose syntax is detailed in Fig. 2.

An application is composed of several modules (*Module* element in Fig. 2). A module is represented by a source code and can have several binary versions (*Binary* element in Fig. 2). A task is an instantiation of one module. Thus, several tasks can correspond to the same module.

Precedence rules between tasks are described by *Task* elements. A task can have several inputs (*Input* element in Fig. 2) but only one output (*Output* element in Fig. 2). The element *cmdLine* indicates arguments/parameters used by the task. This field is optional.

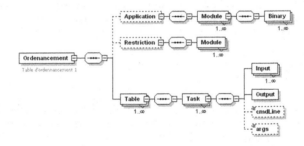

Fig. 2. XML syntax of a parallel/distributed application

A parallel/distributed application is thus, represented by:

- its XML file representing its data flow graph,
- the binary codes of its modules. Let's recall that one module can have several binary codes,
- its input data.

These files are compressed into one file.

XWCH can be perceived as a layer on *XW* that takes into account the communications between tasks belonging to the same parallel/distributed application. In this context, a task belonging to a given parallel/distributed application is considered by *XW* as a standalone application.

A client can submit his application to *XWCH* by uploading its corresponding compressed file. In addition to the three states that a task can have: *ready*, *running* and *saving*, *XWCH* adds a fourth state: *blocked*. Tasks of a given application are initially *blocked* and cannot be assigned to any worker, since their input data are not available. Only tasks whose input data are given by the user are in *ready* state and can be allocated to workers. When a task is assigned to a worker, it moves from *ready* to *running* state. Input data needed by *blocked* tasks are progressively provided by *running* tasks which finish their processing. *XWCH* detects the *blocked* tasks which can pass to ready state and can, thus, be assigned to a worker.

4.2 Granularity and Scheduling

In parallel computing, the grain's size (granularity) depends on the application and the number of processors in the target parallel machine. This number is generally known and fixed before the execution. Thus, the granularity is fixed during the development of the application. In our context, the computer is the network, workers are free to join and/or leave the GC platform whenever they want. The exact number of available workers is known just before the execution and could be varied during the execution. As a consequence, the best granularity can not be fixed before execution time. This section describes how *XWCH* optimize the granularity of tasks and how these tasks are scheduled during execution.

Data flow graph representing an application comprises generally a set of stages $\{S_i\}$. A stage S_i is represented by a set of tasks having the same source code (module in the XML file) and can be executed in parallel on different workers. The precedence rules between two stages S_i and S_{i+1} depends on the application. Tasks belonging to the same stage have no precedence rules. They are fed with different data and are executed according to the Single Program Multiple Data (SPMD) model. Thus, every stage is responsible of processing a "quantity" of data noted Q_i. The number of tasks belonging to stage S_i depends also on application but could be fixed according to the number of workers.

Fig. 3 and 4 show two kinds of parallel/distributed applications experimented on *XWCH*.

In Fig. 3, odd stages contain one task while even stages contain a variable number of tasks. This means that odd stages concentrate results of even stages before sending them to the next stage.

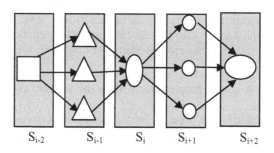

Fig. 3. Phylogenetic applications

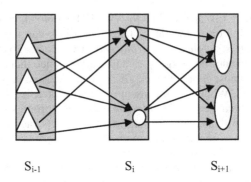

S_{i-1} S_i S_{i+1}

Fig. 4. Numerical application

In Fig. 4, every task of a stage S_i sends its result to all tasks of stage S_{i+1} (multicast operation).

To deploy an application on *XWCH*, three steps are required:

Discovery step: This step consists of searching for a set of available workers W to execute the application (or one stage of the application). The output of this step is a set of workers $W = \{(w_j, p_j)\}$ where p_j is the performance of w_j. p_j can be expressed in term of CPU performance, main memory size, network bandwidth, etc.

Configuration step: Assuming that $|W| = n$, this step dispatches the quantity of data to process by a stage S_i (Q) among the n tasks which compose the given stage. A task t_k, supposed to be executed by worker w_j (with performance p_j), is assigned a quantity of data q_k function of p_j. q_k is called the workload of t_k. The more the worker is powerful, the bigger is q_k. At this point, the system behaves as if the n workers are fully monitored by the coordinator. In another term, granularity of the parallelization and load balancing are fixed according to the number of available workers and the state of the targeted P2P platform.

The output of the configuration step for a given stage S of a given application is a set of couples $\{(q_k, p_j)\}$ where p_j is the performance of the worker that will process the task having the workload q_k.

The XML file, describing the application, is automatically generated at the end of this step.

Execution step: Configuration step assumes that available workers W are fixed and controlled by the coordinator. However, during execution, tasks allocation is not totally controlled by the coordinator. Indeed, tasks are allocated to workers when the coordinator receives work requests from workers. At this point, it is worth going into some details:

- A work request is a remote procedure called by the workers and executed by the coordinator.
- A work request, called by a worker, indicates its current performance p.
- One or several workers selected during discovery step can disappear during execution step.
- One or several new workers can connect and start to send work requests after discovery step.

During execution, the coordinator manages a set of tasks $T = \{t_k\}$ belonging to different applications. Every task t_k has its workload q_k.

Ideally, tasks belonging to a given stage of a given task are executed in parallel on workers selected during configuration step (or new workers with higher performance). Since workers are volatiles, a work request received by the coordinator is not necessarily sent by one of the workers selected during the configuration step. Moreover, arrivals of work requests are unpredictable. For that reasons, the scheduling policy of XWCH is the following: when receiving a work request from a worker w having performance p, the task t allocated to w is the one whose workload q is closer to p. Thus, the scheduler of XWCH allocates task t of T to w if:

$$|q - p| = \min (|q_k - p|) \text{ for all } t_k \text{ belonging to } T.$$

The scheduling algorithm is executed inside the work request call. According to this algorithm, a given task is not executed unless an appropriate worker calls a work request. This means that a task could stay indefinitely in a *ready* state and never assigned to a worker, the application is blocked. In order to avoid this situation, a deadline is affected to each stage of the application: if a task spends in a ready state a time higher than its deadline, it is automatically allocated to the first free worker. A small value of the deadline, means that the user prefers allocate tasks to workers as soon as possible. In this case, tasks could be assigned to a non appropriate worker. A high value of the deadline means that the user prefers wait and allocate tasks to the best appropriate worker. In this case, the task could be blocked indefinitely.

4.3 Direct Communication

Two versions of *XWCH* were developed. The first, called *XWCH-sMs*, manages inter-tasks communications in a centralized way. The second version, called *XWCH-p2p*, allows a direct communication between workers without passing by the coordinator.

In the *XWCH-sMs* (slave-Master-slave) version, workers cannot directly communicate, they cannot "see" each other. Any communications between tasks take place through the coordinator. This architecture overloads the coordinator and could affect the application performances.

In order to cure the gaps of the *XWCH-sMs* version, it is necessary to have direct worker-to-worker communications. In other term, the worker executing module *A* (called *worker A* in Fig. 5) must be able to directly send its results to *workers B* and *C*.

The *XWCH* coordinator can, thus, allocate tasks *B* and *C* to two available workers. Every worker receives the binary code of the module it will execute and the necessary information relating to its input file (IP address, path and name of the input file). Data transfer between workers *A* and *B* (*resp.* C) can thus take place on the initiative of the receiver.

This version called *XWCH-p2p* has two main advantages:

1. it discharges the coordinator from data routing.
2. it avoids the duplication of communications.

In this context, the coordinator keeps only the responsibility of tasks scheduling. *XWCH-p2p* tends towards the Peer-To-Peer concept which one of its principles is to avoid any centralized control.

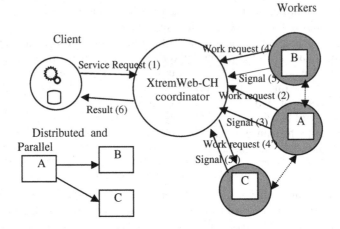

Fig. 5. Execution of an application on a *XWCH-p2p* platform

Direct communication can only take place when the workers can "see" each other. Otherwise (one of the two workers is protected by a firewall or by a NAT address), direct communication is impossible. In this case, it is necessary to pass by an intermediary (*XWCH* coordinator for example). This scenario is similar to *XWCH-sMs* version. However, to avoid overloading the coordinator, one possible solution consists on installing a relay machine, called "data collector" which acts as an intermediary. This machine is used by worker *A* (in our example) to store its results and by workers *B* and *C* to seek their data. "Data collector" machine is chosen by the user when launching the application. This machine must be reachable by all workers contributing to the execution of the concerned application.

4.4 Monitoring Tools

XWCH proposes a package of tools allowing the user to debug and/or visualize the progression of the execution of their applications:

- Tasks allocation: The user can "spy" the execution of his application. He can follow the allocation of tasks (which worker is executing which task).
- Progression of tasks execution: When executing, every task can send progression report to its worker informing it about its state. Currently, this progression report is expressed in term of percentage of execution.
- Step by step execution: It's a debugging mode. When activated, every task sends messages to the worker. These messages are inserted in the source code by the developer.

5   Experimental Measures

The purpose of this section is to assess the performances of *XWCH* in a real case of a CPU time consuming application. *XWCH* was evaluated in the case of a phylogenetic application: *PHYLIP* (the *PHYLogeny Inference Package*) package [12]. The parallelized version of *PHYLIP* is used by the Laboratory of virology at the Geneva Hospital in order to generate phylogenetic tree related to HIV virus.

Phylogenetic is the science which deals with the relationships that could exist between living organisms. It reconstructs the pattern of events that have led to "the distribution and diversity of life". These relationships are extracted from comparing Desoxyribo Nucleic Acid (DNA) sequences of species. An evolutionary tree, termed life tree, is then built to show relationship among species. This tree shows the chronological succession of new species (and/or new characters) appearances.

In a medical context, the generation of a life tree for a family of microbes is particularly useful to trace the changes accumulated in their genomes. These changes are due, inter-alia, to the "reaction" of the virus to the treatments.

A multitude of applications aiming at building evolutionary trees are used by the scientific community [13] [14] [15] [16]. These applications are known to be CPU time consuming, their complexity is exponential (*NP-difficult* problem). Approximate and heuristic methods do not solve the problem since their complexity remains polynomial with an order greater than 5: $O(n^m)$ with m > 5. Parallelization of these methods could be useful in order to reduce the response time of these applications.

PHYLIP is a package of programs for inferring phylogenies (evolutionary trees). It is the most widely-distributed phylogeny package. *PHYLIP* has been used to build the largest number of published trees. It has been in distribution since 1980, and has over 15,000 registered users. *PHYLIP* was ported on *XWCH* platform.

An evolutionary tree is composed of several branches. Each branch is composed of sub-branches and/or leaf nodes (sequences). Two sequences belonging to the same branch are supposed to have the same ancestors. To construct the tree, the application defines a "distance" between all pairs of sequences. Evolutionary tree is then gradually built by sticking to the same branch, the pairs of sequences having the smallest distance between them. Even if the concept is simple, *PHYLIP* is a CPU time consuming application. This complexity is due to two factors:

1. Methods used to group sequences into branches are complex. As an example, the *Fitch* program, one of the most used methods, takes two hours to execute on a Pentium 4 (3 GHz) with 100 sequences.

2. The application constructs not only one tree from the origin data set, but a set of trees generated from a large number of bootstrapped data sets (somewhere between 100 and 1000 is usually adequate). These data are randomly generated from origin data. The final (or consensus) tree is obtained by retaining groups that occur as often as possible. If a group occurs in more than a fraction *l* of all the input trees it will definitely appear in the consensus tree.

The application, as adapted to *XWCH*, is composed of 5 programs: Seqboot, Dnadist, Fitch-Margoliash, Neighbor-Joining and Consensus.

- *Seqboot* is a general bootstrapping and data set translation tool. It is intended to generate multiple data sets that are re-sampled versions of the input data set. It involves creating a new data set by sampling *N* characters randomly with replacement, so that the resulting data set has the same size as the original, but some characters have been left out and others are duplicated.

- *Dnadist* uses sequences to compute a distance matrix. It computes a table of similarity between the sequences. The distance, for each pair of species, estimates the total branch length between the two species. Each distance that is calculated is an estimate, from that particular pair of species, of the divergence time between those two species.
- *Fitch-Margoliash (FITCH)* and *Neighbor-Joining (NJ)*: These two programs generate the evolutionary tree for a given data set. *FITCH* method is a time consuming method and can not be applied to a large number of sequences.
- *Consensus*: This program constructs the consensus tree from the set of trees generated from bootstrapped data sets.

The structure of the obtained parallel/distributed application is shown in Fig. 3. The application, as developed, has two parameters (fed by the user):

- Set of DNA Sequences from species under investigation.
- Number of evolutionary tree to generate: This parameter represents the quantity of data: Q. It's used to produce multiple data sets from original DNA sequences by bootstrap re-sampling. The higher is Q, the finest is the result.
  Two versions of PHYLIP were deployed on *XWCH*:
- The first version (Version 1 in Fig.6) is composed of *Q* tasks in the stage corresponding to the *FITCH* module. Each task processes one data (one tree)
- In the second version (Version 2 in Fig.6), the number of tasks and their workload are processed as explained in paragraph 4.2.

Execution times consumed by the two versions are shown in Fig. 6. *PHYLIP* was executed on an *XWCH* platform composed of more than 100 heterogeneous PC (Pentium 2, 3, 4) with Windows and Linux operating systems.

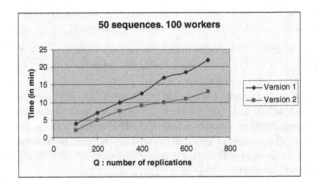

Fig. 6. Execution times of PHYLIP

For both versions, *XWCH* insures that executing codes are transferred from coordinator to workers only at the start of the execution: if the same task is re-executed on the same worker, its code is not downloaded again. The difference of execution times in Fig. 6 is due to the synchronization between the coordinator and workers: When a worker ends the execution of one task it stores the results locally and on the relay, generates a work request call to ask for a new job, and finally generates a data request call to receive input data it needs.

6 Conclusion

This paper presents a new GC environment (*XtremWeb-CH*), used for the execution of high performance applications on a highly heterogeneous distributed environment. *XWCH* can support direct communications between workers, without passing by the coordinator. A scheduling policy is proposed in order to minimize synchronization between coordinator and workers and optimize load balancing of workers. The porting of *PHYLIP* on *XWCH* has demonstrated the feasibility of our solution. Other experiments are in progress to evaluate *XWCH* in other High Performance applications cases.

The current version of *XWCH* allows the decentralization of communications between workers. The next step consists on designing a distributed scheduler. This scheduler shall avoid allocating communicating tasks to workers that can not reach each other. This approach offers a strong basis for the development of distributed and dynamic scheduler and could confirm and reinforce the tendency detailed in section 2.

References

1. http://setiathome.berkeley.edu/
2. http://www.entropia.com/
3. http://www.ud.com/home.htm
4. Parabon Computation, Inc: The Frontier Application. Programming Interface, Version 1.5.2. 2004 (www.parabon.com)
5. Gilles Fedak et al. XtremWeb : A Generic Global Computing System. CCGRID2001, workshop on Global Computing on Personal Devices. Brisbane, Australia. May 2001. http://xtremweb.net
6. KAN G., Peer-to-Peer: harnessing the power of disruptive technologies, Chapter Gnutella, O'Reilly, Mars 2001.
7. Ian Clarke. A Distributed Decentralised Information Storage and Retrieval System. Division of Informatics. Univ. of Edinburgh. 1999. http://freenet.sourceforge.net/
8. Babin, G; P. Kropf; and H. Unger. A two-level communication protocol for a Web Operating System: WOS. Vasteras, Sweden, Aug 1998. In IEEE Euromicro Workshop on Network Computing, 939–944.
9. I. Foster, C. Kesselman, J. Nick, and S. Tuecke. Grid Services for Distributed System Integration. IEEE Computer, pages 37-46, June 2002.
10. Franck Cappello et al. Computing on Large Scale Distributed Systems: XtremWeb Architecture, Programming Models, Security, Tests and Convergence with Grid. In Future Generation Computer Science (FGCS), 2004.
11. Samir Djilali. P2P-RPC: Programming Scientific Applications on Peer-to-Peer Systems with Remote Procedure Call. GP2PC2003 colocated with IEEE/ACM CCGRID2003. Tokyo Japan, May 2003.
12. http://www.phylip.com/
13. http://biowulf.nih.gov/apps/puzzle/tree-puzzle-doc.html
14. http://www.tree-puzzle.de/
15. http://www.dkfz.de/tbi/tree-puzzle/
16. Heiko A. Schmidt, Phylogenetic Trees from Large Datasets, 'Ph.D.' in Computer Science, Düsseldorf, Germany, 2003.

Brokering Multi-grid Workflows
in the P-GRADE Portal*

Attila Kertész[1,2], Gergely Sipos[2], and Péter Kacsuk[2]

[1] Institute of Informatics, University of Szeged
H-6720 Szeged, Arpad ter 2, Hungary
keratt@inf.u-szeged.hu
[2] MTA SZTAKI Computer and Automation Research Institute
H-1518 Budapest, P. O. Box 63, Hungary
CoreGRID Institute on Resource Management and Scheduling
{sipos,kacsuk}@sztaki.hu

Abstract. Grid computing has gone through some generations and as
a result only a few widely used middleware architectures remain. The
Globus Toolkit is the most widespread middleware in most of the current production grid systems, but the LCG-2 middleware dominates in
Europe. The paper describes a brokering solution that enables the interoperability of various Globus and LCG-2 based grids during the execution of workflow applications, and supports users to utilize computing
and storage resources from multiple production grids by a single application. The development and execution of such applications can be
managed by a Web-based Grid portal called P-GRADE Portal, and the
brokering of the workflows is carried out by its integrated GTbroker and
LCG-2 broker component.

Keywords: Grid Computing, Grid Portal, Resource Broker, Workflow
Management, Globus Toolkit.

1   Introduction

The Grid was originally proposed as a global computational infrastructure to
solve grand-challenge, computational intensive problems that cannot be handled
within reasonable time even with state of the art supercomputers and computer
clusters [1]. Grid computing tackles these tasks by aggregating geographically
and architecturally dispersed hardware and software resources into large virtual
super-resources.

Meanwhile grids can be realized relatively easily by building a uniform middleware layer, such as Globus [2], on top of the hardware and software resources,
the programming concept of such distributed systems is not obvious. Complex
problems often require the integration of several existing sequential and parallel
programs into a single application in which these codes are executed according

---

* This research work is carried out under the FP6 Network of Excellence CoreGRID
funded by the European Commission (Contract IST-2002-004265).

W. Lehner et al. (Eds.): Euro-Par 2006 Workshops, LNCS 4375, pp. 138–149, 2007.

to a graph, called workflow. The workflow concept introduces parallelism at two levels. The top level parallelism comes from the graph concept, i.e., codes contained by independent branches can be executed simultaneously. The bottom level parallelism can be applied if some of the workflow nodes are themselves parallel programs. Both top level and bottom level parallelism can be exploited if the parallel branches contain parallel nodes. In such case several supercomputers or clusters can be used simultaneously, and every parallel program would use one of these systems. Consequently, multi-site parallel application execution can be achieved without any performance degradation. The approach combines the benefits of traditional single-site parallel processing and grid-like multi-site processing. Although there are a large number of workflow-oriented grid activities, most of them do not exploit these two possible levels of parallelism [4][5][6].

After the proper parallel processing approach has been selected the next step is to choose a suitable application developer and execution environment. Grids are typically accessed through portals that serve as both grid application developer and executor environments. As grid technology matures the number of production grids dynamically increases. Although sometimes multiple grids are served by the same portal, usually different portals are installed for different grids. Even if a portal is connected to multiple grids, applications that utilize services from these grids simultaneously are not supported.

The P-GRADE Portal [20] is a workflow-oriented portal that supports applications that utilize services from multiple grids simultaneously and demonstrates how Web-based Grid portals can be implemented on top of the Globus middleware [2].

Executing a job in a grid environment requires special skills like how to find out the actual state of the grid, how to reach the resources, etc. Not only computer scientists, but also people from other scientific fields started to deal with this topic, because using the grid resources makes scientific development and research faster and produces better results. As the number of the users are growing and grid services are starting to become commercial, resource brokers are needed to free the users from the cumbersome work of job handling. Though most of the existing grid middlewares give the opportunity to choose the environment for the user's task to run, but originally they are lacking such a tool that automates the discovery and selection. Brokers meant to solve this problem. The Globus middleware does not provide brokering though it has an API that can be used to build such a tool. GTbroker [24] is a proper solution for this toolkit to provide automated job submission for the users.

This paper describes how this broker can be adopted by grid portals to reach Globus resources in an automated way. The following sections introduce the workflow management of the P-GRADE portal and the execution of the workflows with its incorporated brokers. This combination enables multi-grid brokering, even for different middlewares (Globus 2, 3, 4 and LCG-2). Since this portal is already connected to various grids, the automatic multi-grid workflow execution has become reality.

2   The P-GRADE Portal

The P-GRADE Portal is a workflow-oriented grid portal with the main goal to support all stages of grid workflow development and execution processes. It enables the graphical design of workflows created from various types of executable components (sequential, MPI [3] or PVM [17] jobs), executing these workflows in Globus-based [2] computational grids relying on user credentials, and finally, analyzing the monitored trace-data by the built-in visualization facilities. The P-GRADE Portal provides the following functions (see also Fig. 1.): Defining grid environments, creation and modification of workflow applications, managing grid certificates, controlling the execution of workflow applications on grid resources and monitoring and visualizing the progress of workflows and their component jobs.

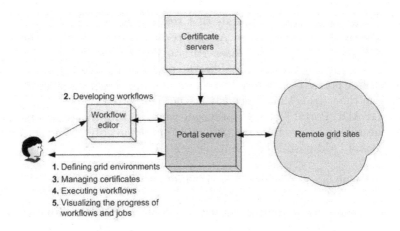

Fig. 1. User activities supported by the P-GRADE Portal

Current portals support only isolated users, i.e., grid users cannot collaborate via the portal either to develop applications together or collaboratively run existing applications. Multi-user portals provide controlled and concurrent access to grid applications for multiple users during both the application development and execution phases. This portal can connect several grids and able to support the simultaneous, collaborative execution of components of a workflow in several connected grids. The P-GRADE portal can give the users all these functionalities, so this portal is a collaborative-grid/user portal. In this paper we are focusing on workflow management. For more information on the portal please refer to [20].

3   Workflow Management in the P-GRADE Portal

Every workflow-oriented portal consists of a workflow GUI and a workflow manager part. While the workflow GUI is the interface that enables the development, submission and steering of workflows and the visualization of results, the workflow manager is responsible for the execution and scheduling of workflow components in the connected grids. It can simultaneously utilize multiple grids to execute different components of a workflow.

3.1   Workflow Notation

Workflow applications can be developed in the P-GRADE Portal by the graphical Workflow Editor. The Editor is implemented as a Java Web-Start application that can be installed on the client machines "on the fly", using a standard Web browser. The Editor communicates only with the Portal Server, and it is completely independent from the grid infrastructures the Server is connected to.

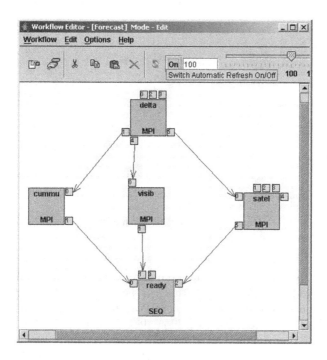

Fig. 2. A workflow graph in the P-GRADE Portal

A P-GRADE Portal workflow is a directed acyclic graph that connects sequential and parallel programs into an interoperating set of jobs. The nodes of such a graph are batch jobs, while the arc connections define data relations among these jobs. Arcs define the execution order of the jobs and the input/output dependencies that must be resolved by the workflow manager during execution (Fig. 2.).

Nodes labeled as delta, cummu, visib, satel and ready represent executable programs. Small squares labeled by numbers around the nodes are called ports and represent input and output data files that the corresponding executables expect or produce. (One port represents one input/output file.) Directed arcs interconnect pairs of input and output ports if an output file serves as an input file for another job. An input file – represented by an input port – can come from three different sources: It can be produced by another job of the workflow, come from the workflow developer's desktop machine or from a storage resource. An output file – represented by an output port – can also have the following three target locations: A computational resource, the Portal server or a storage resource (Fig. 3.).

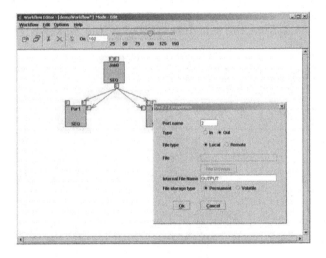

Fig. 3. Input/output file handling

The semantics of the workflow execution means that a node (job) of the workflow can be executed if, and only if all of its input files are available, i.e., all the jobs that produce input files for this job have successfully terminated, and all the other input files are available on the Portal Server and at the pre-defined storage resources. Therefore, the workflow describes both the control-flow and the data-flow of the application. If all the necessary input files are available for a job, then the workflow manager transfers these files - together with the binary executable - to the computational resource where the job is allocated for execution. Managing the transfer of files and recognition of the availability of the necessary files is the task of the workflow manager component of the Portal Server. In case of using the brokering service over Globus-based VO-s, GTbroker handles the necessary file transfers.

3.2 Developing and Editing Workflows

For simplicity let us examine a scenario, when a user works on workflows individually during both the development and execution phases. In the P-GRADE

Portal a workflow can be loaded from the user's private storage space – allocated on the Portal Server – into the client-side Editor, can be edited locally, and the updated version can be loaded back to the Server. The development of a P-GRADE Portal workflow consists of two subtasks: Defining the structure of the graph and specifying the properties of nodes (jobs and ports).

The graph structure can be defined using the drag and drop GUI elements of the Workflow Editor. The properties of nodes can be specified using property windows: by double clicking a job or a port a corresponding property window can be popped up and the attributes of the affected component can be defined.

The job component of a workflow node can be defined in the following way: using the job property window the user must specify the client side location and the type of the binary executable. Optional start-up parameters can also be given here (e.g. command line attributes). The job can be mapped onto a computational resource in the following way: Using the "Grid" and "Resource" dropdown listboxes first a grid, then a computational resource from that grid must be chosen. Jobs can be mapped onto resources of the VO the user has valid certificates to. If a broker is selected for job submission it can be seen from the "Grid" name (SZTAKI_MDS_2_BROKER – means that GTbroker is used for the job in the SZTAKI Grid). The portal can interface with 2 kinds of brokers: GTbroker for Globus 2 or 3 Grids and the broker component of the LHC Grid infrastructure [22].

The job requirements can be set in the Workflow Editor of the portal. GTbroker needs an RSL [2] file, which is created by the portal through the so-called RSL Editor. The other broker needs a JDL [22] file, created by the JDL Editor. They have a similar interface, so the user can use the same data to set the Job attributes. From the job property window the user can select the RSL Editor, when he/she has already chosen GTbroker for the "Grid" field. In this Editor window the redirection of standard streams and brokering options can be set, and a summary of the input/output files for the job can be viewed. The "Broker options" enables selection of resource mapping guidelines and defining minimum disk size, CPU speed and memory size requirements (Fig. 4.). Only this panel requires additional information about the job compared to the JDL Editor. The guidelines tell the broker to order the resources by disk size, CPU speed or memory size, or to use only clusters for execution environment. Clicking on "View" at the bottom of the window the generated RSL file can be viewed.

4   Workflow Execution

Because none of the largest production grids contain workflow manager services, workflow-oriented portals connected to them must incorporate workflow managers, too. The P-GRADE Portal contains a DAGMan-based [11] workflow manager subsystem which is responsible for the scheduling of workflow components in grids. This section discusses the workflow executor subsystems of the P-GRADE Portal with the brokering functions provided by GTbroker and the LCG-2 Broker.

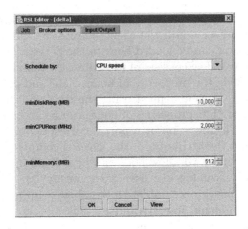

Fig. 4. The RSL Editor

4.1   Workflow Management in Details

One of the main goals of the P-GRADE Portal is to hide the low level details of grid systems with high-level, technology-neutral interfaces that can be easily integrated with different middleware. The GUI of the Portal is built with the GridSphere portal framework [19], thus the various portal functions are implemented as nearly independent portlets.

The "Certificate manager" portlet is responsible for uploading X.509 certificates into MyProxy servers [7] and for downloading short-term GSI proxies [12] into the workflow manager. These proxies are used for authentication. The "Settings" portlet can be used to specify Globus VOs and computational resources for the portal application. The "Workflow" portlet is the graphical interface of the workflow manager and can be used to submit and control workflows, to monitor and visualize execution.

The "Workflow" portlet is interfaced with the Condor DAGMan [11] workflow scheduler. DAGMan degrades workflows into elementary file transfer and job submission tasks and schedules the execution of these tasks. Although DAGMan itself cannot invoke grid services, it supports customized grid service invocations by its pre/post script concept [11]. One pre and one post script can be attached to every node of a DAGMan workflow. DAGMan guarantees, that it first executes the pre script, then the actual content script and finally the post script when it reaches a workflow node. Consequently, the Portal Server automatically generates appropriate pre, content and post scripts for every workflow node when the workflow is saved on the server. These scripts – started by DAGMan according to the graph structure –, invoke the GridFTP and GRAM clients to access files and start up jobs in the connected grids. DAGMan invokes these scripts in the same way in both single- and multi-grid configuration. In general, when a broker is used for job submission, the pre script invokes the broker, and the post script waits till the execution is finished. The broker provides information about the actual job status and the post script notifies the portal about the status changes.

4.2   Multi-grid Workflow Brokering

During workflow editing in the P-GRADE Portal the user has the opportunity to select a resource for each job to run on, or to let a broker choose one. Currently two brokers are used by the portal: the LCG-2 Broker and GTbroker (Fig. 5.).

Regarding Globus Grids, when the right order of the jobs is selected by DAG-Man (according to the dependencies of the jobs), the actual job is given to GT-broker to find a suitable environment and guide the job through the submission process. This broker uses GT2 C API [2] functions to perform interaction with the Globus resources and job submission. For determining the available hosts in the grid it queries the MDS [2]. The job submission to resources is done through GRAM, and a GASS server [2] is used to put the files needed for the job to the remote host and to get back the result files if there are any. These tools enable this broker to work without additional software on Globus Grids. Since most of the current grids use this middleware, the simply adaptation makes this broker relevant.

When a job of the workflow is selected to run with GTbroker, the pre script executes the broker with an RSL file created by the RSL Editor. For QoS, user requirements are taken into account during resource selection. The extended RSL file contains the user requirements and job properties. Static and dynamic information are also used for matchmaking.

In Globus Grids the MDS contains the static properties of the appropriate VO resources. After getting the resources from the MDS, GTbroker orders them by a predefined criterion. In the criteria one can use the following metrics: CPU speed, number of CPUs, free CPUs on the node, disk size and whether a node is a cluster. With these metrics the hosts can be ordered in a way that the ones having the best resources for the actual job get higher priority than the others. The user can modify the priority by selecting the suitable one in the RSL Editor of the portal.

Dynamic information is also used by the broker. For PBS clusters the broker can determine the actual load, right before submitting the job to the selected resource. The pbsnodes command gives back the present availability and load of each node in the selected cluster. This additional piece of information makes the broker able to react for dynamic changes, and reject choosing an overloaded cluster. With this method it automatically finds the best resources and submits only jobs that can actually run.

Fault tolerance is supported by resubmissions. Should a job fail or be pending for too long on a resource (this time interval is set in the broker), the broker cancels and resubmits it to another high priority one. The actual state of the jobs is tracked by the broker, that's why it is possible to cancel and resubmit jobs. After the job is successfully finished, the result files are staged back by the broker and the workflow execution is continuing with the post script of DAGMan. The job states are sent to the portal, so it can visualize the execution phases of the job and therefore of the whole workflow. With this functionality the users are aware of the state of their workflows and notified about each step the execution is going through.

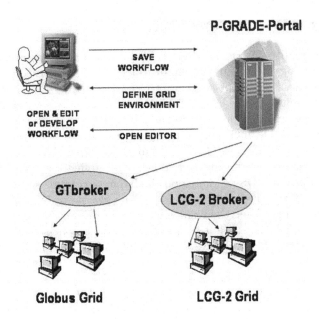

Fig. 5. Multi-grid workflow brokering

The LCG-2 brokering solution is also used by the P-GRADE portal to reach LCG-2-based grids. This kind of broker is built in the LCG-2 architecture: it uses the following parts: a User Interface machine is needed to use the Workload Management System. The Workload Manager is responsible for calling the Matchmaker, interacting with the Information System (BDII), the Replica Location Service, the Log Monitor and the Logging and Bookkeeping server. The Matchmaker gets the job data in a JDL file, and tries to find a "close" host: it takes into account the distance of the physical files on the Storage Elements to the actual Computing Element; finally it submits the job there. The WM can be informed of job failures through the Log Monitor, but automatic resubmission is not provided unlike in the case of GTbroker. Both solutions provide automatic workflow execution, but GTbroker relies only on Globus services and its usage is not limited to grids with LCG-2 architecture.

In the executable workflow the jobs are mapped to resources or brokers. The selected broker determines a VO for the job, from which the broker has to select the executing site. In the case of pure Globus-based grids GTbroker queries the MDS for resource information of the selected VO, and in LCG-2-based grids the LCG-2 broker asks the BDII. The nodes of the workflow that need LCG-2 services can be selected to run on a grid supporting them, and nodes that require only Globus services can be mapped to VOs handled by GTbroker (Fig. 5.). It means that a workflow can be brokered over several VOs, even on different grids. This multi-grid brokering is performed by the brokers connected to the portal. The portal is able to adopt other brokering services, therefore the number of reachable grids is growing.

5   Related Work

Workflows can be managed not only by grid portals, but by other traditional grid user interfaces and problem solving environments (PSE) as well. Unicore [13] and Triana [6] are two of the most well-known workflow-oriented PSEs. They provide neither multi-grid access, nor collaborative user support. Although the server of Triana is built on top of the GAT API [14] - thus it could abstract the underling grid services from their actual implementations - it cannot distinguish security domains from each other, which is a prerequisite of multi-grid access.

Pegasus [15] is a Web-based grid portal, which has the same isolated environment. Based on a special configuration file, filled up by the portal administrator with Globus GRAM and GridFTP [2] site addresses, Pegasus is able to map abstract workflows onto physical resources. At the same time – because of the centrally managed resource list and the single certificate the manager applies during workflow execution – Pegasus cannot be considered a multi-grid portal.

The GridFlow portal [16] applies a more complex workflow executor subsystem than the above discussed environments. The workflow manager of GridFlow handles workflows at two levels. It manages workflows at a global grid level and schedules them at the level of different local grids, but it does not provide collaborative development and execution capabilities.

KOALA [21] is a grid scheduler that uses some of the components of the Globus Toolkit, supports processor and data co-allocation and provides automatic resource selection. Users can interact with KOALA through so-called runners, which are command line tools and require RSL file specifications. KOALA has various runners for submitting and monitoring various kinds of jobs (MPI, Ibis). Currently it is only available on their own multicluster system (DAS 2 – Distributed ASCI Supercomputer 2), but they are planning to make it available on other grids.

Regarding brokers, several solutions have been developed up till now. These solutions usually require other tools to run and the user usually needs to modify its configuration or even additional software needs to be installed to the grid middleware. GTbroker is a broker for the Globus Toolkit, which uses only the APIs and services provided by the toolkit, performs automatic resource discovery and job submission with QoS and fault tolerant features. Since it does not need any other tools, it can be easily incorporated into portals.

6   Summary and Conclusions

With the most advanced portals multiple users can work together to define and execute grid applications that utilize resources of multiple grids. By connecting previously separated grids and previously isolated users together, these portals will revolutionize multidisciplinary research.

The P-GRADE Portal gives a Globus-based implementation for workflow management even for the collaborative-grid, collaborative-user concept [23]. Due to the multi-grid concept a single portal installation can serve user communities of multiple grids. These users can define workflows using the high-level graphical notations of the Workflow Editor, can manage certificates, workflows and jobs through the Web-based interface of the Portal Server. With exploiting the advanced workflow management features of the P-GRADE portal and the brokering functions of GTbroker and LCG-2 Broker users can develop and execute multi-grid workflows in a convenient environment. Users have access to more VOs can create such multi-grid workflows that reach resources from even different grids. Furthermore, the execution of these workflows is carried out in an efficient, brokered way. Since almost every production grid uses Globus middleware today, these grids could all be accessed by the P-GRADE Portal and the workflows created by the portal can produce the expected results.

P-GRADE Portal 2.1 [25] has been already connected to several European grids (LHC Grid [22], EU GridLab testbed [14], UK OGSA test-bed [8], UK NGS [10]) and serves as a graphical interface for several production grids like SEE-GRID [9], HunGrid [18] and UK NGS [10]. While the 2.2 version is already connected to the broker of the LCG middleware [22], the next, upcoming version is connected to GTbroker.

References

1. I. Foster, C. Kesselman, "Computational Grids, The Grid: Blueprint for a New Computing Infrastructure", Morgan Kaufmann, 1998. pp. 15-52.
2. I. Foster C. Kesselman, "The Globus project: A status report", in Proc. of the Heterogeneous Computing Workshop, IEEE Computer Society Press, 1998, pp. 4-18.
3. M. Snir, S. W. Otto, S. Huss-Lederman, D. W. Walker, J. Dongarra, "MPI: The Complete Reference", MIT Press, 1995.
4. Ewa Deelman, et al, "Mapping Abstract Complex Workflows onto Grid Environments", Journal of Grid Computing, Vol.1, no. 1, 2003, pp. 25-39.
5. Matthew Addis, et al: "Experiences with eScience workflow specification and enactment in bioinformatics", in Proc. of UK e-Science All Hands Meeting (Editor: Simon J. Cox), 2003.
6. I. Taylor, et al., "Grid Enabling Applications Using Triana", Workshop on Grid Applications and Programming Tools, Seattle, 2003.
7. J. Novotny, S. Tuecke, V. Welch, "An Online Credential Repository for the Grid: MyProxy", in Proc. of 10th IEEE International. Symposium on High Performance Distributed Computing, 2001
8. UK e-Science OGSA Testbed: http://dsg.port.ac.uk/projects/ogsa-testbed/
9. Southern Eastern European GRid-enabled eInfrastructure Development (SEE-GRID): http://www.see-grid.org/
10. UK National Grid Service: http://www.ngs.ac.uk/
11. D. Thain, T. Tannenbaum, and M. Livny, "Distributed Computing in Practice: The Condor Experience", Concurrency and Computation: Practice and Experience, 2005, pp. 323-356.
12. Butler, R., Engert, D., Foster, I., Kesselman, C., Tuecke, S., Volmer, J. and Welch, V. A National-Scale Authentication Infrastructure. IEEE Computer, 33 (12). 60-66. 2000.

13. D. W. Erwin and D. F. Snelling., "UNICORE: A Grid Computing Environment", In Lecture Notes in Computer Science, volume 2150, Springer, 2001, pp. 825-834.
14. G. Allen et. al., "Enabling Applications on the Grid: A GridLab Overview", International Journal of High Performance Computing Applications, Issue 17, 2003, pp. 449-466.
15. G. Singh et al, "The Pegasus Portal: Web Based Grid Computing" In Proc. of 20th Annual ACM Symposium on Applied Computing, Santa Fe, New Mexico, 2005.
16. J. Cao, S. A. Jarvis, S. Saini, and G. R. Nudd, "GridFlow: WorkFlow Management for Grid Computing", In Proc. of the 3rd IEEE/ACM International Symposium on Cluster Computing and the Grid (CCGRID'03), 2003, pp. 198-205.
17. V. Sunderam, J. Dongarra, "PVM: A framework for parallel distributed computing", Concurrency: Practice and Experience, 2(4), 1990, pp. 315-339.
18. The HunGrid Virtual Organisation: http://www.lcg.kfki.hu/?hungrid&hungrid-general
19. J, Novotny, M. Russell, O. Wehrens: "Grid-Sphere: A Portal Framework for Building Collaborations" in Proc. of the 1st International Workshop on Middleware in Grid Computing, Rio de Janeiro, Brazil, 2003.
20. Csaba Németh, Gábor Dózsa, Róbert Lovas, Péter Kacsuk, "The P-GRADE Grid Portal", Lecture Notes in Computer Science, Volume 3044, Jan 2004, pp. 10-19.
21. KOALA Co-Allocating Grid Scheduler: http://www.st.ewi.tudelft.nl/koala
22. LCG-2 User Guide, 4 August, 2005: https://edms.cern.ch/file/454439/2/LCG-2-UserGuide.html
23. Péter Kacsuk, Gergely Sipos, "Multi-Grid, Multi-User Workflows in the P-GRADE Grid Portal", Journal of Grid Computing, Feb 2006, pp. 1-18.
24. A. Kertész, "Brokering solutions for Grid middlewares", in Pre-proc. of 1st Doctoral Workshop on Mathematical and Engineering Methods in Computer Science, 2005.
25. P-GRADE Grid Portal: http://lpds.sztaki.hu/pgportal

Diet: New Developments and Recent Results[*]

A. Amar[1], R. Bolze[1], A. Bouteiller[1], A. Chis[1], Y. Caniou[1],
E. Caron[1], P.K. Chouhan[1], G. Le Mahec[2], H. Dail[1], B. Depardon[1],
F. Desprez[1], J.-S. Gay[1], and A. Su[1]

[1] LIP Laboratory (UMR CNRS, ENS Lyon, INRIA, UCBL 5668) / GRAAL Project
[2] LPC / PCSV (CNRS / IN2P3 UBP Clermont-Ferrand)
Frederic.Desprez@inria.fr

Abstract. Among existing grid middleware approaches, one simple, powerful, and flexible approach consists of using servers available in different administrative domains through the classic client-server or Remote Procedure Call (RPC) paradigm. Network Enabled Servers (NES) implement this model also called GridRPC. Clients submit computation requests to a scheduler whose goal is to find a server available on the grid. The aim of this paper is to give an overview of an NES middleware developed in the GRAAL team called DIET and to describe recent developments. DIET (Distributed Interactive Engineering Toolbox) is a hierarchical set of components used for the development of applications based on computational servers on the grid.

1 Introduction

Large problems ranging from numerical simulation to life science can now be solved through the Internet using grid middleware. Several approaches exist for porting applications to grid platforms; examples include classic message-passing, batch processing, web portals, and GridRPC systems [31]. This last approach implements a grid version of the classic Remote Procedure Call (RPC) model. Clients submit computation requests to a scheduler that locates one or more servers available on the grid. Scheduling is frequently applied to balance the work among the servers and a list of available servers is sent back to the client; the client is then able to send the data and the request to one of the suggested servers to solve their problem. Thanks to the growth of network bandwidth and the reduction of network latency, relatively small computation requests can now be sent to servers available on the grid. To make effective use of today's scalable resource platforms, it is important to ensure scalability in the middleware layers.

The Distributed Interactive Engineering Toolbox (DIET) [18] project is focused on the development of scalable middleware with initial efforts focused on distributing the scheduling problem across multiple agents. DIET consists of a set of elements that can be used together to build applications using the

---

[*] DIET was developed with financial supports from the French Ministry of Research (RNTL GASP and ACI ASP) and the ANR (Agence Nationale de la Recherche) through the LEGO project referenced ANR-05-CIGC-11.

GridRPC paradigm. This middleware is able to find an appropriate server according to the information given in the client's request (e.g. problem to be solved, size of the data involved), the performance of the target platform (e.g. server load, available memory, communication performance) and the local availability of data stored during previous computations. The scheduler is distributed using several collaborating hierarchies connected either statically or dynamically (in a peer-to-peer fashion). Data management is provided to allow persistent data to stay within the system for future re-use. This feature avoids unnecessary communication when dependencies exist between different requests.

Several other Network Enabled Server (NES) systems have been developed in the past [4,24]. Among them, NetSolve [5], Ninf [25], and OmniRPC [30] have particularly pursued research involving the GridRPC paradigm. NetSolve, developed at the University of Tennessee, Knoxville allows the connection of clients to a centralized agent and requests are sent to servers. This centralized agent maintains a list of available servers along with their capabilities. Servers report information about their status at given intervals, and scheduling is done based on simple models provided by the application developers, LINPACK benchmarks executed on remote servers, and/or information given by the Network Weather Service (NWS). Some fault tolerance is also provided at the agent level. Data management is managed either through request sequencing or using the Internet Backplane Protocol (IBP). Client Proxies ensure portability and interoperability with other systems like Ninf or Globus [6]. Ninf is an NES system developed at the Grid Technology Research Center, AIST in Tsukuba. Close to NetSolve in its initial design choices, it has evolved towards several interesting approaches using either Globus [34,37] or Web Services [32]. Fault tolerance is also provided using Condor and a checkpointing library [26]. The performance of the platform can be studied using a powerful tool called BRICKS. As compared to the NES systems described above, DIET is interesting because of the use distributed scheduling to provide better scalability, the ability to tune behavior using several APIs, and the use of CORBA as a core middleware.

In this paper, we present the last developments done within the DIET project that will provide the user with an efficient, scalable, and fault-tolerant system for the deployment to deploy large scale applications over the net. This paper is organized as follows. In Section 2, we recall the architecture of the DIET middleware and the characteristics that make it scalable over large scale grids. Then in Section 3, we describe our most recent developments in resource and server management. The DIET platform deployment tool is described in Section 4 and fault-tolerance detection and recovery are explained in Section 5. The visualization of DIET's behavior on large scale platforms is described in Section 6. Finally, before a conclusion, we describe two new applications ported over DIET.

2   DIET Architecture

The DIET architecture is hierarchical for better scalability. The architecture provides flexibility and can be adapted to diverse environments including

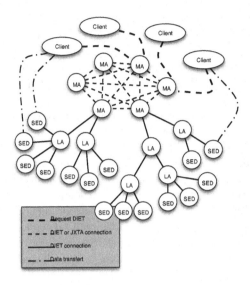

Fig. 1. DIET hierarchical organization

heterogeneous network hierarchies. DIET is implemented in CORBA and thus benefits from the many standardized, stable services provided by freely-available and high performance CORBA implementations. DIET is based on several components. A **Client** is an application that uses DIET to solve problems using an RPC approach. Users can access DIET via different kinds of client interfaces: web portals, PSEs such as Scilab, or from programs written in C or C++. A **SeD**, or server daemon, provides the interface to computational servers and can offer any number of application specific computational services. A SeD can serve as the interface and execution mechanism for a stand-alone interactive machine, or it can serve as the interface to a parallel supercomputer by providing submission services to a batch scheduler.

Agents provide higher-level services such as scheduling and data management. These services are made scalable by distributing them across a hierarchy of agents composed of a single **Master Agent (MA)** and any number of **Local Agents (LAs)**. Each DIET hierarchy is independent but the MA can connect to other MAs either statically or in a peer-to-peer fashion to access resources available via other other hierarchies. Figure 1 shows an example of several DIET hierarchies.

A **Master Agent** is an entry point of our environment. In order to access DIET scheduling services, clients only need a string-based name for the MA (e.g. "MA1") they wish to access; this MA name is matched with a CORBA identifier object via a standard CORBA naming service. Clients submit requests for a specific computational service to the MA. The MA then forwards the request in the DIET hierarchy and the child agents, if any exist, forward the request onwards until the request reaches the SeDs. SeDs then evaluate their own capacity to perform the requested service; capacity can be measured in a variety of ways including an application-specific performance prediction, general server load, or local availability of data-sets specifically needed by the application. SeDs forward their responses back up the agent hierarchy. The agents perform a distributed collation and reduction of server responses until finally the MA returns to the client a list of possible server choices sorted using an objective function such as computation cost, communication cost or machine load. The client program may then submit the request directly to any of the proposed servers, though typically the first server will be preferred as it is predicted to be the most appropriate server. The scheduling strategies used in DIET are described in Section 3.

Finally, NES environments like Ninf and NetSolve use a classic socket communication layer. Nevertheless, several problems with this approach have been pointed out such as the lack of portability or limitations in the number of sockets that can be opened at once. A distributed object environment such as *CORBA* has been proven to be a good base for building applications that manage access to distributed services. It provides transparent communication in heterogeneous networks, but it also offers a framework for the large scale deployment of distributed applications. Moreover, CORBA systems provide a remote method invocation facility with a high level of transparency. This transparency should not dramatically affect the performance since the communication layers have been carefully optimized in most CORBA implementations [17]. Thus, CORBA has been chosen as a communication layer in DIET.

3   DIET Scheduling

3.1   Plug-In Schedulers

DIET provides a special feature for scheduling through its plug-in schedulers. As the applications that are to be deployed on the grid vary greatly in terms of performance demands, the DIET user is provided with the possibility of defining requirements for scheduling of tasks by configuring the appropriate scheduler. The performance estimation values to be used for scheduling are stored in a performance estimation vector by the SeDs as a response to a client call propagated from master agent to local agents and finally to the server level. The values to be stored in this structure can be provided by CoRI (Collector of Resource Information), which will be described in Section 3.2.

The standard values are to be identified based on standard estimation tags given in Table 1. Application developers may also define performance values to be included in a SeD response to a client request. For example, a DIET SeD that provides a service to query particular databases may need to include information about which databases are currently resident in its disk cache so that data transfer times can be minimized.

The application developer can define their own performance estimation routine or function when developing the application-specific portion of the SeD. At this point, any services added to the SeD will be associated with the declared performance estimation routine. In the performance estimation routine, the SeD developer should store in the provided estimation vector any performance data to be used in the server response aggregation methods. At the time a DIET service is defined, an aggregation method - the logical mechanism by which SeD responses are sorted - is associated with the service. If application-specific data are supplied (i.e., the estimation function has been redefined), an alternative method for aggregation is needed. Currently, a basic priority scheduler has been implemented, enabling an application developer to specify a series of performance values that are to be optimized in succession. From the point of view of

Table 1. Standard estimation tags used in DIET

Information tag starts with EST_	multi-value	Explanation
TCOMP		the predicted time to solve a problem
TIMESINCELASTSOLVE		time since last solve started (seconds)
FREECPU		amount of free CPU (fraction between 0 and 1)
LOADAVG		CPU load average
FREEMEM		amount of free memory (Mb)
NBCPU		number of available processors
CPUSPEED	x	frequency of CPUs (MHz)
TOTALMEM		total memory size (Mb)
BOGOMIPS	x	the BogoMips
CACHECPU	x	cache size CPUs (Kb)
TOTALSIZEDISK		size of the partition (Mb)
FREESIZEDISK		amount of free space on partition (Mb)
DISKACCESREAD		average time to read from disk (Mb/sec)
DISKACCESWRITE		average time to write to disk (Mb/sec)
ALLINFOS	x	[empty] fill all possible fields

an agent, the aggregation phase is essentially a sorting of the server responses from its children. A priority scheduler logically uses a series of user-specified tags to perform the pairwise server comparisons needed to construct the sorted list of server responses.

3.2 Collectors of Resource Information

As we have seen in the previous section, to make a good decision the scheduler requires a measurement tool. In particular, DIET needs reliable resource information from grid resource information services. In this section, we introduce the requirements of DIET for a grid information service and the architecture of a new tool called Collectors of Resource Information (CoRI).

For some time DIET has depended on a performance prediction tool called FAST [29]. In this paper, we add new functionality to DIET. We are now able to add any new monitoring tool interface or even any new prediction tool within DIET. It could be dangerous to rely on a single prediction tool for all resource information needs. For example, the prediction tool may not be available on a given architecture and the software dependencies may fail or be too difficult to satisfy in a particular environment. In this case, the scheduler does not receive enough information. We propose a new feature which provides a basic set of performance measurements that can satisfy basic scheduler needs. This tool must always provide an answer in order to avoid the blockage of the grid system. If the tool is not able to provide a measurement, a generic response must be provided. Finally, the tool must provide one single interface for all kinds of resource information services.

The new tool has to solve two main problems. First, it must provide basic measurements that are available regardless of the execution environment. The service developer can then rely on this collector of resource information even if no other resource services like FAST or NWS are available . Secondly, the tool must manage the use of different collectors at the same time and in a similar way. We offer two solutions to these problems: the **CoRI-Easy** collector for the first problem, namely the collector, and the **CoRI Manager** for the second problem, namely management of different collectors. In general, we refer to these two solutions together as the **CoRI** tool.

CoRI-Easy is a set of simple requests for basic resource information, and CoRI Manager allows developer teams to add other resource information services. As CoRI-Easy is a resource information service, it is logical to add it as a collector in the new CoRI Manager. FAST is also available as a collector in the Manager. In addition, it is possible to add new collectors.

The CoRI Manager allows access to different **modules** (also referred to as **collectors**). A module is any kind of element that can provide information about the system. This **modularity** allows the separation of measurement sources and the selection of a module. Even if the manager should unify the different resource information services, the trace of data remains, and so the origin can be determined. For example, it could be important to distinguish the data coming from the CoRI-Easy module and the FAST module, because the information from FAST may give a more accurate estimation of the real value. The **extensibility** of the system is also ensured by the modular design. Because the interface of the manager allows the addition of a new module in some steps, additional modules like Ganglia or NWS can easily be added.

To conclude this section and as a proof of concept, Figure 2 shows an experiment using two kinds of scheduler. The first scheduler uses a simple round robin algorithm wherein we have six servers and round robin works on a rotating basis so that one server is assigned some work, then moves to the back of the list. The second scheduler is a CPU scheduler that maximizes the ratio of $\frac{BOGOMIPS}{1+load\_average}$.

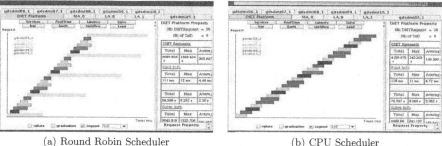

(a) Round Robin Scheduler   (b) CPU Scheduler

Fig. 2. Comparison between the taskflows for 25 consecutive requests with task inter-arrival time equal to 1 minute

The behavior of both schedulers was also studied for requests with different inter-arrival times on an heterogeneous cluster. In this paper we focus on 1 minute for the request inter-arrival time in order to see how the CPU scheduler performs when sufficient time is provided for an accurate estimation of the load average. The distribution of the tasks for the CPU scheduler was performed only on the four fastest nodes resulting in quasi-equal small times for all the tasks. In the case of the Round Robin scheduler, some tasks were privileged by being assigned to the fastest servers while others required longer computing times because all servers were used and some were slower. The total computation time on the platform is smaller with the CPU scheduler due to the fact that the tasks are assigned on the fastest servers. The overlapping of tasks observed in the case of the Round Robin scheduler on the slowest processor resulted in larger computing times.

3.3   DIET Batch Scheduler Management

Parallel grid resources, parallel machines or clusters of workstations are generally managed by a reservation batch system such as Loadleveler[1], PBS[2], or OAR[3]). Such a system is responsible for managing the submitted jobs and locating and allocating the required resources. It accepts user submission scripts which must normally contain a variety of information including the requested number of resources and the amount of time (walltime) needed for the reservation. In order to get the job completed as soon as possible, users can take into account the hardware (walltime), the software he can rely on (NFS for copying some data) and the actual workload on the system (number of resources to use in order to receive resources as soon as possible).

An efficient grid middleware should provide transparent access to parallel resources for the user. It must choose the best parallel resources that suit the request, tune the parallel task to the right number of processors, provide the corresponding walltime, and submit this information to the batch system in an automatically built script in the language of the reservation system.

Simbatch[4] is a C API which relies on the grid simulator Simgrid[20] to provide models of clusters and their batch systems for multi-site grid simulations of parallel tasks. Some batch systems have already been implemented, like PBS and OAR (which respectively rely on FCFS and Conservative Backfilling), but the API is defined to easily integrate new ones.

Simbatch has been designed to fulfill numerous goals such as facilitating the conception and evaluation of grid scheduling algorithms using batch systems. Experiments have been undertaken to validate the batch simulator. Figure 3 shows representative results for an experiment composed of 100 tasks, whose input data and output data sizes are drawn uniformly between 1 and 20 Mbytes,

---

[1] http://www-03.ibm.com/servers/eserver/clusters/software/loadleveler.html

[2] http://www.clusterresources.com/pages/products/torque-resource-manager.php

[3] http://oar.imag.fr/

[4] http://graal.ens-lyon.fr/Simbatch

whose computation time is drawn uniformly between 600 and 800 seconds, and where the number of processors required is between 1 and 5. The platform is composed of 7 machines, interconnected by a star topology and managed by an OAR batch system. The experiment has been run on a real architecture and simulated with the Simbatch tool and one can observe the error between the measured flow in a real OAR batch system and the flow obtained for the same experiment within Simbatch as a function of the submission date of the tasks.

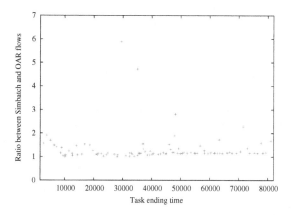

Fig. 3. Flow comparison between Simbatch and OAR results.

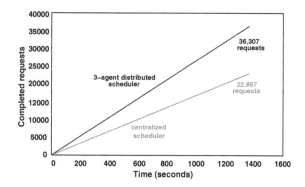

Fig. 4. Comparison of requests completed by a centralized DIET scheduler versus a three agent distributed DIET scheduler

The high precision of Simbatch simulation results makes it possible to use it as a performance prediction module in the DIET environment. Assuming that a performance prediction function is given by the application programmer in the SeD server, Simbatch can simulate several scenarios to choose the best number of resources to use for the application to finish the soonest, as well as the corresponding walltime.

A grid environment must dialogue with batch schedulers to get information on the parallel resources in order to make performance predictions and to submit tuned parallel applications with the correct semantics.

Our work relies on the Elagi[5] library. It provides in particular the possibility to submit jobs to batch systems including Loadlever, Sun Grid Engine[6], and PBS. We have extended the recognized systems list with the OAR system and we plan to complete integration for the WMS system used in the EGEE[7] project (*Enabling Grids for E-science in Europe*).

The DIET batch API provides several functions. A client can explicitly ask for a parallel job, but otherwise, whenever possible DIET will choose the best available allocation to minimize a given objective function. On the server side, the resolution of the application must end with the `diet_submit_call()` which builds and submits the script to the batch scheduler.

3.4   DIET Workflow Management

A large number of scientific applications are represented by interconnected tasks which are structured based on their control and data dependencies. The workflow paradigm on grids is well adapted for representing such applications and the development of several workflow engines [3,27,33,35] illustrate significant and growing interest in workflows within the grid community. The success of this paradigm in complex scientific applications can be explained by the ability to describe such applications in high levels of abstraction and in a way that makes it easy to understand, change, and execute them.

Several techniques have been established in the grid community for defining workflows. The most commonly used model is the graph and especially the directed acyclic graph (DAG). Since there is no standard language to describe scientific workflows, the description language is environment dependent and usually XML based, though some environments use scripts. In order to support workflow applications in the DIET environment, we have developed and integrated a workflow engine. Our approach has a simple and a high level API, the ability to use different advanced scheduling algorithms, and it should allow the management of multi-workflow.

DIET users have traditionally submitted individual tasks, but we have extended the agent hierarchy by adding a new special agent to handle workflow submissions. This special agent, called a MA_{DAG}, manages the different workflow submissions. An overview of the new DIET architecture is shown in Figure 5.

The two architectures presented in the above figure can be used within the same DIET platform. The use of the MA_{DAG} is based on the user choice to use his own scheduling strategy or to use the global one provided by the MA_{DAG}. It is obvious that when the user decides not to use the MA_{DAG}, there is no collaboration between the different clients but he can use and test easily a new

---

[5] http://grail.sdsc.edu/projects/elagi/

[6] http://www.sun.qassociates.co.uk/software-grid-engine.htm

[7] http://public.eu-egee.org/

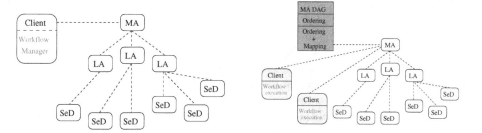

Fig. 5. Software architecture of DIET workflow engine

scheduling algorithm by plugging it in the client code. On the other hand, when the MA$_{DAG}$ is used, the workflow submissions go through this special agent and the multi-workflow can be handled more efficiently using core heuristics. To avoid overloading due to multiple workflow submissions from different clients, the MA$_{DAG}$ is not responsible for workflow execution but it only manages the scheduling aspects. Two working modes can be used in the MA$_{DAG}$: in the first a complete schedule (which assign priority and mapping to each task) is provided to the client, while in the second only task priorities are returned to the client.

4   DIET Deployment

An important factor that influences the efficiency of DIET is the mapping of its components on available resources. We call mapping of the components on available resources "deployment". GoDIET [13] is designed to automate the deployment of DIET platforms and associated services for diverse grid environments. Key goals of GoDIET included portability, the ability to integrate GoDIET in a graphically-based user tool for DIET management, and the ability to communicate in CORBA with LogService [14]. GoDIET automatically generates configuration files for each DIET element while taking into account user configuration preferences and the hierarchy defined by the user, launches complimentary services (such as a name service and logging services), provides an ordered launch of components based on dependencies defined by the hierarchy, and provides remote cleanup of launched processes when the deployed platform is to be destroyed.

To show that the efficiency of an NES environment can depend on the arrangement, or deployment, of its components on available resources we performed several experiments using DIET and shown in Figure 4. These experiments were performed using 151 nodes of the Orsay cluster of the Grid'5000 testbed [8]. Depending on the number of nodes available and the number and sizes of requests, several deployments are possible. For example, we tested two deployments with 150 nodes. In the first deployment, one node is a centralized scheduler that is

---

[8] http://www.grid5000.org

used to manage scheduling for the remaining 150 nodes, which are dedicated computational nodes servicing requests. In the second deployment, three nodes are dedicated to scheduling and are used to manage scheduling for the remaining 148 nodes, which are dedicated to servicing computational requests. In this test the centralized scheduler is able to complete 22,867 requests in the allotted time of about 1400 seconds, while the hierarchical scheduler is able to complete 36,307 requests in the same amount of time.

The distributed configuration performed significantly better, despite the fact that two of the computational servers are dedicated to scheduling and are not available to service computational requests. The deployment plan of components is a very important factor that influences the throughput of the environment. Thus a good planning approach is needed to arrange the resources in such a manner that when the components are deployed on the resources, the maximum number of requests can be processed in a time step. We called this process *deployment planning*. We have shown in [16] that the optimal deployment on a cluster is a **C**omplete **S**panning **d**-ary (CSD) tree; a CSD tree is a tree that is both a complete d-ary tree and a spanning tree. This result conforms with results from the scheduling literature. More importantly, we have presented an approach for determining the optimal degree d for the tree. Finding the best deployment among heterogeneous resources is a hard problem since it amounts to finding the best broadcast tree on a general graph, which is known to be NP-complete. So we presented a deployment heuristic that predicts the maximum throughput that can be achieved by the use of available nodes. The main focus of the heuristic is to construct an hierarchy so as to maximize the throughput of each node, where the throughput depends on the number of children attached as children to the node in the hierarchy. The given heuristic provides a deployment that can meet the user request demand, if user demand is at most equal to the maximum throughput.

Finally, we gave a mathematical model [12] that can analyze an existing deployment and can improve the performance of the deployment by finding and then removing the bottlenecks. This is an heuristic approach for improving deployments of NES environments in heterogeneous grid environments. The heuristic is used to improve the throughput of a deployment that has been defined by other means. The approach is iterative: in each iteration, mathematical models are used to analyze the existing deployment to identify the primary bottleneck, and the bottleneck is then removed by adding resources in the appropriate area of the hierarchy. Using this model we can evaluate a virtual deployment before making a real deployment, provide a decision builder tool (i.e., designed to compare different hierarchies or add new resources) and take into account the hierarchies' scalability.

5   DIET Fault-Tolerance

Grids are composed of many geographically distributed resources, each having its own administrative domain. These resources are gathered using a WAN or even the Internet. Those characteristics lead grids to be more error prone than other computing environments, raising the issue of fault tolerance. NETSOLVE and Ninf includes some fault tolerant mechanisms based on a centralized design. In this section we describe fault tolerant mechanisms incorporated in DIET that specifically target decentralized designs.

5.1   Fault Detection

Most common failure scenarios include both intermittent network failures between sites and node crashes. When considering unreliable networks, ensuring application correctness requires fault detection approaches. Failure detectors can be classified based on two criteria: time to detect a failure and accuracy. Detection time represents the time between a failure and definitive suspicion by the failure detector. Accuracy is the probability of a correct answer from the failure detector when queried at a random time. Classical failure detection systems, like TCP, are based on heartbeats and timeouts. Maximum time to detect a failure depends both on the arrival date of the previous heartbeat and on the maximum delivery time. Considering a WAN or larger network, maximum delivery time is hard to bound, leading to a tradeoff between long failure detection time and poor accuracy.

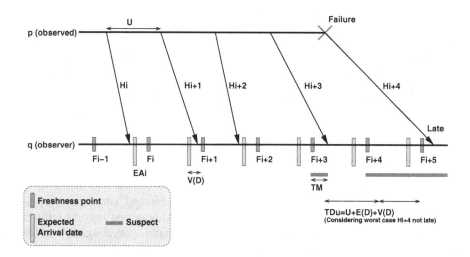

Fig. 6. Sample execution of the DIET fault detector

DIET implements the Chandra, Toueg and Aguilera failure detector [15]. Considering a given heartbeat frequency and maximal time to detect a failure, this detector has optimal accuracy. Another benefit of this fault detector is its ability to adapt to detected network parameters and reconfigure itself according to changing network delay or message loss probability. To our knowledge, this is the first implementation of this algorithm.

Figure 6 presents a sample execution of the fault detector. The basic idea of the algorithm is as follows. Given some QoS parameters, compute a heartbeat period U to be sent by the observed process p, and some time intervals $[F_i, F_{i+1})$ on the observing process q (with respect to the local clock). At time F_i, if q has received a heartbeat $H_j, j \geqslant i$, then q trusts p for the entire time interval $[F_i, F_{i+1})$. If not, q starts suspecting p until it receives a heartbeat $H_j, j \geqslant i$. F_i is called the freshness point: any outdated message with respect to F_i is ignored. Thus maximum time to detect a failure does not depend on maximum message delay, but on average message delay E(D) (worst case when failure happens just after heartbeat emission). Parameters of the detector are the expected quality of service, namely TDu the upper bound on time to detect a failure, TMu the upper bound on average mistake duration, and $TMRl$ the lower bound on average mistake recurrence time. As freshness point F_{i+1} does not depend on the reception date of heartbeat H_i, F_{i+1} has to be set based on network parameters and expected QoS. Once a large enough sample of previous heartbeats is collected, we compute three values: EAi - the expected arrival date of Hi on q, V(D) - the variance of message delay, and Pl - the probability that a message will be lost. The variance is added to the estimated arrival date on q to set F_{i+1}. When network parameters are changing, the observing process may reconfigure the heartbeat period on p according to newly computed parameters.

DIET FD is a part of the DIET library included in every DIET entity. There is no centralized fault detector: each entity is in charge of observing its neighbors. Each observed server costs 750B of memory on the observer. As most computation is triggered by the reception of a heartbeat, and the typical heartbeat period is less than one heartbeat per 5s, the computational cost per observed service is marginal. Heartbeats are very small UDP messages (40 bytes including UDP headers), thus the impact on the network of the fault detection service is small. DIET FD is connected to DIET LogService and VizDIET and may be used to collect statistics on network parameters and grid reliability.

5.2 MA Topology Recovery

Compared to other GridRPC systems, DIET uses a decentralized hierarchical infrastructure. Master Agents and Local Agents are organized using a tree topology. Each Agent is responsible for monitoring its neighbors and reports failures to the GoDIET component. The consequence of the failure of an agent is a disconnection of the tree topology: some available services are not found by client service requests. In order to reconnect the tree, each agent keeps the list of its f ancestors. When detecting failure of its parent, the agent tries to reconnect to the nearest (in the tree) alive ancestor. If this ancestor has also failed, it then tries with the second nearest until it is able to reconnect the tree. Thus, the algorithm is able to recover without central coordination from $f - 1$ simultaneous failures. During recovery, some available services may not be found. This is the same property as in auto-stabilization (except that we are considering only the crash of nodes and message loss). When reconnected, the agent updates its ancestor list from its new parent. When the root of a tree has failed, GoDIET is in charge of replacing the MA and notifies any client to use the new MA instead of the failed one.

5.3   Checkpoint/Restart Mechanism

Unexpectedly, the most intrusive failures are not those hitting infrastructure such as the MA and LAs but the computing nodes [19]. This is mainly due to the large amount of lost computation time. Process replication and process checkpointing are two well known techniques to decrease the amount of lost computation in case of crash failure. In replication, the same program is running on several hosts. Any input of the program is atomically broadcasted to all the replicas. When a failure hits the main process, one of the replicas is promoted as the main process. Whenever a failure occurs, a new replica is created to replace the missing one. Thus, having f replicas is sufficient to tolerate $f - 1$ simultaneous failures, but divides the available computing power by f. Checkpoint based fault tolerance relies on taking periodic snapshot of the state of a process, saving it to another (safe) place. In case of a failure, the process state is recovered from the checkpoint. This is the approach used in NETSOLVE and Ninf [2,26].

   In Ninf the Condor checkpoint library is used. Checkpoint data are stored on a stable checkpoint server holding all recovery data. If a failure hits the checkpoint server, it is no longer possible to recover from any other failure. In order to solve this issue the checkpoint data have to be replicated. [28] suggests the use of computing nodes to host replicated checkpoint data. In DIET, checkpoint data are considered equivalent to persistent data and distributed on computing nodes using JUXMEM. JUXMEM manages data persistence across failures by replicating it on several computing nodes.

   Another important issue in checkpointing is software development cost. On the one hand automatic checkpointing has low software development cost but high overhead. On the other hand application checkpointing is tightly adapted to the algorithm but requires new development for any new algorithm. The DIET checkpoint API is designed to allow both simplicity and performance. From the client point of view, checkpointing and fault tolerance are fully automatic. The fault tolerance management is silently included in the usual RPC mapping layer of the DIET client library. On the service side, each computational service can choose whether it provides its own checkpoint mechanism to DIET (it only has to register checkpoint files and notify the SeD when a checkpoint is ready to be stored), or it can rely on DIET automatic checkpointing. In this case, the service is linked with the Condor Stand-alone Checkpoint Library [21] to periodically create a checkpoint file that can be restarted on any compatible architecture.

6   DIET Visualization and Large Scale Validation

DIET uses an event monitoring system called LogService [22]. This monitoring service offers the capability to monitor information that must be gathered from a distributed platform. *LogComponent* attaches to a component and relays information and messages to *LogCentral*. *LogCentral* collects messages received from *LogComponents*, then it stores or sends these messages to *LogTools*. *LogTools* connect themselves to *LogCentral* and wait for messages. The main interest of LogService is that information are collected by a central point *LogCentral* that receives *LogEvents* from *LogComponents* that are attached to the component that need to be monitored. *LogCentral* offers the service of re-sending this information to several tools (*LogTools*) which are responsible for analyzing these messages and offering a comprehensive view of the system to the user. On each component of DIET there is a *LogComponent* that send information to *LogCentral*. VIZDIET [8] has been developed to offer visualization of the DIET hierarchy. VIZDIET implements the *LogTools* library to be able to connect to *LogCentral* and collect all events from the DIET hierarchy (Figure 7). VIZDIET gives a graphical representation of the platform as well as some quantitative and qualitative information (Figure 8) about the performance and behavior of DIET. VIZDIET is very useful to understanding the behavior and performance

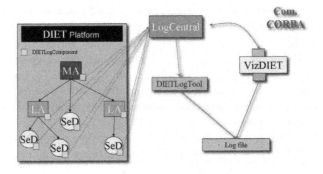

Fig. 7. The LogService mechanism in DIET

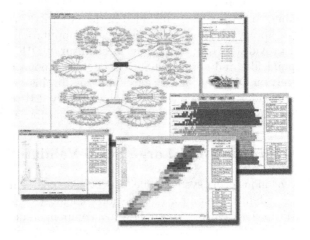

Fig. 8. VIZDIET screenshots

of DIET as it dynamically displays the activity of the platform. Moreover it can also read a log file of a previous DIET run and replay it.

Recently some tests have been done with DIET on Grid'5000 [11]. We have evaluated the scalability of DIET over more than one thousand processors distributed in a nation-wide grid. The experimental process used for performing these tests involved four main steps. (1) Reserve with OAR [10] 550 dual-processor nodes that will be used to run the DIET components. (2) Generate an XML file describing the reserved nodes and the desired deployment. (3) Use GoDIET [13] to deploy the hierarchy of DIET components. (4) Launch 1040 clients which continuously submit requests to the hierarchy for solving a matrix multiplication problem (DGEMM).

During this experiment, there was one local agent managing each cluster. There were a total of 540 SeDs running the same service (DGEMM), eight local agents, and one master agent. 13761 requests were computed by the DIET hierarchy. The scheduling heuristic was a simple round robin approach that used the time since last solve for each SeD to coordinate round-robin behavior amongst the distributed SeDs. Figure 9 shows that the time taken by agents to schedule requests depends on the computing power of the nodes on the cluster. There are some huge variations of response time except at *Lyon* and *Paraci*. This can be explained by the fact that the nodes used by the SeDs and agents were shared with other users. At *Lyon* and *Paraci* the nodes were reserved in an exclusive mode so that response time remains relatively constant. The main goal of this experiment was to prove that DIET can be used on large scale grids while maintaining a low response time (average of 1.9 s) despite a heavy load (1040 clients). Further experiments will be done to test and improve DIET features and performance.

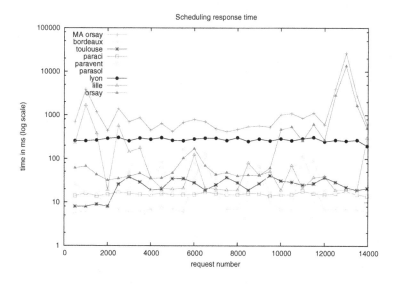

Fig. 9. Response time of agents for scheduling client's requests

7   DIET Applications

7.1   A BLAST Application Using DIET

BLAST is a popular application in bioinformatics for comparing biological sequences such as nucleotides or amino-acid sequences. The aim of such comparisons is to try to determine the function of a new sequence by finding *homologies* with known sequences. A typical use of BLAST is to compare one or more sequences to one or more biological databases. Many approaches to parallelizing BLAST have been investigated [9,1,7,36,23] and three levels of parallelization can be identified. In the fine grained parallelization approach, alignment searches are performed in parallel on a single sequence pair. For the medium grained parallelization, databases are partitioned so that alignments between a sequence and each part of the database can be searched at once. With coarse grained parallelization, the input is partitioned so that multiple sequences can be compared against one or more databases at once.

Using the DIET middleware, we developed an "N-sequences versus one database" service for BLAST queries. On the client side, the multi-request files are partitioned into several smaller requests according to a user strategy (decided by choosing an existing input plug-in or a self-made one). Then, the requests are distributed over the available SED's, which treat them independently and send back the results to the client which merges them after choosing an output plug-in. On the server side, a scheduling strategy can be applied to choose the most appropriate server to execute a request. The server then sends back the result of the execution of a BLAST implementation (including mpiBLAST when a server manages a cluster).

The next focus of the work on DIET BLAST is to introduce data replication in the database management. In the current version, we assume that every server declaring the DIET BLAST service owns the databases used for client requests. We also want to introduce a generic data partitioning information system into the DIET architecture to develop an "intelligent" request decomposition plug-in for the client. This system should be used when the input data of a problem can be divided into multi-size parts and could benefit other applications that use DIET as middleware. The last step is to implement a database partition system acting like mpiBLAST to improve the performance of the single sequence versus one large database requests.

7.2   Cosmological Simulation with RAMSES and Galics

RAMSES[9] is a grid-based hydro solver with adaptive mesh refinement. This code is used to study large scale structure and galaxy formation: from the early universe's structure, the evolution of the position, mass, and velocity of the different particles is followed until now. The raw data produced by RAMSES are then processed using the GALICS[10] software (HaloMaker, TreeMaker and GalaxyMaker)

---

[9] http://www-dapnia.cea.fr/Phocea/Vie_des_labos/Ast/
ast_sstechnique.php?id_ast=904

[10] http://galics.iap.fr/

to extract the halos of matter (gathering of particles), to build the evolution tree (how each particle has evolved), and finally to build galaxies.

The experiments are done on the Grid'5000 platform. RAMSES is suited for this platform as it is an MPI code. It is then convenient to use the DIET middleware to provide a simpler, transparent way of using this cosmological simulator.

The simulation we're working on is called a zoom simulation, wherein the goal is to study in detail the evolution of the distribution of dark matter in the universe. The first part consists of using RAMSES on low resolution initial conditions (few particles) to have a global map of the different particle clusters formed from the primordial universe until now. These data are then post-processed using HaloMaker, the halos' descriptions are sent back to the user, who decides which parts may be interesting to analyze more precisely. The simulation is then rerun on all these different parts at a higher resolution (lots of particles), and on specific locations of the universe. The post-processing uses HaloMaker, TreeMaker and GalaxyMaker sequentially and the final results are sent back to the user for further interpretation.

The structure of this experiment is divided in three parts: the client who sends the request and analyzes the data; the servers that run the simulation; and a database containing the initial conditions. The two parts of the simulation are basically the same: they run RAMSES on initial conditions, post-process the data, and return them to the client. Therefore, many SeDs capable of managing the whole simulation may be deployed (each SeD offering two services: one for each part), allowing DIET to chose the most accurate one at a given time, and bringing total transparency to the user. The user will only have to send a request through DIET, which will ask its hierarchy for the service, and run it. The access to the database will also be transparent, as only the SeDs will have to extract the initial conditions from it. Data management is one of our concerns as the amount of transferred data may be large: we may have file sizes up to 1 Gb. We intend to use the JUXMEM[11] software for data management, which provides location transparency as well as data persistence in a dynamic environment. However, this part is not yet implemented in our prototype and we still use DIET for communications between the SeDs and the client, and `scp` for communications between the SeD and the database.

8   Conclusion and Future Work

In this paper we have presented the overall architecture of DIET, a scalable environment for the deployment on the grid of applications based on the Network Enabled Server paradigm as well as its most recent developments. Like NetSolve and Ninf, DIET provides an interface to the GridRPC API defined within the Global Grid Forum.

Our main objective is to improve the scalability of the platform using a distributed set of agents managing a large set of servers available through the network. By being able to modify the number of schedulers, we are able to ensure

---

[11] http://juxmem.gforge.inria.fr/

a level of performance adapted to the characteristics of the platform (number of clients, number and frequency of requests, performance of the target platform). The management of the platform is handled by several tools like GoDIET for the automatic deployment of the different components, LogService for monitoring, and VizDIET for the visualization of the behavior of DIET's internals. Scheduling is of course one of the main research issue addressed within our tool. Thanks to several APIs, we are able to tune the scheduler itself to either best fit the needs of specific users or to test new heuristics for particular problems.

In our future work we plan to improve the flexibility of the plug-in schedulers, improve the performance evaluation feature, port new applications, and finally to test several DIET platforms at a large scale within the Grid'5000 project [11].

References

1. L. Carey A. E. Darling and W. chun Feng. The design, implementation, and evaluation of mpiblast. In *ClusterWorld 2003*, 2003.
2. A. Agbaria and J.S. Plank. Design, implementation, and performance of checkpointing in netsolve. *dsn*, 00:49, 2000.
3. K. Amin, G. von Laszewski, M. Hategan, N.J. Zaluzec, S. Hampton, and A. Rossi. GridAnt: A Client-Controllable Grid Workflow System. *hicss*, 07:70210c, 2004.
4. P. Arbenz, W. Gander, and J. Mori. The Remote Computational System. *Parallel Computing*, 23(10):1421–1428, 1997.
5. D. Arnold, S. Agrawal, S. Blackford, J. Dongarra, M. Miller, K. Sagi, Z. Shi, and S. Vadhiyar. Users' Guide to NetSolve V1.4. Computer Science Dept. Technical Report CS-01-467, University of Tennessee, Knoxville, TN, July 2001. http://www.cs.utk.edu/netsolve/.
6. D.C. Arnold, H. Casanova, and J. Dongarra. Innovations of the NetSolve Grid Computing System. *Concurrency And Computation: Practice And Experience*, 14:1–23, 2002.
7. R.D. Bjornson, A.H. Sherman, S.B. Weston, N. Willard, and J. Wing. Turboblast: A parallel implementation of blast based on the turbohub process integration architecture. In *Parallel and Distributed Processing Symposium., Proceedings International, IPDPS*, pages 183–190. TurboGenomics, Inc., 2002.
8. R. Bolze, E. Caron, F. Desprez, G. Hoesch, and C. Pontvieux. A monitoring and visualization tool and its application for a network enabled server platform. In M. Gavrilova et al., editor, *Computational Science and Its Applications - ICCSA 2006*, volume 3984 of *LNCS*, pages 202–213, Glasgow, UK., 8-11 May 2006. Springer.
9. R. C. Braun, K. T. Pedretti, T. L. Casavant, T. E. Scheetz, C. L. Birkett, and C. A. Roberts. Parallelization of local BLAST service on workstation clusters. *FGCS*, 17(6):745–754, 2001.
10. N. Capit, G. Da Costa, Y. Georgiou, G. Huard, C. Martin, G. Mouni, P. Neyron, and O. Richard. A batch scheduler with high level components. In *Cluster computing and Grid 2005 (CCGrid05)*, 2005.
11. F. Cappello, E. Caron, M. Dayde, F. Desprez, E. Jeannot, Y. Jegou, S. Lanteri, J. Leduc, N. Melab, G. Mornet, R. Namyst, P. Primet, and O. Richard. Grid'5000: a large scale, reconfigurable, controlable and monitorable Grid platform. In *SC'05: Proc. The 6th IEEE/ACM International Workshop on Grid Computing Grid'2005*, pages 99–106, Seattle, USA, November 13-14 2005. IEEE/ACM.

12. E. Caron, P. K. Chouhan, and A. Legrand. Automatic Deployment for Hierarchical Network Enabled Server. In *The 13th Heterogeneous Computing Workshop (HCW 2004)*, Santa Fe. New Mexico, April 2004.

13. E. Caron, P. Kaur Chouhan, and H. Dail. Godiet: A deployment tool for distributed middleware on grid'5000. In IEEE, editor, *EXPGRID workshop. Experimental Grid Testbeds for the Assessment of Large-Scale Distributed Apllications and Tools. In conjunction with HPDC-15.*, pages 1–8, Paris, France, June 19th 2006.

14. Eddy Caron and Frédéric Desprez. Diet: A scalable toolbox to build network enabled servers on the grid. *International Journal of High Performance Computing Applications*, 20(3):335–352, 2006.

15. W. Chen, S. Toueg, and M. Kawazoe Aguilera. On the quality of service of failure detectors. *IEEE Transactions on Computing*, 51(1):13–32, 2002.

16. P. K. Chouhan, H. Dail, E. Caron, and F. Vivien. Automatic Middleware Deployment Planning on Clusters. *International Journal of High Performance Computing Applications*, 2007. To appear.

17. A. Denis, C. Perez, and T. Priol. Towards high performance CORBA and MPI middlewares for grid computing. In Craig A. Lee, editor, *Proc. of the 2nd International Workshop on Grid Computing*, number 2242 in LNCS, pages 14–25, Denver, Colorado, USA, November 2001. Springer-Verlag.

18. DIET. Distributed Interactive Engineering Toolbox. http://graal.ens-lyon.fr/ DIET.

19. S. Djilali, T. Herault, O. Lodygensky, T. Morlier, G. Fedak, and F. Cappello. Rpc-v: Toward fault-tolerant rpc for internet connected desktop grids with volatile nodes. In *SC '04: Proceedings of the 2004 ACM/IEEE conference on Supercomputing*, page 39, Washington, DC, USA, 2004. IEEE Computer Society.

20. A. Legrand, L. Marchal, and H. Casanova. Scheduling distributed applications: the simgrid simulation framework. In IEEE Computer Society, editor, *3rd International Symposium on Cluster Computing and the Grid*, page 138. IEEE Computer Society, May 2003.

21. M. Litzkow, T. Tannenbaum, J. Basney, and M. Livny. Checkpoint and migration of UNIX processes in the condor distributed processing system. Technical Report 1346, University of Wisconsin-Madison, 1997.

22. LogService. http://graal.ens-lyon.fr/DIET/logservice.html.

23. D.R. Mathog. Parallel blast on split databases. *Bioinformatics*, 19(14):1865–1866, September 2003.

24. S. Matsuoka, H. Nakada, M. Sato, , and S. Sekiguchi. Design Issues of Network Enabled Server Systems for the Grid, 2000. Grid Forum, Advanced Programming Models Working Group whitepaper.

25. H. Nakada, M. Sato, and S. Sekiguchi. Design and Implementations of Ninf: towards a Global Computing Infrastructure. *Future Generation Computing Systems, Metacomputing Issue*, 15(5-6):649–658, 1999. http://ninf.apgrid.org/papers/ papers.shtml.

26. H. Nakada, Y. Tanaka, S. Matsuoka, and S. Sekiguchi. The Design and Implementation of a Fault-Tolerant RPC System: Ninf-C. In *Proceeding of HPC Asia 2004*, pages 9–18, 2004.

27. T. M. Oinn, M. Addis, J. Ferris, D. Marvin, R. M. Greenwood, T. Carver, M. R. Pocock, A. Wipat, and P. Li. Taverna: a tool for the composition and enactment of bioinformatics workflow. *Bioinformatics*, 20(17):3045–3054, nov 2004.

28. J. S. Plank, K. Li, and M. A. Puening. Diskless checkpointing. *IEEE Transactions on Parallel and Distributed Systems*, 9(10):972–980, 1998.

29. M. Quinson. Dynamic Performance Forecasting for Network-Enabled Servers in a Metacomputing Environment. In *International Workshop on Performance Modeling, Evaluation, and Optimization of Parallel and Distributed Systems (PMEO-PDS'02), in conjunction with IPDPS'02*, Apr 2002.

30. M. Sato, T. Boku, and D. Takahasi. OmniRPC: a Grid RPC System for Parallel Programming in Cluster and Grid Environment. In *Proceedings of CCGrid2003*, pages 206–213, Tokyo, May 2003.

31. K. Seymour, C. Lee, F. Desprez, H. Nakada, and Y. Tanaka. The End-User and Middleware APIs for GridRPC. In *Workshop on Grid Application Programming Interfaces, In conjunction with GGF12*, Brussels, Belgium, September 2004.

32. S. Shirasuna, H. Nakada, S. Matsuoka, and S. Sekiguchi. Evaluating Web Services Based Implementations of GridRPC. In *Proceedings of the 11th IEEE International Symposium on High Performance Distributed Computing (HPDC-11 2002)*, pages 237–245, July 2002. http://matsu-www.is.titech.ac.jp/~sirasuna/research/hpdc2002/hpdc2002.pdf.

33. G. Singh, E. Deelman, G. Mehta, K. Vahi, M.-H.i Su, G.B. Berriman, J. Good, J.C. Jacob, D.S. Katz, A. Lazzarini, K. Blackburn, and S. Koranda. The pegasus portal: web based grid computing. In *SAC '05: Proceedings of the 2005 ACM symposium on Applied computing*, pages 680–686, New York, NY, USA, 2005. ACM Press.

34. Y. Tanaka, N. Nakada, S. Sekiguchi, T. Suzumura, and S. Matsuoka. Ninf-G: A Reference Implementation of RPC-based Programming Middleware for Grid Computing. *J. of Grid Comput.*, 1:41–51, 2003.

35. Condor Team. The directed acyclic graph manager. http://www.cs.wisc.edu/condor/dagman.

36. C. Wang, B.A. Alqaralleh, B.B. Zhou, M. Till, and A.Y. Zomaya. A blast service built on data indexed overlay network. In *e-Science*, pages 16–23, 2005.

37. Y. Tanaka and H. Takemiya and H. Nakada and S. Sekiguchi. Design, Implementation and Performance Evaluation of GridRPC Programming Middleware for a Large-Scale Computational Grid. In *Proceedings of 5th IEEE/ACM International Workshop on Grid Computing*, pages 298–305, 2005.

Execution Support of High Performance Heterogeneous Component-Based Applications on the Grid⋆

Massimo Coppola[1,2], Marco Danelutto[2], Nicola Tonellotto[1,3],
Marco Vanneschi[2], and Corrado Zoccolo[4]

[1] Information Science and Technologies Institute, National Research Council
Via G. Moruzzi 1, 56124 Pisa, Italy
[2] Computer Science Department, University of Pisa
Largo B. Pontecorvo 3, 56127 Pisa, Italy
[3] Information Engineering Department, University of Pisa
Via G. Caruso 16, 56122 Pisa, Italy
[4] IAC Search & Media Italia S.r.l.
Corso Italia 58, 56100 Pisa, Italy

Abstract. Application deployment is becoming an increasingly hard task, as complex, component-based Grid applications have to be deployed on heterogeneous and dynamic Grids, interfacing to several different component frameworks and Grid middlewares. We describe the architecture of the Grid Execution Agent (GEA), the deployment and resource brokering tool of the *Grid.it* project. GEA has been designed to ease the deployment of complex Grid applications written in a high-level, structured way. To easily handle different component models over heterogeneous Grid resources, the GEA design exploits multiple levels of abstraction. Our approach allows consistent translation of the high-level requirements from heterogeneous, multi-component applications, to low-level operations over different middlewares. GEA architecture provides a unified interface with services to locate resources, devise initial mapping, and instantiate applications, and it is extensible to new component models. It supports dynamically reconfiguring, self-adapting applications by allowing execution-time resource allocation changes.

1 Introduction

The vision of Computational Grids set forth at the end of last century is becoming reality, at least from the point of view of the raw capability of coordinating Grid resources into executing applications. However, standardization of middleware and practical and efficient programming models for the Grid are still to be

---

⋆ This work has been supported by: the Italian MIUR FIRB Grid.it project, No. RBNE01KNFP, on High-performance Grid platforms and tools, and the European CoreGRID NoE (European Research Network on Foundations, Software Infrastructures and Applications for Large Scale, Distributed, GRID and Peer-to-Peer Technologies, contract no. IST-2002-004265).

W. Lehner et al. (Eds.): Euro-Par 2006 Workshops, LNCS 4375, pp. 171–185, 2007.

achieved. Thus, the advantages of large Grid computing platforms for several tasks, including collaborative engineering, data exploration, high-throughput computing, and of course distributed super-computing, are still hindered by the difficulty in writing truly portable applications able to exploit dynamic, heterogeneous platforms, as well as to integrate legacy code.

While portals and graphical interfaces allow to manage simple applications and to expose legacy ones as publicly available services, more complex applications designed to benefit from the nature of the Grid platforms still have to be developed exploiting direct interaction with Grid middleware.

Beside the efforts spent in developing middleware systems, the tools provided to deploy and manage the elements of the application do not offer yet a high level of abstraction. Nowadays, the vast majority of applications exploiting Grids are structured as bags of independent jobs, or workflows with simple, file-transfer based interactions.

In the future, complex, multi-disciplinary applications will have to provide an agreed QoS with respect to their fundamental characteristics, e.g. performance, fault tolerance, security. In order to support these requirements more complex and flexible programming models are needed, and applications will have to be able to dynamically alter the set of resources allocated during the execution, and to support multiple interaction protocols with resource management middlewares.

There is general consensus on the adoption of the software component abstraction to simplify the task of programming high performance and distributed applications, especially on Grids. Early examples of this trend are the CCA [1] and GridCCM [2] approaches. Large, international research projects on Grid-aware component models, like CoreGRID and GridCOMP, are a more recent outcome of this trend.

Within the *Grid.it* project we developed the ASSIST [3] structured parallel programming environment to produce software components, and we addressed in the runtime tools the problem of composite application deployment on heterogeneous clusters and Grid resources. The ASSIST environment also supports mixing different kinds of components in the same application (*Grid.it* components, Web services, CCM). These abilities generate the need to integrate different protocols in the run-time for communications, resource query and deployment activities.

In this work we describe the architecture of the *Grid Execution Agent* (GEA), which provides resource brokering and management functionalities for the ASSIST environment. GEA insulates the run-time support of components from the actual Grid middleware. Preliminary versions of GEA have already been introduced in previous works [4,5].

Our main contribution is the design of a Grid application execution framework where a very high-level, abstract description of applications, which is based on software components, is translated into deploy actions using multiple levels of interpretation. The proposed solution exploits plug-in classes to encode the peculiarities, at the different levels of interpretation, of deployment protocols w.r.t. component frameworks, computing processes and supporting middlewares.

The multi-level design of GEA allows easy and seamless configuration of resources and component infrastructures, generally done at run-time, in order (1) to host components from different frameworks, (2) to host components interacting by means of different middlewares, supporting multi-framework integration, (3) to add support for user-transparent deployment of applications on different Grid-middlewares. GEA is thus customizable to support different high-level abstractions interacting with different existing middlewares, supporting HPC Grid applications in large-scale Virtual Organizations, and providing the functionalities of an *Invisible Grid* [3].

The rest of this paper is structured as follows. In Sect. 2 we give a general definition of the Deployment process, with special regard to hierarchical and component-based applications. In Sect. 3 we discuss related work w.r.t. Grid deployment. In Sect. 4 we describe the ASSIST programming environment and the *Grid.it* component model. In Sect. 5 we discuss the approach to application and requirement analysis and translation adopted by GEA, and the overall architecture of the deployment system that results. In Sect. 6 we sum up our contributions and illustrate future work directions.

2   Component Deployment in a Multi-middleware Heterogenous Environment

Our aim is quite general: we want to be able to deploy applications made up of *distributed and parallel components*, which can possibly belong to *different component frameworks*, over *a set of Grid resources* that span a Virtual Organization, possibly encompassing resources managed by *different middleware systems*.

According to our approach, the input of the deployment process includes the application structure, a set of *resource requirements* (fixed constraints on the execution of a single process or components), a set of *QoS models* (analytical expressions of Quality of Service, relating it to the execution parameters of a component) and a set of *contracts* (constraints that the free variables in the application or component model shall satisfy). The initial application deployment will usually involve assembling an overall application QoS model, to balance resource allocation, and decomposing global application contracts into contracts suitable for the single components and modules. Merging contracts and requirements for each component with static knowledge about its implementation, we obtain the information for its initial deployment.

We have to map a non trivial amount of high level information, concerning application structure, deployment requirements and user-expected QoS into a large amount of low-level actions about resource reservation and configuration, process/job mapping and scheduling. Moreover, at run-time more sophisticate models can be used, leading to dynamic changes to the initial deployment choices.

The general problem almost naturally breaks down into levels corresponding to levels of abstraction in the application structure (see Fig. 1).

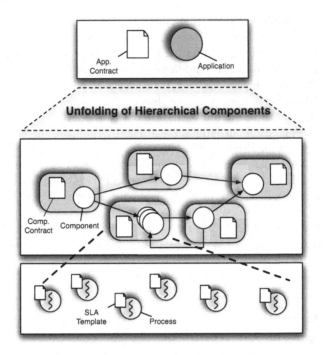

Fig. 1. Abstraction layers for component-based applications execution

Application. An application is a hierarchical composition of components interacting through communication patterns. We need to run it on a Grid while enforcing a user-agreed performance contract, that is a user-dependent specification of the expected behavior of the application at runtime.

Component. The hierarchical composition of components can be unfolded until the whole application is "exploded" to a complex graph of atomic components. We need a standardized way to convey information about the structure and characteristics of every component, as well as models of the runtime behavior of the components and their interactions. Such information is exploited to characterize every component with its own contract, in order to select Grid resources that will host the component, deploy it and control its behavior at runtime.

Process. Each component is made up of several processes (including functional processes and support services). Every process will need its own mapping and scheduling over concrete resources. A Service Level Agreement (SLA) template has to be derived from the component contract for each implementation process, in order to negotiate an agreement with the Grid resources and to globally enforce the QoS required by the user. Moreover, to deploy a component, its processes need to be properly configured to interact.

Middleware. For each single process or job, the Grid middleware used to access the selected resource (e.g. Globus) will generally need a specific set of actions in order to successfully deploy the process.

The deployment activities at each abstraction layer below the first one (component, process, middleware) can be arranged in a workflow, that encodes their dependencies (a partial order over activities), the parameters and the configuration information that each activity needs to transmit to the depending ones.

If we consider each deployment level as the satisfaction of a dependency graph of actions, the overall application deployment is actually the product of the three dependency graphs over components, over processes within them, and over Grid middlewares exploited to access the resources.

To avoid generating large optimization problems, whose solutions would be anyway approximate, we chose not to unroll and flatten the whole application deployment to a single dependency graph of elementary activities.

Our approach instead exploits the hierarchical structure of the application to split the deployment problem into smaller and smaller subproblems. This way we can more easily devise deployment heuristics to reach good initial resource mappings, and it is possible to reuse the same deployment system to perform application adaptation, by deploying locally optimized additional entities (components, processes).

3   Related Work

First-generation deployment mechanisms based on Globus [6] can deploy only sequential jobs and "bag of tasks", that is parallel "uniform" (SPMD) jobs to be executed by homogeneous clusters. Condor-G [7] is a typical example of this kind of approach, as deployment requirements are specified in detail at a very low level of abstraction. Deployment is defined by elementary actions which depend both on the application process structure, and on the middleware. Another obstacle to the aggregation of heterogeneous Grid resources is that Grid middlewares provide in general different APIs, functionalities and servers for resource location, access and management.

These are clearly fundamental issues, which we must solve in order to enhance support for component-based applications, applications with non-trivial structure, e.g. exploiting different frameworks and middlewares, and dynamically adaptive applications, which need execution-time reconfiguration and (re)deployment. A number of systems are currently being developed for the Grid, which aim at solving the mentioned issues and supporting high-level programming languages and environments.

Adage [8] is a tool for Grid deployment whose approach is based on the translation of different kinds of application descriptions, both flat message passing and component-oriented ones, into a common XML format called GADe (Generic Application Description). GADe represents an application as a graph of computing entities, each one made up of processes, and each process containing a set of code entities (components and DLLs). Currently CCM, and MPI translators have been developed which feed with GADe description the deployment planning and execution modules of Adage. GridCCM and CCA translators are in development.

Grid middleware interface in Adage is based on the GAT toolkit [9]. Adage and GEA share a common approach [10] in adopting a core deployment engine independent of application and middleware details, and exploiting a high-level application description language. We differentiate from Adage as our description language (ALDL) allows applications to mix processes and components from different frameworks, and can express dynamic adaptive process networks. While support for automatic translation of descriptions is less developed, GEA can exploit ALDL to manage coallocation of resources over multiple middlewares and frameworks.

The Proactive library [11] is a Java-based solution for parallel and distributed programming. This library provides a programming model and a set of API to develop complex Grid applications. Parallelism in Proactive applications is defined by *Active Objects*, which host application control threads.

To obtain seamless deployment on different runtime environments, Proactive exploits a descriptor-based approach. The *Descriptor Deployment Model* [12] of Proactive is based on three levels of abstraction, (1) Virtual Nodes (VNs) hosting the application specific Active Objects, (2) Java Virtual Machines (JVMs), hosting the application runtime environment, (3) processes, to create and/or acquire JVMs. Virtual nodes are defined in the application code. They are possibly replicated, and instantiated (as Nodes) to run on actual JVMs. JVMs are recruited exploiting information provided by processes.

These mappings are coded in XML Deployment Descriptors, with the target of completely abstracting away from each other the hardware and software runtime configuration, and decoupling application logic from deployment logic.

The approach results in a highly configurable deployment mechanism, which can start new JVMs as needed on local and remote resources. Configuration is left to the Proactive runtime, avoiding any reference to concrete resources in the application code. The drawbacks, and main differences with respect to GEA approach, are that it is difficult to extend the approach to non-Java components, that mapping of application objects to resources is not automatic, and that a first mapping step has to be performed by the user in designing the application, by specifying the mapping of Active Objects to Virtual Nodes.

On the contrary, GEA's Virtual Nodes represent compiler-generated sets of processes, which are not related to the programmer's view of the application, but whose existence is suggested by the implementation of the run-time support.

To the best of our knowledge, KOALA [13] is the only other high-level tool to provide extensive support for coallocation over Grids. KOALA manages reservation and deployment of multi-job applications over the Grid. The job model in KOALA is an executable to be run over a specified number of nodes, taking as input a single data file. The Grid model consists of a set of clusters with homogeneous nodes, interconnected by a known network and each one managed by a compatible local resource manager. These assumptions allow to develop algorithms for multi-site job scheduling taking into account problem parameters like job sizes, input data sizes, available resource loads, network transfer bandwidths and job priority. With respect to KOALA, GEA is less integrated with resource reservation mechanisms, and it would need more complex job scheduling algorithms to find optimal allocation, due to the Grid model adopted. Nevertheless, GEA can deal with much more detailed resource constraints, allowing to exploit a much more heterogeneous Grid and to satisfy compiler-generated resource requirements in a dynamically evolving environment.

4   The ASSIST Environment Architecture

ASSIST is a high-level parallel programming environment: it provides a structured parallel programming language and a compiler to develop QoS enabled parallel components [3,14]. Basically, applications are described by means of a coordination language, which can express arbitrary graphs of (possibly) parallel modules, which are the basic structural units of applications. ASSIST modules are interconnected by typed streams of data, host portions of sequential code (C, C++, Fortran) and can explicitly define even complex parallel semantics at the module level.

A parallel module (*parmod*) coordinates a set of concurrent activities which are performed by *Virtual Processes* (VPs). We do not fully describe here the ASSIST parallel coordination language [15] or its implementation, we only underline that parmods allow to express parallel computations which are reconfigurable during execution. User-defined code sections within the VPs are seen as the set of atomic computations in the application by the reconfiguration run-time support.

The environment is designed to allow execution of parallel programs over resources that are heterogeneous w.r.t. many characteristics, including CPU architecture, operating system and middleware interface. The compiling tools can also generate *Grid.it* components containing ASSIST code as well as *alien* software resources (e.g. software components from other frameworks).

ASSIST/*Grid.it* components [3] are graphs of modules that are explicitly declared as deployment units, and export information and control ports to allow coordinated execution of several of them. *Grid.it* Components expose, among others, *Non-Functional* ports related to QoS control. As shown in Fig. 2a, *Grid.it* native components have a sophisticated internal structure, including different classes of processes devoted to application execution and run-time support,

(1) computational processes (ASSIST processes in the figure),
(2) processes supporting the shared memory abstraction (HOC processes),
(3) manager processes implementing autonomic component behavior (CAM and DCServer), i.e. steering and managing adaptation at the component level,
(4) proxy processes allowing inter-component communication (component bridges).

To deploy even a single component, a workflow of deployment actions is needed, of which we show an example in Fig. 2b. Moreover, *Grid.it* native component can interoperate with Corba/CCM ones (via IIOP-based RPC) [16] and with Web Services (via HTTP/SOAP) [17], forming composite, multi-framework applications. Whether a wrapping approach is adopted, or component bridges

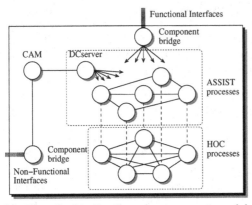
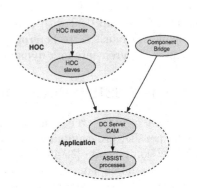

(a) The *Grid.it* distributed component model implementation as a set of processes.

(b) Dependencies at the component level among sets of processes.

Fig. 2. Process structure of a typical Grid.it component, and example of the deployment dependencies among it composing processes

are created, the overall deployment gets more complex. Running such a multi-framework application requires the ability to devise and set up proper support processes for any combination of resource, supporting middleware and supported component framework.

Super-component have been introduced in [18] to describe higher-order components, which can manage parametric graphs of arbitrary components according to a parallel skeleton (i.e. well-known, parametric pattern of parallelism). They provide a fully compositional structure for self-managing Grid applications.

Super-components interact with the resource management and deployment system to manage the life-cycle of their controlled components, and coordinate their overall dynamic reconfiguration. To accomplish this task, they leverage on a compositional self-adapting infrastructure, and on suitable behavioral models corresponding to the parallel skeletons the specific super-component implements.

At any moment during an ASSIST application run, modules and components can be assigned a new QoS contract, e.g. specifying a performance, security or fault tolerance requirement. In order to fulfill the contracts, the component framework continuously adapts component configurations, in terms of parallelism degree, and process mapping [19]. This means having a **progressive, dynamic deployment process** where portions of the application are re-deployed in order to meet a specific QoS target.

The adaptation mechanism relies on automatic user code instrumentation, and on a **hierarchy of Application Managers** [3] exploiting knowledge about the application structure and the run-time implementation. The hierarchy of managers operating at different levels in the application structure is reflected in the connections among the Non-functional ports of modules, components, and super-components. Eventually, the whole execution is steered by the top-level Application Manager (AM) component. Semantics and protocols of these interactions are out of the scope of the paper, but we point out that some dynamic effects of resource management have non-local impact which need proper handling in the management hierarchy (e.g. load balancing may need computing resources in excess to be re-allocated to some seemingly unrelated part of the application).

From the ASSIST point of view, the GEA is a component of the environment run-time support, the Grid Abstract Machine, as it manages the resource allocation at all levels. ASSIST super-component managers leverage the GEA when deploying new component(s), and the component and module run-time support for reconfiguration (CAM and MAM entities) can contact GEA whenever resources are needed to spawn new processes in order to satisfy performance contracts at the module level.

As a final remark, compositional, hierarchical component models (e.g. an implementation of Fractal in Java or C++) also need an algorithmic way to break down the overall application description into the set of descriptions of its components, and in particular to **project the application QoS specification over that of components**, in order to find appropriate resources to deploy each component. This phase of "requirement unfolding" can happen outside of, or as part of the deployment workflow. Current approach in *Grid.it* exploits the model embedded in super-components. Transformation from application to component-level requirements is recursively performed by the Application Manager, which directly controls the unfolding step of deployment. As a different approach, a compositional performance model for launch-time mapping has also been developed [20], which is suitable to devise a good initial resource allocation and speed up the deployment of the whole application. Such a model can be produced and evaluated during deployment, for instance implementing it within a particular Component Translation Engine Plug-in (see Sect. 5.2).

5 Grid Execution Agent Design

The Grid Execution Agent (GEA) is the automatic tool developed within the *Grid.it* project to seamlessly search resources for, deploy, and run complex component-based Grid applications. The ASSIST/*Grid.it* environment targets high-performance, data/computation intensive, and distributed applications. GEA is designed to be a high-level resource management system, handling all the low-level interaction with multiple Grid middlewares and with the code providing dynamic adaptation. The Core of the Deployment cycle, as shown in Fig. 3 in the inner box, follows the outline presented in [4]. We recall it in short in the next section, before discussing its extension to a dynamic and multi-middleware scenario.

5.1 Core Deployment Cycle

The input of the cycle is a description of the entities to deploy in a general format, the Application Level Description Language (ALDL). This XML dialect can encode the requirements for all the deployment entities, ranging from high level performance specifications for components to concrete constraints on target architectures for processes.

ALDL descriptions contain different types of information:

- static resource constraints and dynamic constraints – e.g. constraints related to quantities that do not vary over time, like peak performance of a resource, as opposed to those related to varying features, like available computation bandwidth, which depends on resource load.
- hardware and software constraints – expressing the specific need of architectures, operating systems, support or application libraries to be available to an entity at its execution site.
- aggregate constraints – specifying constraints on groups of entities. Most notably, constraints over communication networks (security, reachability) and over sets of processes which employ specific common resources or launch protocols (e.g. name services or fault-tolerant communication schemes).

GEA, starting from the ALDL description, automatically performs resource discovery and selection, handles data and executable file transfers. Different GEA modules perform successive steps of translation of high-level specifications into deploy actions.

1. The ALDL description is parsed and an internal representation of the graph of tasks is generated, annotated with specific requirements and constraints.
2. From the internal representation, resource queries are computed, which aim at locating a set of resources satisfying all the static constraints.
3. Resource queries are executed exploiting the middleware.
4. A subset of the resources is selected, also exploiting information related to dynamic constraints.
5. The graph of processes is mapped over the resources. This can result in mapping cooperating entities over independently managed resources, thus triggering coallocation in the following phase.
6. Finally, each entity is executed on the corresponding resource through the middleware. This means e.g. staging and executing process code, configuring its execution environment, or deploying a specific network configuration to ensure a stated goal of communication QoS.

Actually, the discovery and mapping phases can loop until a set of resources is found that allow to map the whole entity according to its execution constraints. As a result of the deployment process, different parts of the ALDL description are filtered, instantiated and translated into the appropriate forms to be enacted on the middleware used by each part of the Grid computing platform.

5.2   A Modular Multi-middleware Architecture

The core cycle described in the previous section deploys applications over a heterogeneous Grid, can exploit different middleware to access resources, but it coordinates them in the execution of a single program.

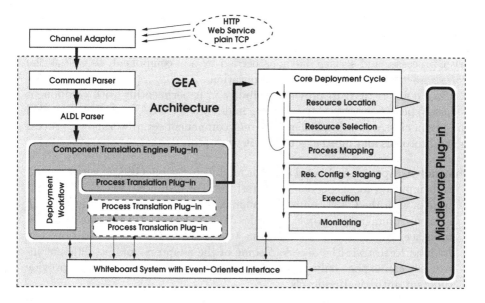

Fig. 3. GEA high-level architecture

As ASSIST was being developed into a component model, additional requirements were put onto GEA.

Separate/Dynamic Deployment. Applications are made up of multiple components, and components are separate units of deployment. GEA has to fully behave like a server, allowing to launch multiple components (eventually from different applications) and to manage each one separately over possibly overlapping portions of the Grid.

Dynamic adaptation. *Grid.it* components can ask for and free resources (e.g. processing nodes) at run-time, so they need access to deployment functions.

Access Protocols. GEA functionalities have to be accessible through different protocols (plain TCP sockets, HTTP, Web Services), in order to exploit them easily across the Grid.

Flexibility w.r.t. Component Models. *Grid.it* applications can exploit components and services from other frameworks (CCM, Web Services), which need to be deployed and accessed with their own protocols. Moreover, it is a key feature to ease experimentation with various implementations of a component model.

Higher Scalability. Provision to the user of a single point of access to the whole Grid must not bring unneeded centralization and impair deployment scalability.

Crossing of Domain Boundaries. GEA should be able to operate at the boundary of a network and provide to the outside a unified access abstraction, independently of any Grid middleware present within the network.

The resulting extended architecture in Fig. 3 takes into account all these issues and builds upon the core cycle used in the first design of GEA. It plans and enacts the deployment workflow of *Grid.it* components, starting in the proper order the server processes and service daemons needed by any component, as well as the processes actually performing the computation.

A key point is that, to provide flexibility in experimenting with component models, types of processes and diverse middlewares, GEA has been extended via plug-in classes that implement different component setup workflows, process launch protocols and interfaces to middleware primitives.

Channel Adaptor. A channel adaptor module is used in GEA to support multiple input protocols and control interfaces. For instance, the GEA server interacts with the CAM component manager via TCP, with the user through a set of command-line tools, provides Web Services / HTTP as a standard interface. Authorization mechanisms (e.g. local or GSI authentication) can be used to restrain the access to some of the adaptors, and, as different interfaces can expose only part of the full set of GEA commands, to provide different authorization levels.

Command Parser. The GEA command parser supports commands to manage the life-cycle of deployed entities (from providing them in archival form, to monitoring their termination) and to control the configuration of the GEA server. provide components in form of archives, to deploy component instances, to dynamically add new resources, to monitor component termination, to quit the GEA server or to reload its static configuration information (e.g. addresses of dynamic Grid information services). The command parser is also in charge of managing multiple sessions: each component's ALDL is kept linked to a session identifier, which is a handle to monitor and steer the component deployment and the set of allocated resources. The ALDL representation of each component must be kept (1) to allow caching of component archives and easily instantiate multiple copies of a component, (2) to simplify handling of dynamic adaptation, as we can deploy additional processes within a component by referring to their identifiers in the component description. The parser manages multiple command session and internally caches ALDL representation of components, in order to ease creating multiple instances of a component, and to allow partial redeployment of a running component.

ALDL Parser. The ALDL parser has been extended (w.r.t. [4]) to allow expressing explicit co-allocation and superposition of processes. A concept of *virtual node* is used (close to that of Proactive) which is the abstraction of a physical resource. Processes of different types are mapped to virtual nodes, which are the units of low-level resource mapping. In order to subsequently map virtual nodes on resources, process constraints are gathered and summed up with the proper aggregation function (e.g. sum, for memory requirements, union, for requirements over available libraries, and so on).

Component Translation Engine. The Component Translation Engine is actually the highest level plug-in, transforming a component ALDL specification into a network of process dependencies. The deployment work-flow, hard-coded as a Java class in the plug-in, fulfills the dependency graph among different types of processes in a given component model,

Process Translation Plug-in. Process Translation Plug-ins are a set of mid-level translator classes, associated with the types of processes we need to start. Each of these plug-ins can add special constraints over the resources to select (e.g. having a public IP address), it can exploit information from previously configured/deployed processes and service daemons, and it knows the protocols to configure and deploy one type of processes (e.g. ASSIST DCserver and ASSIST application processes). Translated requirements of all processes within a component are produced by the appropriate plug-ins, before starting the actual deployment process.

Core Deployment Cycle. As reported, the core deployment cycle has been adapted from the previous GEA architecture, with all requirements gathered from all virtual nodes first, and then satisfied all at the same time as described in Sect. 5.1. The actual deploy order of processes (configuration, staging and execution phases) in governed by the dependencies explicitly introduced by the Component Engine and Process Translation plug-ins.

Middleware Plug-in. This is a low-level set of classes, one for each middle-ware supported, exposing a set of primitives for the basic operations of all the steps defined in the Core Deployment Cycle (resource location, selection, mapping, staging and so on). Existing plug-ins support Globus managed re-sources (using MDS as information repository), as well as SSL-based access to clusters and local networks (XML static configuration files are used in place of MDS). In some cases GEA extends the functionalities of the mid-dleware, e.g. to provide status monitoring for resources accessed via simple SSL.

Event System and GEA Whiteboard. A communication module is used as a scratch-pad interface to allow uniform parameter management to plug-ins of all levels. Different process types and different instances within a compo-nent in general need to exchange and propagate synchronization informa-tion, service addresses and other execution parameters (e.g. CCM processes are run after their naming service is known and it is up). The whiteboard implementation manages several kinds of events (TCP and HTTP commu-nications, process termination, monitoring information) and uses them to trigger deployment dependencies by means of callback functions that notify registered plug-ins.

6   Conclusion and Future Work

We have presented the architecture of the GEA deployment tool developed within the *Grid.it* project. The presented extension to the deployment cycle developed in the previous implementation [4] allows greater flexibility and easier customization of the application model, with respect to the types of component we can actually deploy. The ASSIST application model, the *Grid.it* component model and the launch and configuration procedures for component support elements are all boxed into separate plug-ins.

We are currently working on the integration of different component and application models within GEA, including CCM components and Web Services (within the *Grid.it* project) and POP C++ applications (within the CoreGRID NoE). The new implementation matches with the extensions of the ALDL language, which we just mentioned in this work, to cope with new mapping constraints and with contract specifications.

Currently, we are still working on the ALDL language to improve expressiveness w.r.t. QoS contract specification for applications and components. A technique that allows to derive component constraints from application ones has already been devised [20], which can be used to develop a Component Translation Plug-in to deal with a fully hierarchical component model (e.g. Fractal) to devise an optimal initial application mapping starting from the ALDL specification and the description of the available resources.

Both the implementation of a hierarchy of GEA servers for distributed deployment, and of more scalable deploy protocols for very large Grids, are in our future plans. We have made preliminary experiments in this direction, exploiting hierarchical and distributed communication schemes in overlay networks of servers. We are going to exploit the flexibility of the GEA architecture to integrate these prototypes as a new channel adaptor and dedicated low-level plug-ins.

References

1. Armstrong, R., Geist, A., Keahey, K., Kohn, S., McInnes, L., Parker, S., Smolinski, B.: Toward a common component architecture for high-performance scientific computing. In: 8th IEEE International Symposium on High-Performance Distributed Computing. (1999)
2. Pérez, C., Priol, T., Ribes, A.: A parallel CORBA component model for numerical code coupling. In: GRID 2002 : Third International Workshop, Springer (2002)
3. Aldinucci, M., Campa, S., Coppola, M., Danelutto, M., Laforenza, D., Puppin, D., Scarponi, L., Vanneschi, M., Zoccolo, C.: Components for high performance Grid programming in Grid.it. In: Proc. of the Workshop on Component Models and Systems for Grid Applications. CoreGRID series. Springer (2005)
4. Danelutto, M., Vanneschi, M., Zoccolo, C., Tonellotto, N., Baraglia, R., Fagni, T., Laforenza, D., Paccosi, A.: HPC Application Execution on Grids. In Getov, V., Laforenza, D., Reinefeld, A., eds.: Future Generation Grids. CoreGRID. Springer (2006) Dagstuhl Seminar 04451 – Nov. 2004.
5. Adami, D., Giordano, S., Repeti, M., Coppola, M., aforenza, D.L., Tonellotto, N.: Design and Implementation of a Grid Network-Aware Resource Br oker. In Fahringer, T., ed.: Parallel and Distributed Computing and Networking 2006. ACTA Press (2006)
6. Foster, I., Kesselman, C.: Globus: A Metacomputing Infrastructure Toolkit. Int. J. of Supercomputer Applications and High Performance Computing **11** (1997) 115–128
7. Frey, J., Tannenbaum, T., Foster, I., Livny, M., Tuecke, S.: Condor-G: A Computation Management Agent for Multi-Institutional Grids. In: Proceedings of the 10th IEEE Symp. on High Performance Distributed Computing (HPDC10), San Francisco, California, IEEE (2001)

8. Lacour, S., Pérez, C., Priol, T.: Generic application description model: Toward automatic deployment of applications on computational grids. In: 6th IEEE/ACM International Workshop on Grid Computing (Grid2005), Seattle, WA, USA, Springer-Verlag (2005)

9. Allen, G., et al.: Enabling Applications on the Grid – A GridLab Overview. International Journal of High Performance Computing Applications **17** (2003) 449 – 466 Special issue on Grid Computing: Infrastructure and Applications.

10. Coppola, M., Danelutto, M., Lacour, S., Pérez, C., Priol, T., Tonellotto, N., Zoccolo, C.: Towards a Common Deployment Model for Grid systems. To appear in CoreGRID series. (2005)

11. Baduel, L., Baude, F., Caromel, D., Contes, A., Huet, F., Morel, M., Quilici, R.: Programming, Deploying, Composing, for the Grid. In: Grid Computing: Software Environments and Tools. Springer-Verlag (2006)

12. Baude, F., Caromel, D., Mestre, L., Huet, F., Vayssière, J.: Interactive and descriptor-based deployment of object-oriented grid applications. In: Proceedings of the 11th IEEE International Symposium on High Performance Distributed Computing, Edinburgh, Scotland, IEEE Computer Society (2002) 93–102

13. Mohamed, H., Epema, D.: The Design and Implementation of the KOALA Co-allocating Grid Scheduler. In Sloot, P.M.A., Hoekstra, A.G., Priol, T., Reinefeld, A., Bubak, M., eds.: Advances in Grid Computing - EGC 2005: European Grid Conference. Volume 3470 of LNCS. (2005) 640–650

14. Aldinucci, M., Coppola, M., Danelutto, M., Vanneschi, M.V., Zoccolo, C.: ASSIST as a research framework for high-performance Grid programming environments. In Cunha, J.C., Rana, O.F., eds.: Grid Computing: Software environments and Tools. Springer (2005) 1–32 (To appear, draft available as TR-04-09, Dept. of Computer Science, University of Pisa, Italy, 2004).

15. Vanneschi, M.: The programming model of ASSIST, an environment for parallel and distributed portable applications. Parallel Computing **28** (2002) 1709–1732

16. Magini, S., Pesciullesi, P., Zoccolo, C.: Parallel software interoperability by means of CORBA in the ASSIST programming environment. In: Kosch, H., Böszörmeny, L., Hellwagner, H.(eds.) Euro-Par 2003. LNCS, vol. 3648, pp. 679–688. Springer, Heidelberg (2004)

17. Aldinucci, M., Danelutto, M., Paternesi, A., Ravazzolo, R., Vanneschi, M.: Building interoperable grid-aware ASSIST applications via WebServi ces. In: PARCO 2005: Parallel Computing, Malaga, Spain (2005)

18. Aldinucci, M., Bertolli, C., Campa, S., Coppola, M., Vanneschi, M., Veraldi, L., Zoccolo, C.: Self-Configuring and Self-Optimising Grid Components in the GCM model and their ASSIST Implementation. In: Joint Workshop on HPC Grid Programming Environments and Components (HPC-GECO/CompFrame). (2006)

19. Aldinucci, M., Petrocelli, A., Pistoletti, E., Torquati, M., Vanneschi, M., Veraldi, L., Zoccolo, C.: Dynamic reconfiguration of grid-aware applications in ASSIST. In Cunha, J.C., Medeiros, P.D., eds.: 11th Intl Euro-Par: Parallel and Distributed Computing. Volume 3648 of LNCS., Lisboa, Portugal, Springer (2005) 771–781

20. Tonellotto, N., Zoccolo, C.: Characterization of the performance of ASSIST programs. Technical Report TR-0007, CoreGRID - Network of Excellence (2005)

Towards a Grid Information Knowledge Base*

Wei Xing[1], Marios D. Dikaiakos[1], and Rizos Sakellariou[2]

[1] Department of Computer Science, University of Cyprus, Cyprus
[2] School of Computer Science, University of Manchester, UK
xing@ucy.ac.cy, mdd@ucy.ac.cy, rizos@cs.man.ac.uk

Abstract. In this paper, we present our work on building a Grid information knowledge base, which is a key component of a semantic Grid information system. A Core Grid Ontology (CGO) is developed for building a Grid knowledge base; and the SPARQL query language is adopted to query the knowledge base.

1 Introduction

Information Services are regarded as a vital component of the Grid infrastructure. They address the problem of the discovery and ongoing monitoring of the existence and characteristics of resources, services, computations and other entities of value to the Grid [1]. As Grids grow larger and gain widespread use, there is an increasing need for Grid information systems to support complex queries, such as:

1. Is there a VO providing exclusive access to a shared-memory multiprocessor system with at least 16 processor, 8 GB of main memory, and a usage charge of not more than 100 euros per CPU time?
2. Find services running Quantum Chromo-Dynamics calculations (QCD) using F90 and MPI?
3. Locate Grid-sites that offer access to a LAPACK software library installed on a shared-memory multiprocessor with 16 to 64 processors?
4. Find the pricing and prior clientele of Grid service that provide access to the XYZ workflow for real-time oil refinery simulations?

However currently no information system can answer the above queries. The main problem of the existing information systems (e.g. MDS, RGMA, BDII) is that they are actually designed and developed for providing particular information to specific Grid sub-systems [2]. For instance, MDS of Globus is designed and developed to support resource discovery; BDII of LCG is used mainly for job scheduling. This makes them not adequate to provide information about whole aspects of Grid systems, such as Grid entities, capability of Grids, Grid resources, Grid middleware, Grid services, Grid applications, and Grid users.

---

* Work supported in part by the European Commission under the CoreGrid project.

W. Lehner et al. (Eds.): Euro-Par 2006 Workshops, LNCS 4375, pp. 186–190, 2007.

In brief, the main limitations of those systems are: (1) limitations on absorbing heterogeneous information sources; (2) inadequate in supporting information management, retrieval, and sharing in a large-scale multi-Grid systems.

To tackle these issues, we propose a semantic approach, which builds Grid information knowledge bases for Gris information services. The knowledge bases contain semantic metadata of Grid entities, resources, middleware, services, applications, and users. In this paper, we present the work of building a Grid information knowledge base. By using a Core Grid Ontology, we propose a ontology-driven method to create a Grid knowledge base [3]. We also adopt SPARQL query language to support the complex query to the Grid knowledge base [4].

The remaining of this paper is organized as follows. In Section 2, we describe the Core Grid Ontology. In Section 3, we introduce our work of building a Grid information knowledge base. Then, we describe how to query the knowledge base using SPARQL in Section 4. Finally, we conclude our work in Section 5.

2   A Core Grid Ontology Framework

The main issue is how to build, update and manage the Grid knowledge base. A Grid knowledge base may be built based on Grid ontologies, which define fundamental Grid-specific concepts, and the relationships between them. Hence, a Grid ontology is needed in order to build a Grid knowledge base. The main problem for building an ontology for Grids is that there is currently a multitude of proposed Grid architectures and Grid implementations, which are comprised Grid entities, services, components, and applications. It is thus very difficult, if at all feasible, to develop a complete Grid ontology that will include all aspects of Grids. Furthermore, different Grid sub-domains, such as Grid resource discovery and Grid job scheduling, normally have different views of, or interests about a Grid entity and its properties. This makes the definition of Grid entities and the relationships between them very hard. To tackle these issues, we propose a Core Grid Ontology (CGO) that defines fundamental Grid-specific concepts, and relationships. One of our main goals was to make this Core Grid Ontology general enough and easily extensible to be used by different Grid architectures or Grid middleware, so that the CGO can provide a common basis for representing Grid knowledge about Grid systems, including resources, middleware, services, applications.

A Core Grid Ontology (CGO) is proposed to define fundamental Grid-specific concepts, and the relationships between them. One of the key goals is to make this Core Grid Ontology general enough and easily extensible to be used by different Grid architectures or Grid middleware, so that the CGO can provide a common basis for representing Grid knowledge about Grid systems, including Grid resources, Grid middleware, services, applications, and Grid users.

The Core Grid Ontology is designed and developed based on a general model of Grid infrastructures, and described in the Web Ontology Language OWL [3]. Such an ontology can play an important role in building Grid-related Knowledge bases and in supporting the realization of the Semantic Grid. We adopt the CGO as the key building block for the GriSen. It is used for both the creation of a Grid knowledge base and knowledge-based query.

3   Building a Grid Knowledge Base Using CGO

A Grid knowledge base is normally comprised of a set of Grid Ontology classes, the relationships and constraints among those ontology classes, and instances of the classes (i.e. Individuals). In reality, the knowledge base may contain a large number of instances of different CGO classes. To build a Grid knowledge base, creation and updating of the instances of the Grid knowledge base is most important and difficult work. Traditionally, the instances of a knowledge base are created by a manual process with ontology editor. However, the manual process to build and maintain the instances is impossible or difficult for a Grid system. First of all, Grids contain large number of different Grid entities. Hence, the number of their instances is large. It is impossible to generate those instances manually. Secondly, Grids are characterized as dynamic. Consequently, the metadata information about them is also changed frequently. To catch up those changes by hand is very difficult. To cope with these issues, we design an ontology-driven approach that can fetch the information from grid information sources, and represent heterogeneous information about Grids in OWL format.

4   Querying a Grid Knowledge Base

Another important consideration is how to query the Grid knowledge base. Grids are complicate distributed system, which may comprise of a set of interacted components and massive heterogeneous resources. Hence, a Grid user normally does not know exactly what to ask about to the Grid knowledge base. To this end, we design a ontology based query service that supports Grid information navigation based on the definitions and relationships of the Grid entities in the CGO. It can process user requests, and generate queries according to the knowledge of CGO and users' willing.

We adopt the SPARQL as the query language to query the metadata in the Grid knowledge base [4]. SPARQL is a query language for getting information from RDF graphs. It provides facilities to: 1) extract information in the form of URIs, blank nodes, plain and typed literals; 2) extract RDF sub-graphs; 3) construct new RDF graphs based on information in the queries graphs.

The SPARQL query language is based on matching graph patterns. The simplest graph pattern is the triple pattern, which is like an RDF triple , but with the possibility of a variable instead of an RDF term in the subject, predicate or object positions. The query consists of two parts, the SELECT clause and the WHERE clause. The SELECT clause identifies the variables to appear in the query results, and the WHERE clause has one triple pattern.

We design a OntoQuery service, which can help users make a SPARQL query according to user's questions. We illustrate how to query Grid knowledge base using SPARQL with examples as follows:

(1) Is there a VO providing exclusive access to a shared-memory multiprocessor system with at least 16 processors, 8 GB of main memory, and a usage charge of not more than 100 euros per CPU time?

```
PREFIX  cgo: <http://grisen.grid.ucy.ac.cy/cgo/0.1/>
FROM    <grisen.owl>
SELECT  ?VO
WHERE   { ?x cgo:hasName ?VO .
            OPTIONAL { ?x cgo:hasCPUnum ?number . FILTER (?number > 64)}
            OPTIONAL { ?x cgo:hasCPUType ? .    FILTER (?number > 8)}
            OPTIONAL { ?x cgo:price  ?price  . FILTER (?price =< 100)}
```

(2) Find services running Quantum Chromo-Dynamics calculations (QCD) using F90 and MPI.

```
PREFIX  cgo: <http://grisen.grid.ucy.ac.cy/cgo/0.1/>
FROM    <grisen.owl>
SELECT  ?Service
WHERE   { ?x cgo:runningService ?service   .
  OPTIONAL { ?x cgo:hasName "QCD"   .
                    ?y cgo:installedOn    .
                    ?z cgo:hostsSoftware  . }
```

(3) Find the pricing and prior clientelle of Grid services that provide access to the XYZ workflow for real-time oil refinery simulations.

```
PREFIX  cgo:  <http://grisen.grid.ucy.ac.cy/cgo/0.1/>
FROM    <grisen.owl>
SELECT  ?Service ?Workflow
WHERE   { ?x cgo:runsOn ?service .
  OPTIONAL { ?x cgo:hasName "PPC" . }
        UNION    { ?y cgo:access ?Workflow .
        OPTIONAL { ?z cgo:hasName ''XYZ'' .
                    ?s cgo:simulation   ''Oil'' .}
        }
```

The above examples show that the SPARQL is capable of querying a Grid system, in particular, supporting complex queries about "the particular properties&values of the Grid entities". Since SPARQL is a RDF-based query language, we currently also investigate how it can be used to query the complicate relationships among the CGO classes which are represented in OWL.

5   Conclusions

In this paper, we present our on-going work on building a Grid Information knowledge base, and querying the knowledge of Grid resources, Grid middleware, services, applications, and Grid users based on a Core Grid Ontology.

Next step, we plan to implement a set of Grid services that can be used to build and update a Grid information knowledge base automatically and dynamically.

References

1. M. D. Dikaiakos, R. Sakellariou, and Y. Ioannidis, "Information Services for Large-scale Grids: A Case for a Grid Search Engine," in *Engineering the Grid: status and perspective*, J. Dongarra, H. Zima, A. Hoisie, L. Yang, and B. DiMartino, Eds. American Scientific Publishers, January 2006.
2. S. Campana and A. S. M. Litmaath, "LCG-2 Middleware Overview," LCG Technical Document, https://edms.cern.ch/file/498079//LCG-mw.pdf.
3. W. Xing, M. Dikaiakos, and R. Sakellariou, "A Core Grid Ontology for the Semantic Grid," in *Proceedings of 6th IEEE International Symposium on Cluster Computing and the Grid (CCGrid 2006)*. Singapore: IEEE Computer Society., May 2006, pp. 178–184.
4. E. Prud'hommeaux and A. Seaborne, *SPARQL Query Language for RDF*, W3C Working Draft, July 2005.

UNICORE Summit 2006

Introduction

Achim Streit and Wolfgang Ziegler

Workshop Chairs

The UNICORE Grid technology provides a seamless, secure, and intuitive access to distributed Grid resources. UNICORE is a full-grown and well-tested Grid middleware system, which today is used in daily production worldwide. Beyond this production usage, the UNICORE technology serves as a solid basis in many European and International projects. In order to foster these ongoing developments, UNICORE is available as open source under BSD licence at http://www.unicore.eu.

The UNICORE Summit is a unique opportunity for Grid users, developers, administrators, researchers, and service providers to meet. The first UNICORE Summit was held in conjunction with "Grids@work - 2nd Grid Plugtests," October 11–12, 2005 in Sophia Antipolis, France. In 2006 the style of the UNICORE Summit was changed by establishing a Program Committee and publishing a Call for Papers. The UNICORE Summit 2006 was held in conjunction with the Euro-Par 2006 conference in Dresden, Germany, August 30–31, 2006. Although it was a workshop at the Euro-Par 2006 conference, all papers were conference-reviewed by at least four members of the Program Committee. The acceptance rate was 38%.

We would like to thank the Program Committee members Agnes Ansari, Rosa Badia, John Brooke, Anton Fank, Edgar Gabriel, Alfred Geiger, Odej Kao, Paolo Malfetti, Ralf Ratering, Johannes Reetz, Mathilde Romberg, Bernd Schuller, Dave Snelling, Stefan Wesner, and Ramin Yahyapour for their excellent job. Special thanks go to Graham Fagg and Björn Hagemeier for providing additional reviews.

Finally, we would like to thank all authors for their submissions, camera-ready versions, and presentations at the UNICORE Summit 2006 in Dresden as well as Dave Snelling for giving the opening talk.

The next UNICORE Summit will again take place in conjunction with the Euro-Par conference, this time in Rennes, France, on August 28, 2007. More information can be found at http://summit.unicore.org/2007/. We are looking forward to the next UNICORE Summit!

W. Lehner et al. (Eds.): Euro-Par 2006 Workshops, LNCS 4375, p. 193, 2007.
© Springer-Verlag Berlin Heidelberg 2007

A Versatile Execution Management System for Next-Generation UNICORE Grids

Bernd Schuller, Roger Menday, and Achim Streit

Research Center Jülich, Central Institute for Applied Mathematics, Jülich, Germany
{b.schuller,r.menday,a.streit}@fz-juelich.de

Abstract. This paper builds on extensive experience with the UNI-CORE middleware to derive requirements for the next generation of Grid execution management systems. We present some well-known architectural ideas and design principles that allow building Grid servers that are adaptable to any type of target systems, from single workstations or PCs to huge supercomputers, and flexible enough for the novel usage scenarios and business models that are coming up in next-generation Grid systems. These ideas are used to implement an execution management system similar in scope to the UNICORE NJS.

1 Introduction

Compute resources available in present-day Grids range from small systems, such as single PCs, to very large systems such as supercomputers (for example in the DEISA project [1]) or PC farms as in EGEE[2].

These resources are made accessible through Grid middleware, specifically *execution management systems* (EMS). They serve a variety of functions in the areas authentication, authorisation and accounting (AAA), data management and execution management.

Grid execution management systems have to serve a wide range of compute resource capabilities, number of concurrent client connections, number of concurrent jobs, amount of data transferred and so on.

Additionally, in next-generation Grids new requirements are emerging [4]. In traditional scenarios such as scientific computing Grids, business rules such as billing or auditing procedures are simple, and usually hardcoded. However, business concerns such as service level agreements play an increasingly prominent role, as investigated for example in the NextGrid project[3]. To accommodate these needs, EMSs in next-generation Grids have to be highly flexible and reconfigurable.

The remainder of this paper is organised as follows: in the next section we present and review some experiences with the UNICORE Grid middleware made in the course of project and production use. From the capabilities and more importantly the shortcomings of this mature system, we derive a set of requirements for next-generation Grid servers. The remaining sections are devoted to design and partial implementation of a system called XNJS, respecting these requirements.

W. Lehner et al. (Eds.): Euro-Par 2006 Workshops, LNCS 4375, pp. 195–204, 2007.

2   Experiences with UNICORE

UNICORE, developed in the course of several German and European projects since 1997, is a mature Grid middleware that is deployed and used in a variety of settings, from small projects to large (multi-site) infrastructures involving high-performance computing resources. UNICORE can be characterised as a vertically integrated Grid system, that comprises a graphical client and various server and target system components. The communication is based on a proprietary protocol using serialised Java objects (*abstract job objects, AJOs*). An overview on the history and usage scenarios is given in [6]. UNICORE is being used in various projects and production environments such as DEISA [1]. In the EU FP6 project UniGrids [7], it has evolved into a web services based Grid environment compliant with the web service resource framework (WSRF)[8], which is the prime candidate for realising the Open Grid Services Architecture (OGSA)[9] vision.

The UNICORE software is available open-source under a liberal, BSD-type license from the SourceForge repository [5].

The server side components of the current version of the UNICORE middleware (UNICORE 4) are organised into three tiers, the Gateway, NJS and TSI, that usually run on separate machines (Fig. 1).

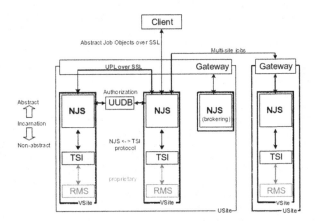

Fig. 1. The UNICORE 4 architecture

They serve distinct functions. The *gateway* is the primary point of entry, and can be considered a software firewall. It authenticates client requests and forwards them on to the next tier. The *target system interface* (TSI) is a stateless component talking directly to the underlying batch system. It offers a simple, text-based protocol to the batch system, and is used to execute scripts, submit batch jobs, request job status, get or write files and perform some common file system operations such as "list directory" or "copy". The main component in a UNICORE server installation is the *network job supervisor* (NJS), which will be discussed in detail in the next section.

2.1   The UNICORE NJS: A Gap Analysis

The central component of the UNICORE server side is the NJS (network job supervisor). The NJS is a multithreaded Java application that offers a variety of features, such as

- authorising users using the UNICORE user database (UUDB),
- translating the incoming abstract jobs into concrete jobs for the target system using a process called incarnation,
- submitting the concrete jobs to the TSI and monitoring their status,
- managing the outcome,
- communicating with the gateway,
- submitting sub-jobs to other Grid sites,
- keeping job state.

In UNICORE, abstract jobs can be arbitrarily complex and may involve workflows spanning multiple Grid sites. The NJS is a combination of a workflow processing engine and an execution management system for "atomic" jobs such as executing a script on the target system associated with the NJS.

Furthermore, the NJS offers several interfaces for add-on functionality such as brokering, resource reservation, and alternative file transfer mechanisms.

While the NJS (and thus UNICORE as a whole) offers a lot of functionality, there are some shortcomings as well, in the areas of flexibility, scalability and fault resilience.

Flexible processing and business rules. One limitation of the NJS is the fact that the processing rules as well as the business rules are hardcoded. Therefore, new requirements are only implementable by changing the core NJS code.

Currently, the processing rules are encoded into an object model, thus to add new types of actions or to modify the processing for certain types of action needs modifications of existing Java code.

Business rules are currently fixed as well (and mostly implicit). As an example scenario for the need for flexible business rules, one might think of different billing schemes based on the current user, such as pay-per-use for User A and a computing time budget for User B. Another business rule might be related to providing different resource views for different users. For example, User A should be able to use at most 10 nodes of a cluster, while User B should be allowed to use the full cluster. Currently, all users have the same set of available resources, and UNICORE relies on the underlying batch system to enforce policies such as the ones mentioned.

Scalability. As with any single software component, there are scalability issues with the NJS as well. There is no possiblity for clustering groups of NJSs. Furthermore, the current implementation of the NJS keeps a lot of state information in-memory, so during long-term operation, out of memory errors may occur.

Fault resilience. Fault resilience has many facets, but as an example scenario, consider the following. UNICORE was the Grid middleware of choice in the OpenMolGRID project [10][11], that sucessfully targeted Grid-based drug design. Often, complex multi-step jobs involving many Grid sites were run. However, sometimes job parts failed due to networking problems, or failure of one site, etc. This led to a failure of the whole job, and often to loss of results from other job parts, because the users did not take any precautions such as saving intermediary results. Here, the need for improved fault handling was felt, which should be based on a flexible set of business rules.

Limiting the scope. The UNICORE NJS does both atomic jobs (plain execution tasks) and multi-step, multisite workflows. We believe the workflow functionality should be provided elsewhere, in the interest of simplicity and modularity, and thus maintainability.

2.2   Requirements for a Next-Generation NJS

Analysing the experiences in using the current UNICORE implementation, we can identify a set of principles for designing a next-generation Grid execution management system. These can also be seen as non-functional requirements.

- *Highly modular, reconfigurable and scalable architecture:* The system must be composed of building blocks with well-defined functionality and well-defined responsibilities. These components must interact only using interfaces. The components must not make any assumptions about their environment.
  This ensures that an implementation of a subsystem can be replaced by another implementation without breaking other parts of the system.
- *No hardcoded processing rules:* The actions executed to run a job, or to transfer a file must be easy to modify, extend or even switch off. Additional processing steps (such as encryption / decryption) should be pluggable into the processing. It must be possible to add new types of actions without having to change the system core.
- *No hardcoded business rules:* There must be no static business rules in the system core. All business rules should be explicitly defined and configurable, ideally even on a per client request basis.
- *Scope limited to single-site actions:* The system should only deal with actions on a single site. Multi-site workflow functionality should be provided elsewhere.

A system that respects these principles will be easy to extend and adapt to new requirements.

To keep such a highly flexible system transparent and manageable, one has to take special care to give system administrators a detailed view on the current configuration, and to allow them to monitor the system closely.

3   The XNJS: Design and Implementation of a Next-Generation UNICORE NJS

In this section we outline the design and partial implemention of a Grid execution management system with we have named XNJS, that respects the basic requirements outlined in section 2.2.

A high-level depiction of a typical Grid node is shown in Fig. 2. The execution server resides in the " Grid tier".

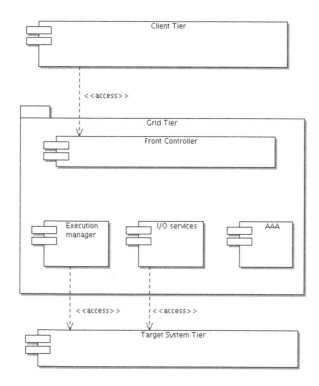

Fig. 2. Three-tier architecture representation of a Grid node

The front-controller subsystem takes care of the client communication. In present-day Grid systems this will usually be a set of WSRF compliant web services or components talking a proprietary protocol.

The services offered by the EMS can be grouped into execution services, file I/O services, and security services (authentication, authorization and accounting).

The actual resource, such as a batch system or database, is in the separate target system tier. Here, there are two scenarios: the target system and the EMS are running on the same machine, or, the EMS is running on a separate machine.

Additionally, there are aspects such as persistence, management, logging, etc., that are not shown in the figure but have to be taken into account in the design and implementation as well.

3.1   Core Architecture

It is well known that one fundamental principle for designing modular systems is the separation of interface from implementation [12]. When some component uses some other component of the system, it must not refer to a concrete implementation, but to the abstract interface of that service only.

For the XNJS, we have chosen an architecture similar to the "Microkernel" pattern [13] to maximise reconfigurability.

To realise this architecture, a component repository or component *container* is necessary, allowing storage and retrieval of concrete realisations of needed interfaces.

Using Java, this can be implemented using simple, lightweight containers such as the PicoContainer [16] or more complex frameworks such as Spring [17]. These containers offer convenient methods for dealing with component lifecycle, i.e. starting and stopping components. For our implementation we have chosen Pico-Container because of its easy embeddability and low memory footprint. Most services will in turn be dependent on other services. To resolve these dependencies, it is convenient to let the container take care of this task, and let it *inject* the dependencies using setter methods or constructor parameters[15]. A very important non-functional aspect of using dependency-injection is the improved testability of the individual components. For tests, "mock" dependencies can be used, allowing unit testing.

The actual runtime system configuration is defined at deployment time using configuration files.

3.2   Execution Management

In this section we focus on the execution management engine. Its functionality is the ability to execute a set of actions, keep track of action state, and offer some interfaces to the outside world for adding new actions and querying their status. Keeping such an engine flexible involves an extensible set of basic executable "building blocks" and a reconfigurable and extensible processing scheme for these executables.

Actions. Actions are the basic execution units in the XNJS. Actions are usually created within the front-end controller of the XNJS, and submitted for processing to the core execution engine. Their state chart is shown in Fig. 3.

If needed, the actions will communicate with the target system, invoke I/O services, start sub-actions etc.

The Action includes an XML work description, for example, a JSDL [18] document. The type of XML that is used defines the "type" of Action. The XNJS can be extended easily by adding support for new types of actions, specifically by adding processing code as outlined in the next section.

Processing Chains. The design of the processing itself should be flexible and adaptable to various deployment and usage scenarios. For this purpose, we have chosen to adopt a design pattern similar to the "chain of responsibility" from

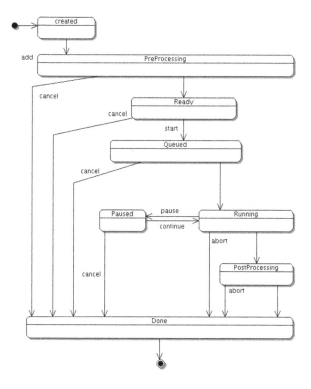

Fig. 3. The statechart for actions

[14]. The processing of an Action instance is done by a finite chain of processing elements (processors), that are called one after the other as depicted in Fig. 4. Each processor may perform arbitrary operations, change the action state, etc. Context information can be stored in the Action object itself that is passed along the processing chain. The processing chain for a given Action type is configurable, even at runtime if needed.

Processors can be used for any type of operation relevant during action execution, such as running an application, data encryption and decryption, , accounting and billing, or user notification. In this way, we realise the requirement that there are no hardcoded business or processing rules.

One disadvantage of this concept may be noted: processing can become quite complex, especially when different processors share context information, as is common during processing of workflows. Here, the processor responsible for workflow processing will start sub-actions, and will have to monitor these in order to decide when to start any dependent parts of the workflow.

3.3 Security

In Grids, many different trust and security policies are used, which may also change, thus it is vital that the EMS is flexible and does not have any hardcoded

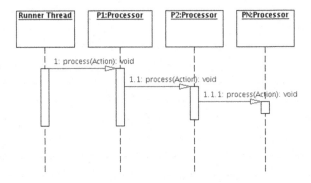

Fig. 4. A chain of processing elements

security settings. Thus, the core XNJS just provides a set of classes and patterns that can be used to build a solution that meets the security needs of the Grid it is deployed into.

Client information. A Client class can be used to capture information about the party that executes Actions on the XNJS, such as their name, authorization tokens. The Client object is usually generated in the front-end component, using information from the transport layer or the original message. This Client object is part of the Action during the Action's lifecycle.

Method call interception. We have employed the method call interception technique to allow very fine grained security checks. In the XNJS, the core method calls include authentication information in the form of a "Client" object. These method calls can be intercepted, and security checks can be executed. We have used aspect-oriented programming techniques [19] using AspectJ [20]. Rules and policies used to enforce security are pluggable, and easily extensible.

3.4   Status of the XNJS

The current status of the XNJS implementation is as follows.

Status of JSDL Processing. The most important functionality available in the XNJS is processing of atomic jobs, as defined by a JSDL document. The XNJS can execute JSDL jobs, including file staging in and out through local copy or HTTP. It supports important UNICORE concepts, such as abstract Application resources and abstract Filespaces (such as Root, Home, or Uspace).

Target System Support. Two target system interfaces currently exist. To support the use of the XNJS as "embedded" execution engine, a Java target system interface is available. This executes jobs locally by spawning a sub-process. Alternatively, an interface to a conventional Unicore 4.x TSI is available. Thus the XNJS can be used as execution management system for all those batch systems that can be accessed using Unicore 4.x, such as IBM LoadLeveler or PBS.

Management Interface. A running XNJS instance may be managed through the standard Java Management Extensions (JMX) interface. This allows to monitor the status of the Java virtual machine, to modify operational parameters, and to cleanly shutdown the XNJS.[21]

4 Conclusions and Outlook

Starting from an analysis of the Unicore NJS Grid execution server, we have derived some principles we believe to be indispensable for the next generation of Grid execution management servers. Using several well known design principles and patterns, we have designed and partly implemented a versatile, highly modular system that can be configured to suit various deployment needs and usage scenarios.

The use of a microkernel architecture with dependency injection allows easy configuration and simple testing and deployment of the system. The use of the "chain of responsibility" pattern within the execution engine allows building arbitrarily complex processing and business logic without modifying the core software.

The inherent flexibility and reconfigurability of the XNJS makes it useful in a variety of scenarios, for example

- as the backend behind a set of WSRF services implementing the UniGrids Atomic Services or OGSA-BES interfaces, with the XNJS embedded into the web services hosting,
- as a execution engine behind a web-application front end or a Representational State Transfer (REST)[22] interface,
- as part of a dynamic cluster of simple standalone worker nodes.

Future work will focus on ways to make the business rules of the system (including terms of use, billing, access rights and permissions) more explicit and dynamic.

References

1. DEISA: Distributed European Infrastructure for Supercomputing Applications http://www.deisa.org
2. EGEE: Enabling Grids for e-Science http://public.eu-egee.org/
3. NextGrid: Next-Generation Grids http://www.nextgrid.org
4. Third report of the "Next Generation Grids" Expert Group ftp://ftp.cordis.lu/pub/ist/docs/grids/ngg3_eg_final.pdf
5. UNICORE at SourceForge: http://unicore.sourceforge.net
6. A. Streit, D. Erwin, Th. Lippert, D. Mallmann, R. Menday, M. Rambadt, M. Riedel, M. Romberg, B. Schuller, and Ph. Wieder: UNICORE - From Project Results to Production Grids. L. Grandinetti (Edt.) "Grid Computing: The New Frontiers of High Performance Processing", pp. 357-376, Elsevier 2005

7. UniGrids homepage:
   http://www.unigrids.org
8. Web Service Resource Framework:
   http://www.oasis-open.org/committees/tc_home.php?wg_abbrev=wsrf
9. The Open Grid Services Architecture, version 1:
   http://www.ggf.org/documents/GFD.30.pdf
10. OpenMolGRID homepage:
   http://www.openmolgrid.org
11. Dubitzky, W., McCourt, D., Galushka, M., Romberg, M., Schuller, B. Grid-enabled data warehousing for molecular engineering; Parallel Computing **30** (2004), 1019–1035
12. David L. Parnas: "On the criteria to be used in decomposing systems into modules", Communications of the ACM 15(2), Dec. 1972,1053-1058.
13. Buschmann, F., Meunier, R., Rohnert, H., Sommerlad, P.: "A System of Patterns: Pattern-Oriented Software Architecture, Volume 1", Wiley, 1996
14. E. Gamma, R. Helm, R. Johnson, J. Vlissides: "Design Patterns", Addison-Wesley Publishing Company, 1995
15. Dependency Injection: http://www.martinfowler.com/articles/injection.html
16. PicoContainer:
   http://picocontainer.codehaus.org
17. The Spring framework:
   http://www.springframework.org
18. Job submission description language (JSDL):
   http://forge.gridforum.org/projects/jsdl-wg/
19. Elrad, T., Filman, R.E., Bader, A.: "Aspect-oriented programming: Introduction", Communications of the ACM, **44** (2001), p. 29-32
20. AspectJ:
   http://www.eclipse.org/aspectj
21. Java Management Extensions (JMX):
   http://java.sun.com/products/JavaManagement
22. Fielding, R. Th.: "Architectural Styles and the Design of Network-based Software Architectures." Doctoral dissertation, University of California, Irvine, 2000.

Towards More Flexible and Increased Security and Privacy in Grids

Willy Weisz

University of Vienna, Institute for Scientific Computing, VCPC
weisz@vcpc.univie.ac.at

Abstract. The development of UNICORE started as a Grid-enabling middleware with a monolithic security policy that restricted Grid activities to a set of users whose credentials (X.509 certificates) are pre-recorded in a UNICORE User Database (UUDB), and to a task distribution completely defined at job-submission time because the sub-jobs have to be signed by the user with his private key. Later on projects aiming at allowing a restricted interoperability with other Grid middleware lead to the adoption of more flexible approaches like the the Explicit Trust Delegation (ETD). ETD involves implicitly a more general concept: That of an attribute or role which is attached to an identified and authenticated entity and which defines the extent of the authorisations granted to that entity by the target resource. Extending this concept to other authorisation-related aspects of Grid computing is today an area of intensive research, that should also be taken up by the UNICORE developers in order to enable the creation of Virtual Organisations (VOs) that are able to take security as seriously as necessary, and to opt for flexibility as much as possible.

1 General Remarks

Virtual Organisations (VOs) that make up the organisational units which use Grid resources have two almost contradictory requirements: (1) Security that is the prime requirement for the establishment of the trust required when allowing the interoperation of resources belonging to different administration domains, and (2) Flexibility that enable VOs to easily adapt to structures in user membership and resources changing during their lifetime.

The initial decision of the UNICORE design was to give security an overriding primacy that resulted in a very strict and rather inflexible Security Model [1], that originally didn't foresee any interoperability with other Grid middleware. Nevertheless the modular design of UNICORE eases the implementation of new UNICORE Security Models that are more suitable to the security and working requirements of VOs as seen as result of the continuing Grid research.

Departing from the traditional OS views on security, and analysing security and authorisation models in real organisations recent projects came up with new approaches to secure and flexible authorisation schemes.

W. Lehner et al. (Eds.): Euro-Par 2006 Workshops, LNCS 4375, pp. 205–214, 2007.
© Springer-Verlag Berlin Heidelberg 2007

2   Identification and Authorisation in an Organisation

2.1   Identification

In any organisation the security of internal and external operations relies on the identification of the actors and the authorisations granted to them in any possible action scenario, including (manual or automated) information processing. Virtual Organisations on the Grid have the same requirements.

Employees and collaborators as well as resources must be uniquely identifiable in order to allow a well co-ordinated and optimisable running of the operation. In plans and reports their respective tasks, rights and responsibilities are attached to their identifiers; for people this is generally their common name possibly extended by e.g. a function title or an affiliation with a department.

2.2   Attributes of Entities

In bigger organisations a comprehensive list of individually named human and non-human resources may not be practical. In this case functions or roles with their rights, privileges and duties may be defined and attributed in a many-to-many relationship to individual resources. Overall work can thus be defined, planned, carried out and reported upon as a function of these attributes. The details including the assignments can be left to the possibly dislocated departments. These smaller units are also better suited to track promotions (or demotions), changes of responsibilities and privileges of their local personel, and new or modified non-human resources. Only modifications in terms of changed organigrams or roles need to be passed on to the higher company echelons.

2.3   Authorisation

When it comes to empowerments and thus responsibilities, the company policies should be defined as a function of roles and attributes of the entity (e.g. clearance), not of the name of individuals. If a person may assume different roles within the organisation its empowerment should be defined with regard to the role he is actually assuming when performing a certain task. In analogy the security levels of computer systems (including their environment) have to be at the basis of decisions on which applications are handed over to which hosts.

When organisations are co-operating in projects new authorisation challenges arise:

• each partner provides collaborators which have certain roles and capabilities,
• within the project and even the project phases project roles are defined.

For some actions some capabilities for the project as well as some defined within an individual partner organisation may be required. Projects spanning boundaries will thus base their authorisation decisions on the direct product of the authorisation attributes of the individuals in their home organisations and those defined within the projects.

3   Organisations in the Grid

The co-operative use of resources connected by the Internet (or any other network of local networks) and belonging to independent administration domains — *the Grid* — requires the formation of Virtual Organisations which must define a matrix of authorisations based on the policies regulating the authorisations within a domain and those agreed upon in the projects leading to the establishment of a VO.

The communication power of the Grid makes the creation of flexible inter-organisational projects very attractive. The flexibility of the VOs being such an asset also means that frequent changes at short notice may happen, be it on the user side or on the side of the resources. These changes generally bring about modifications of the authorisation matrix.

4   Identification and Authentication on the Grid

The multi-administration structure of the Grid requires that the identity of consumers and resources be stated unambiguously despite the many different organisations responsible for them. This is made possible by the establishment of X.509 Public Key Infrastructures (PKI) [4] where the public key of a cryptographic key pair is embedded in a certificate which i.a. contains a unique identifier for the entity owning the key pair, and is digitally signed by a "trusted third party", a trusted Certification Authority (CA). This certificate, for which the CA declares that its identification item (the Distinguished Subject Name, Subject DN) is uniquely attributed to this single entity, identifies the entity to resource consumers and providers on the Grid. The private key of the pair is used for authentication purposes and for signing digital documents and messages; the public key is used by the communication partners to send the entity encrypted messages.

The world-wide distribution of Grid consumers and resources makes it necessary to also have CAs distributed over the world. The agreement on minimal rules of operation to establish mutual trust has lead in 2005 to the International Grid Trust Federation [5]. Nevertheless organisations or VOs may establish CAs with special trust requirements, e.g. the UNICORE CA. Such a VO-centered CA has the disadvantage that is doesn't scale when the VO expands its user community or resource pool.

Even more than the trust that can be put in the CAs the storage quality of the private key determines the security level of the PKI. If its repository is a computer disk, then the security level is a product of the user protecting the file containing the key and the system administrator providing an overall secure system: Nobody than the owner is allowed to access and be able to use the private key. Much better security is achieved when the key pair is generated in a secure cryptographic token (SmartCard or USB token) and the private key, that is never allowed to leave the token is encrypted by a PIN only known to the owner. Since the private key is only available on the token, it can only be used by the person who physically owns the token; and even in the case of theft only the person knowing the PIN can activate the key, i.e. make it usable. When the use of such tokens will become the rule, PKIs will reach a really trustable security level.

5   Authorisation Based on Identity

Like the underlying operating systems, the most widely used Grid middlewares, Globus and UNICORE, base their authorisation infrastructure solely on the identity of the user (Discretionary Access Control, DAC). After the requestor has been authenticated his identity is mapped to an OS-based identity and he is welcome to an "almost help-yourself party".

This lack of fine-grain authorisation in most operating systems has led the creators of database systems to define their own access schemes mostly independent of the OS-related identities. They define roles and access rights, and manage them their own way. The lack of authorisation beyond the user identity makes the Grid for the time being unfit for the use of federated databases. Neither the OGSA-DAI project [6] nor the GGF DAIS-WG [7] have tackled the security aspects of a gridified database access. But databases are but one resource that needs fine-grain, role-based access rules, individuals' health records in any kind of container are another example.

This tradition of reducing the authorisation policy to the mere identification and authentication of the entity requesting a resource has determined how secure authorised accesses have been perceived for the Grid. It is clearly insufficient at the level of VOs.

6   Authorisations Based on Properties of the Entities

In high security environments, entities (consumers and resources) are classified according to security clearance levels. The corresponding authorisation scheme (Mandatory Access Control, MAC) allows read access only to objects of the same or a lower clearance level (Read Down) and write objects of the same or a higher clearance level (Write Up). In most organisational environments MAC is too restrictive a scheme.

The Role-Based Access Control (RBAC) has become the preferred authorisation scheme when DAC is too weak and MAC is too restrictive. It allows policies that are more fine-grained than identity-based access rights. Changes of the position in an organisation generally incur changes in roles for the entity. Like with MAC hierarchies of roles can be constructed leading to hierachies of authorisations.

Attribute-Based Access Control (ABAC) provides even more flexibility as the attribute relations are less static than the consumer-resource authorisation relations in RBAC.

7 User Database at the Grid Resource Site

The UNICORE and Globus assumption is that secure operations in Grids require comprehensive lists of entities that are allowed to access resources at a local administrative domain (e.g. the UUDB at a UNICORE V-site or the gridmap-file for a Globus host). This of course doesn't scale well and isn't appropriate for Virtual Organisations which may be short-lived and/or allow compositions of users and network-attached hardware varying over the lifetime of the VO.

The Globus approach is more flexible as it allows a remote management of the gridmap-file, e.g. by a VO management system like VOMS [8], whereas the UNICORE User Database (UUDB) can only be maintained from the site where it is located. UNICORE also requires that the X.509 Certificate be stored in the UUDB which must therefore be continuously updated since, for security reasons, the certificates have a limited life time und must thus be regularly renewed. The certificates for Globus are stored at the user's site where they have their first home after they have been issued by the CA.

8 Managing Authorisation for VOs

As has been described in Sect. 2 the authorisation structure in bigger organisations and for inter-organisational projects should move from concentrating on identities and their rights to access resources to policies based on roles and attributes of human and non-human resources. This is also true for Virtual Organisations.

8.1 The Attributes of Requestors in Their Organisation

In his own organisation a user may assume roles or have certain attributes. These attributes are signed by an Attribute Authority (AA) that is legitimised by this organisation and recognised by the resource providers in the VO. Its statements concerning attributes of an entity must bear the proof of its origin and a digital signature verifiable by a recognised certificate.

8.2 The Attributes of Requestors in the VO

Similarly the VO itself may need an Attribute Authority that issues digital documents stating the attributes of the requestor within the VO. These attributes may be functions of the identity of the requestor and/or of his roles and attributes as defined by his home organisation, or be just defined by his role in the VO.

8.3 Privacy — Anonymity

In certain applications it may be important (or even required by law) that the identity of the requestor be anonymised for the time of the resource usage, but nevertheless be traceable at some point in time, e.g. for accounting purposes or for feedback. This can be realised by mapping the identity to a "general user" which is given attributes allowing traceability on a need-to-know basis.

8.4   Attributes of the Resources

Likewise there must be Attribute Authorities that issue information document-ing attributes of the resources. They may come from their administration domain or be VO-related, e.g. availability to or costs for the VO.

8.5   Authorisation

In big real organisations a complex set of rules defines who is entitled to take which decisions and who is to implement them. The company policies thus de-fined are generally expressed as functions of roles and levels in the hierarchy, not of individuals.

Likewise in VOs policies govern the authorisation decisions. The complete set of information on the identities and attributes of consumer and resources triggers a policy decision to grant (or deny) the requestor a set of privileges and access rights that the policy enforcement engine will have to use in order to grant access to resources.

The policy may even require a third party permission: The right to access a person's Electronic Health Record (or identifiable parts of it) that may be distributed over a national Health Grid will require the patient's consent (at least in Austria). This third party will also need to be authenticated and its role or attributes taken into account.

9   Consequences for the UNICORE Development

The need for a more flexible but nevertheless improved security and privacy protection must trigger major changes in the UNICORE security infrastructure. The integration of such developments is facilitated by the modular architecture.

9.1   Authentication

The UUDB is too inflexible for future Grid environments. Without sacrificing security concerns an authentication mechanism is needed that doesn't store all potential users at the resource site, but rather security policies.

The Security Model of UNICORE doesn't allow the use of Proxy Certificates [9]. Without this facility no message-level security is possible. And the extension to allow a limited interoperability with Globus transmits a private key over the communication lines! Even so it is included in an encrypted blob, this is against the proxy concept that the private key corrsponding to a proxy certificate is only used on the system where it has been generated, and never leaves that system.

9.2   A First Use of Attributes: The Explicit Trust Delegation

The requirement to build dynamic Grid jobs for which an agent (e.g. a por-tal) decides on the distribution of tasks after the end-user has submitted his job description, require that other instances than the job signer (the end-user) get authority to request actions on behalf of the end-user.

Since the UNICORE Security Model doesn't allow the use of proxies this delegation of rights of the end-user to UNICORE agents is managed by the definition of a trust attribute that the end-user issues for that agent, the Explicit Trust Delegation (ETD) [10].

Even so it is not presented as such, ETD can be seen as the first(?) introduction of a formal policy based on attributes (trust) conferred to a Grid entity (the agent) by an AA (the end-user).

9.3   The Proposed Authorisation Architecture

The authorisation arcitecture to be developed for UNICORE should provide the following functions:

For any subject and target of a request a complete policy or set of policies must be defined by a Source of Authority (SOA); this collection will be used to derive decisions whether to accept or deny requests.

After being authenticated the request for use of resources (including all the identity/role/attribute information provided by the client agent) is handed over to the Policy Enforcing Engine (PEP).

The PEP hands the request over to the Policy Decision Point (PDP) which applies the rules taking into consideration the identity/roles/attributes included in the request, and if needed, requesting further information from Policy Information Points (PIP), like e.g. AAs.

The decision to accept or deny derived by the PDP is then handed over to the PEP which has to enforce it. The PEP should be provided with a default rule (accept or deny) that it must enforce when the PDP is unable to decide (e.g. due to insufficient information from PIPs).

9.4   Attribute Authorities

The collection of attributes of the requestor can be orchestrated by the User Client or it can be initiated by the Policy Decision Point at the resource provider. The former solution seems to be more scalable, at least for the attributes of the requestor in his own organisation.

Since the attributes may be stored in different kinds of databases with differing interfaces, it will be necessary to define a standardised protocol for transporting the attributes over the Grid and an interface for plugins to be developed for the individual underlying databases.

The transport protocols will be based on X.509 Attribute Certificates [11] using the ASN.1 format [12] or the XML-coded Security Assertion MarkupLanguage (SAML) [13].

9.5   Authorisation

Plugins replacing the monolithic UUDB have to be developed that implement the authorisation architecture described in 9.3. The Explicit Trust Delegation will have rules in the policy and will be decided by the PDP.

For the formulation of the policies standard languages will be used, like the eXtended Access Control Markup Language (XACML) [14]. They must be able to describe in easy to learn ways simple policy models as well as complex requirements.

10   Authorisation in the Non-UNICORE World

10.1   VOMS

VOMS manages VOs and their constituency. Users can request to be added to the VO and VOMS managers will accept or deny the request. Users can be assigned attributes and capabilities. At the lowest sophistication level VOMS generates for the Globus middleware on each of the systems available to the VO the gridmap-file which contains the mapping of DNs to user identifications known to the OS.

VOMS performs only PIP functions. The PDP function is left to the Globus Gatekeeper.

10.2   Shibboleth

Shibboleth [15] is a middleware that provides a federated authorisation infrastructure for Web Single SignOn across organisational boundaries. It uses SAML v1.1 for the exchange of attributes.

Shibboleth passes the authorisation information in form of opaque handles which provide anonymity of users without loosing the capability to trace them back, if necessary.

10.3   GridShib

The project GridShib [16] integrates Shibboleth with Grid technology as provided by the Globus Toolkit version 4 (GTK 4). One of the major challenges is the efficient mapping of Shibboleth's opaque handle with the DN of the certificates used in GTK 4.

10.4   PERMIS

PERMIS [17] is a "Privilege Management Infrastructure" that provides a complete policy-based authorisation service. Policies are written in XML to support an RBAC paradigm.

10.5 GridShibPERMIS

The GridShibPERMIS project [18] combines the strengths of Shibboleth as an Identity and Attributes Provider, the Grid Infrastructure of GTK 4 and the PDP provided by PERMIS.

The authentication based on the X.509 certificates is performed by GTK, GridShib provides the PIP, PERMIS provides the policy-based authorisation system with its interface called "GridShibPERMIS Context Handler" acting as the PDP in the GTK authorisation framework.

11 Conclusion

UNICORE provides a solid framework for Grid computing that has already started to inter-operate with other Grid middleware like Globus, has a solid security infrastructure for a rather small, not too mobile user and resource community without the need to leave the UNICORE environment. When it comes to communications with other security schemes the isolation of the approach precludes really secure connections and information transmissions.

Since the inception of UNICORE the understanding of security on the Grid has evolved towards more flexibility while providing more control over integrity and privacy of information and usage of resources. UNICORE/GS, the follow-up to the UNICORE framework used today, must provide a completly overhauled security infrastructure. A look into developments in and surrounding the Globus Toolkit provides guidelines and ideas for the development of a new UNICORE Security Infrastructure Model, based on policies that take into consideration the identity as well as attributes of users and resources.

The existence of standards for the expression and communication of attributes and rules will make the inter-operation with other Grid middleware easier than in the past. Even the problem of delegation of trust, which is the big hurdle for a bi-directional UNICORE-Globus inter-operation should be solvable.

References

1. Goss-Walter, T., Letz, R., Kentemich, T., Hoppe, H.-C. and Wieder, P.: An Analysis of the UNICORE Security Model, Global Grid Forum, Grid Forum Document - Informational 18 (GFD-I 18), 2003,
   http://www.gridforum.org/documents/GFD.18.pdf
2. Erwin, D. (ed.): UNICORE Plus Final Report, 2003, ISBN-3-00-011592-7,
   http://www.unicore.org/documents/UNICOREPlus-Final-Report.pdf
3. Grimm, Ch. and Pattloch, M. (coord.): Analyse von AA-Infrastrukturen in Grid-Middleware, Version 1.1, March 2006
   http://www.d-grid.de/fileadmin/user_upload/documents/DGI-FG3-4/Analyse-AAI_v1_1.pdf
4. Housley, R., Polk, W., Ford, W. and Solo, D.: Internet X.509 Public Key Infrastructure — Certificate and Certificate Revocation List (CRL) Profile, IETF RFC 3280, April 2002, http://www.ietf.org/rfc/rfc3280.txt
5. http://www.gridpma.org

6. http://www.ogsadai.org.uk
7. http://forge.gridforum.org/projects/dais-wg/
8. Alfieri, R. et al.: From gridmap-file to VOMS: managing authorization in a GRID environment, April 2004,
   http://infnforge.cnaf.infn.it/docman/view.php/7/61/voms-FCGS.pdf,
9. Tuecke, S., Welch, V., Engert, D., Pearlman, L. and Thompson, M.: Internet X.509 Public Key Infrastructure (PKI) Proxy Certificate Profile, June 2004, IETF RFC 3820, http://www.ietf.org/rfc/rfc3820.txt
10. Snelling, D., van den Berghe, S. and Li, V. Q.: Explicit Trust Delegation: Security for Dynamic Grids, Fujitsu Sci. Tech. J., 40,2,pp.282-294, December 2004, http://www.fujitsu.com/downloads/MAG/vol40-2/paper12.pdf
11. Farrell, S. and Housley, R.: An Internet Attribute Certificate Profile for Authorization, April 2002, IETF RFC 3281
    http://www.ietf.org/rfc/rfc3281.txt
12. CCITT Recommendation X.208: Specification of Abstract Syntax Notation One (ASN.1), 1988
13. Security Assertion Markup Language (SAML) v2.0, OASIS Standard, 2005, http://docs.oasis-open.org/security/saml/v2.0/saml-2.0-os.zip
14. eXtensible Access Control Markup Language (XACML) 21.0, OASIS Standard, 2005
    http://docs.oasis-open.org/xacml/2.0/access_contrpl-xacml-2.0-core-spec-os.pdf
15. http://shibboleth.internet2.edu
16. http://gridshib.globus.org
17. http://www.permis.org
18. Chadwick, D.W., Novikov, A. and Otenko, O.: GridShib and PERMIS Integration, Terena Networking Conference 2006, 15-16 May 2006, Catania (Sicily), Italy http://www.terena.nl/events/tnc2006/core/getfile.php?file_id=753

Integration of Grid Cost Model into ISS/VIOLA Meta-scheduler Environment

Ralf Gruber[1,6], Vincent Keller[1,6], Michela Thiémard[2,6], Oliver Wäldrich[3,6], Philipp Wieder[4,6], Wolfgang Ziegler[3,6], and Pierre Manneback[5,6]

[1] École Polytechnique Fédérale de Lausanne, LIN-STI, Switzerland
Ralf.Gruber@epfl.ch, Vincent.Keller@epfl.ch
[2] École Polytechnique Fédérale de Lausanne, DIT-EX, Switzerland
Michela.Thiemard@epfl.ch
[3] Fraunhofer Gesellschaft, SCAI, St. Augustin, Germany
Oliver.Waeldrich@scai.fraunhofer.de, Wolfgang.Ziegler@scai.fraunhofer.de
[4] Froschungszentrum Jülich GmbH, D-52425, Germany
ph.wieder@fz-juelich.de
[5] Faculté Polytechnique de Mons and CETIC, Mons, Belgium
Pierre.Manneback@fpms.ac.be
[6] members of CoreGRID

Abstract. The Broker with the cost function model of the ISS/VIOLA Meta-Scheduling System implementation is described in details. The Broker includes all the algorithmic steps needed to determine a well suited machine for an application component. This judicious choice is based on a deterministic cost function model including a set of parameters that can be adapted to policies set up by computing centres or application owners. All the quantities needed for the cost function can be found in the DataWarehouse, or are available through the schedulers of the different machines forming the Grid. An ISS-Simulator has been designed to simulate the real-life scheduling of existent clusters and to virtually include new parallel machines. It will be used to validate the cost model and to tune the different free parameters.

1 Introduction

The Intelligent Grid Scheduling System (ISS) [3] has been proposed to submit n components C_k ($1 \leq k \leq n$) of an application A to a computational Grid composed of r resources R_i ($1 \leq i \leq r$) each being a parallel machine with p_i nodes and m_i main memory [9]. A component is executed on $p_k \leq p_i$ processors. It is planned to apply ISS first to the HPC applications in Switzerland that are executed on the parallel machines that form the SwissGrid. These machines are located at the Swiss National Supercomputing Centre (CSCS) in Manno, at the EPFL in Lausanne, at the ETHZ in Zürich, and at other Universities and research institutions in Switzerland. The aim of ISS is to submit the components of the different applications to well suited machines according to a deterministic cost function. This cost function is presented in this paper.

W. Lehner et al. (Eds.): Euro-Par 2006 Workshops, LNCS 4375, pp. 215–224, 2007.

The ISS cost function includes terms that represent the investment costs per sustained Tflops/s rate, energy consumption due to power supply and cooling, maintenance, licences, and management costs, the infrastructure (room and cooling system) and expenses for personnel taking care of the resources. The sustained Tflops/s rate strongly depends on the usage of the machines and on the type of application components that are executed. Machines that are not used all the time can be expensive. Ecological arguments are now more and more considered when deciding on the purchase of a machine. These can be taken care of by the energy price per kWh, and by the cooling installation costs. With all those characteristics, the overall costs of a component can vary by up to an order of magnitude on the different machines. ISS should optimize the usage of the different machines and help to decide on the policy when purchasing future machines.

In a previous paper [1], the integration of the ISS into the VIOLA Meta-Scheduler environment has been described. Specifically, the roles of the Meta-Scheduler and UNICORE clients have been detailed. The whole scheduling of a component C_k on machine i has been decomposed in three major steps: Prologue (starts at time t_k^0) and Decision (made just after t_k^0), Execution (starts at time $t_{k,i}^s$), and Epilogue (starts at time $t_{k,i}^e$ and ends after data collection at $t_{k,i}^d$). In the Prologue phase, data needed to construct the cost model are collected and used to minimize the cost function. The resulting choice is then forwarded to the UNICORE client for submission to the chosen machine. The information about the execution phase is treated in the Epilogue phase to create a file that can be reused in the next job submission.

Different modules help to decide. Besides the UNICORE [4] and the Meta-Scheduling Services (MSS) [5] there are three new modules: The DataWarehouse (DW), the System Information (SI), and the Broker. Before execution of a component, all the cost model relevant data collected during previous executions on different machines can be found in the DW. They are accessed through the SI and transferred to the Broker. SI also collects the data after execution that is prepared by the VAMOS system [6] and sent to the Broker. The VAMOS system maps Ganglia and accounting data into application relevent ones. This Epilogue data are interpreted according to the Γ model [2]. This model characterizes parallel machines and applications. These data on the behaviour of the component during execution are then stored by the SI in the DW, prepared to be reused in the Prologue phase for the next execution.

The Broker includes all the algorithmic steps for the evaluation of the cost model and for the preparation of the Epilogue data. The needed data are requested from the SI and the MSS. All features of the Broker are described in detail in this paper.

The application components C_k are parametrized by the Γ model [2] in which the needs in processor performance, main memory bandwidth, network communication bandwidth and latency are estimated. Together with similar parameters describing the parallel machines, a Γ value is computed that measures the computation over communication needs of C_k. Γ model relevant parameters such as the number of operations O, the number of transferred data S, or the effective processor performance r_a can be measured on a machine that includes PAPI [10]. For those machines that do not have PAPI, these parameters can be estimated according to the input parameters of the next execution, or with the Γ model using measurements on one machine and by transforming them to the other.

The cost function model includes free parameters that have to be tuned for each Grid. For this purpose, a Simulator has been designed.

2 Application Component Characteristics

The major reason for the development of an Intelligent Grid Scheduling System (ISS) is the different needs of the application components in point of view of processing performance, main memory bandwidth, communication bandwidth and network latency. These characteristics have been parametrized in the Γ model [2]. Similarly, the computer architectures have also been parametrized in the same paper. Some important parameters can be directly measured using Ganglia data [11]. This enables predicting to which machine an application component should be submitted. For the Swiss HPC community the following type of applications consume the major part of the HPC resources:

2.1 Embarrassingly Parallel Application Components

These applications do not demand inter-node communications. As a consequence, the Γ parameter is huge. A big number of cases have to be executed, the results collected and handled by a server. Typical applications are the immense amount of independent data in high energy physics that has to be interpreted, the sequencing algorithms in proteomics, parameter studies in plasma physics to predict optimal magnetic fusion configurations, or a huge number of data base accesses for statistical reasons.

Embarrassingly parallel applications need master/slave computer architectures with a sufficiently powerful connectivity between the frontend server and the different slaves. There is no communication needed between the slaves. Thus, a weekly connected workstation farm can offer a sufficient computing performance. Such clusters can for instance be formed of individual machines connected through a bus-based network or even through the internet to a master. As a consequence, the costs of such application components can be small.

2.2 Application Components with Point-to-Point Communications

Point-to-point communications typically appear in finite element or finite volume methods when a huge 3D domain is decomposed in subdomains [7] and an explicit time stepping method or an iterative matrix solver is applied. If the number of processors grows with the problem size, and the size of a subdomain is fixed, γ_a (=number of operations O over amount of data S sent over the network) is independent of the problem size, and, consequently, Γ does not change. The per processor performance is determined by the main memory bandwidth. The number O of operations per step is directly related to the number of variables in a subdomain times the number of operations per variable, whereas the amount of data S transferred to the neighboring subdomains is directly related to the number of variables on the subdomain surface. For huge point-to-point applications using many processing nodes, $\Gamma << 1$ for a bus (inadequate) $2 < \Gamma < 10$ for the Pentium 4 cluster with a Fast Ethernet switch, $10 < \Gamma < 50$ for the Xeon cluster with a GbE switch, and the Opteron cluster with a Myrinet switch, and $\Gamma >> 100$ for a Cray XT3. Application components with $\Gamma > 1$ can run well on a cluster with a relatively slow and cost-effective communication network.

2.3   Application Components with Multicast Communication Needs

The parallel 3D FFT algorithm is a typical example with important multicast communication needs. Here, γ_a decreases when the problem size is increased, and the communication network has to become faster. Since $r_a = R_\infty$, FFT reaches close to peak performance. Thus, γ_M is big, and, as a consequence, the communication parameter b must be big to satisfy $\Gamma > 1$. Such an application has been discussed in [2]. It has been showed that with a Fast Ethernet based switched network, the communication time is several times bigger than the computing time, even when the problem size is small. Such an application needs a faster switched network such as an efficient GbE, a Myrinet, a Quadrics, or an Infiniband network.

2.4   Components Demanding Shared Memory

There are a few application components that demand a shared memory computer architecture. A typical example are the direct simulation applications to study turbulence phenomena applying a spectral method to the Navier-Stokes equations. The main memory needs are small, the component can be run on single processor machine. Very small phenomena have to be studied, needing very small time steps. Typically, a million of time steps are needed for one simulation, one step takes a few seconds on one processor. This leads to a few months of CPU time per case. The user likes to distribute one case among a number of processors. It can be seen that a distributed memory architecture is not well suited for such a problem [8]. The reason comes from the fixed overall size of the application. If the number of processors is increased, the per processor size reduces, and the scalability is very poor. A real shared memory architecture is more adequate. To reduce the turn-around times of these application components, the Swiss computational Grid should include a few shared memory nodes.

3   Meta-scheduling Features

The Meta-Scheduling Service (MSS) delivers data on the availability of the different eligible machines in a Grid as a function of the number p_k of processors reserved for a component. By means of these data it will be possible to estimate $t^s_{k,i}$, i.e. the starting time of the application component C_k on machine i. The time difference $t^s_{k,i} - t^0_k$ is the time during which the component will sit in the input queue. In fact, if p_k is high the execution time $t^e_{k,i} - t^s_{k,i}$ is small, but the waiting time $t^s_{k,i} - t^0_k$ can become big.

4 The Broker

4.1 Action List

The Broker is active during the Prologue, the Decision, and the Epilogue phases. It computes all data related to the cost model. The tasks of the Broker are:

1. Interpret job input data received from the UNICORE client
2. Collect application related data from DW through SI
3. Collect data on machine availabilities through MSS
4. Evaluate cost function and chooses a well suited machine
5. Send decision to MSS, after reservation to UNICORE client
6. Collect data on execution through SI
7. Prepare Γ model related data
8. Send epilogue data to DW through SI.

4.2 Decision: Grid Cost Model

The choice of a well suited machine depends on user requisites. Some users would like to obtain the result of their application execution as soon as possible, regardless of costs, some others would like to obtain results for a given maximum cost, but in a reasonable time, and some others for a minimum cost, regardless of time.

We will describe here in a few words the various elements that compose a cost function z being able to satisfy users' requests. This cost function depends on costs due to machine usage, denoted by K_e, license fees, denoted by K_l, energy consumption and cooling, denoted by K_{eco}, waiting results time, denoted by K_w, and amount of data transferred, denoted by K_d. Introducing two more quantities, $KMAX$ and $TMAX$, respectively maximum cost and maximum time, from users point of view, we can formulate the following optimization problem:

$$\min z = \beta K_w(C_k, R_i, p_k) + \sum_{k=1}^{n} \mathcal{F}_{C_k}(R_i, p_k)$$

$$\text{such that } \sum_{k=1}^{n} \Big(K_e(C_k, R_i, p_k) + K_l(C_k, R_i, p_k)$$

$$+ K_{eco}(C_k, R_i, p_k) + K_d(C_k, R_i, p_k) \Big) \leq KMAX$$

$$max(t^d_{k,i}) - min(t^0_k) \leq TMAX$$

$$(R_i, p_k) \in \mathcal{R}(C_k),$$

$\forall\, 1 \leq k \leq n$, where

$$\begin{aligned}
\mathcal{F}_{C_k}(R_i, p_k) = {} & \alpha_k \Big(K_e(C_k, R_i, p_k) + K_l(C_k, R_i, p_k) \Big) \\
& + \beta_k \Big(K_w(C_k, R_i, p_k) \Big) + \gamma_k \Big(K_{eco}(C_k, R_i, p_k) \Big) \\
& + \delta_k \Big(K_d(C_k, R_i, p_k) \Big) \quad [\text{ECU}] ,
\end{aligned}$$

$$\alpha_k,\ \beta,\ \gamma_k,\ \delta_k \geq 0,$$

$$\alpha_k + \beta + \gamma_k + \delta_k > 0,$$

and $\mathcal{R}(C_k), k = 1, ..., n$ is the eligible set of machines for component C_k. We express the money quantity as Electronic Cost Unit ([ECU]).

In our model, the parameters α_k, β, γ_k, and δ_k are used to weight the different terms. They can be fixed by the users and/or by a simulator. For instance, by fixing $\alpha_k = \gamma_k = \delta_k = 0$ and $\beta \neq 0$, one can get the result as rapidly as possible, independent of cost. By fixing $\beta = 0$ and $\alpha_k, \gamma_k, \delta_k \neq 0$, one can get the result for minimum cost, independent of time. These four parameters have to be tuned according to the policies of the computing centres and user's demands. In the case of the Swiss HPC Grid, the overall usage of the machines should be high. For instance, increasing β will increase usage of underused machines. One recognizes that a simulator is needed to estimate these parameters.

The quantities K_e and K_l have the same weight α_k. The reason is that license fees are either paid directly by the user, then $K_l = 0$, fully paid by the computing centre, then, the license fee is part of K_e, and $K_l = 0$ again, or it is invoiced per unit CPU time, then, $K_l > 0$.

The K_d quantity depends on the amount of data transferred. The other K values are:

Costs Due to Machine Usage: K_e

$$K_e(C_k, R_i, p_k) = \int_{t_{k,i}^s}^{t_{k,i}^e} k_e(C_k, R_i, p_k, \varphi, t)\, dt \ [\text{ECU}].$$

Each computing center has its specific accounting, but generally they just bill the execution time, depending on the number of CPU time used. Figure 1 shows an example of $k_e(t)$ when day time, night time and weekends have different CPU costs. The φ parameter introduces the **priority** notion.

Costs Due to License Fees: K_l

$$K_l(C_k, R_i, p_k) = \int_{t_{k,i}^s}^{t_{k,i}^e} k_l(C_k, R_i, p_k, t)\, dt \ [\text{ECU}].$$

As costs due to machine usage, costs due to licenses may simply depend on execution time and the number of CPUs used, independent of day time.

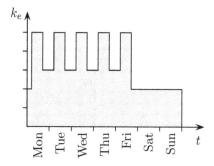

Fig. 1. Example of CPU costs as a function of daytime

Costs Due to Turn-Around Time: K_w

$$K_w(C_k, R_i, p_k) = \int_{min(t_k^0)}^{max(t_{k,i}^d)} k_w(t)\, dt\; [\text{ECU}].$$

This cost is machine and application component dependent since $t_{k,i}^e$ is machine and component dependent. It could be engineer's salary or a critical time-to-market product waiting cost.

Figure 2 shows an example of k_w concerning engineer's salary. Here, it is supposed that the engineer looses his time only during working hours. A more sophisticated function could be yearly graphs also including unproductive periods like vacations.

Figure 2 shows an example of k_w of a critical time-to-market product. This parameter could be also be used to tune the overall usage of the whole machine park of a user community. Increasing β will activate machines that are underused. Putting $\beta = 0$ in the simulator offers the opportunity to recognize overused machines that should in addition be installed.

Costs Due to Energy Consumption and Cooling: K_{eco}

$$K_{eco}(C_k, R_i, p_k) = \int_{t_{k,i}^s}^{t_{k,i}^e} k_{eco}(C_k, R_i, p_k, t)\, dt\; [\text{ECU}].$$

This cost can become relatively important if low-cost PCs are used in clusters. For components that are memory bandwidth bound, the frequency of the processor could be lowered. The energy consumption grows with the second power of the frequency, a reduction by a factor of 2.5 of the processor frequency reduces its energy consumption by a factor of 6. This has been tested with a laptop computer. When reducing frequency from 2 GHz to 800 MHz, the overall performance of the memory bandwidth bound application only was reduced by 10%. We have to mention here that for low-cost PCs energy costs are comparable to investment costs. Thus, in future it is crucial to be able to underclock the processor, adapting its frequency to the application component needs. This would reduce the

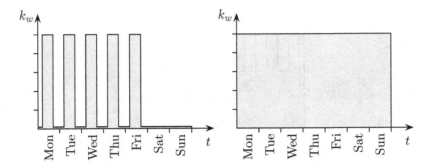

Fig. 2. Left: Engineer's salary cost function $k_w(t)$ due to waiting on the result. Right: Time-to-market arguments can push up priority of the job.

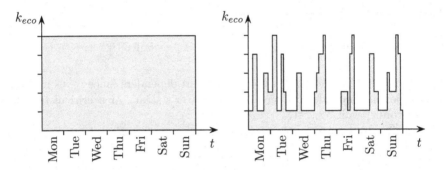

Fig. 3. Today: Excessive costs of energy consumption and cooling. Future: Energy consumption reduction due to frequency adaptation to application component needs. Computer manufacturers are invited to open for on-line frequency underclocking.

worldwide PC energy consumption by at least 75% and would free in the near future 30 nuclear power plants. Computer manufacturers must be convinced to be able to have energy consumption graphs as the one depicted in Figure 3.

4.3   Epilogue: Prepare Data for Next Execution

After execution, the SI collects the application component data from all the databases build up by Ganglia on each node. The Γ model relevant information is extracted and sent to the Broker which computes the Γ model [2] data used for the next execution of the same application component. This data is then sent to the DW through SI.

In addition to the execution data, information on the choice of the machine is also stored on the DW such that in an ulterior step a statistical study can be performed on the adequacy of the machines in the Grid for the set of application components submitted during a certain period in time. This study can then be used to get some insight on which machine should in future be purchased. After a few years it will then be possible to tend towards a well adapted set of machines that form the Grid for a given user community.

5   Simulator to Tune Parameters

The cost model equations include the free variables α, β, γ and δ. It is not clear how they have to be chosen. The ISS-Simulator has been designed to validate the choices of these parameters. In fact, the ISS-Simulator aims to understand the learning process of the Grid system with real machine parameters and real monitored data (see [3]) coming from real applications. It can also be used to predict an imaginary, well suited, set of machines adapted to the applications of a real or a virtual user community. This prediction could also be used in the future to help computer centers to buy new machines. This simulator will be described in more details in another paper.

6   Conclusions

The Broker with the cost function model of the ISS/VIOLA Meta-Scheduling System implementation has been described in detail. The Broker includes all the algorithmic steps needed to determine a well suited machine for an application component. This judicious choice is based on a deterministic cost function model including a set of parameters that can be adapted to policies set up by computing centres or application owners. All the quantities needed for the cost function can be found in the DW, or are available through the schedulers of the different machines in the Grid. An ISS-Simulator has been designed to simulate the real-life scheduling of existent clusters and to virtually include new parallel machines. It is used to validate the cost model and to tune the different free parameters.

Acknowledgement

The development of ISS is a joint project between LIN-EPFL, EIF (Ecole d'in-génieurs et d'architectes de Fribourg) and CSCS (Swiss National Computing Centre), and is part of the Swiss Grid Initiative. Some of the work reported in this paper is funded by the German Federal Ministry of Education and Research through the VIOLA project under grant #01AK605L. This paper also includes work carried out jointly within the CoreGRID Network of Excellence funded by the European Commission's IST program under grant #004265.

References

1. Keller, V., Cristiano, K., Gruber, R., Spada, M., Tran, T.-M., Kuonen, P., Wieder, P., Ziegler, W., Maffioletti, S., Nellari, N., Sawley, M.-C.: Integration of ISS into the VIOLA Meta-Scheduler environment. Pisa, 28-30 November 2005.
2. Gruber, R., Volgers, P., De Vita, A., Stengel, M., Tran, T.-M.: Parameterisation to tailor commodity clusters to applications. *Future Generation Computer Systems*, **19**:111–120, 2003.

3. Gruber, R., Keller, V., Kuonen, P., Sawley, M.-C., Schaeli, B., Tolou, A., Torruella, M., and Tran, T.-M., Towards an Intelligent Grid Scheduling System, Proc. of 6th Int. Conf. PPAM 2005, Poznan, Poland, Lecture Notes in Computer Science 3911 (Springer, 2006) 751-757

4. Erwin, D.,UNICORE plus final report – uniform interface to computing resource,Forschungszentrum Jülich,2003,ISBN 3-00-011592-7

5. Wäldrich, O., Wieder, P., and Ziegler, W., A Meta-Scheduling Service for Co-allocating Arbitrary Types of Resource, In *Proc. of Conference on Parallel Processing and Applied Mathematics PPAM 2005*, Poznan, Poland, 2005, to appear.

6. Gruber, R. and Keller,V. ,Towards an Eco-Grid architecture, submitted to Future Generation Computer Systems

7. Gruber, R. and Tran, T.-M. Scalability aspects on commodity clusters, EPFL Supercomputing Review, **14**, 12-17 (2004)

8. Gruber, R., Keller, V., Leriche, E., and Habisreutinger, M.A., Can a Helmholtz solver run on a cluster?, accepted to appear in Procs. of Cluster 2006

9. Manneback, P., Bergére, G., Emad, N., Gruber, R., Keller, V., Kuonen, P., Noël, S., and Petiton, S., Proposal of a scheduling policy for hybrid methods on computationsl Grids, CoreGRID workshop (Pisa, 2005)

10. Dongarra, J., London, K., Moore, S., Mucci, P., and Terpstra, D., Using PAPI for hardware performance monitoring on Linux systems, www.netlib.org/utk/people/JackDongarra/PAPERS/papi-linux.pdf

11. http://ganglia.sourceforge.net/

A One-Stop, Fire-and-(Almost)Forget, Dropping-Off and Rendezvous Point⋆

R. Menday[1], B. Hagemeier[1], B. Schuller[1], D. Snelling[2], S. van den Berghe[2], C. Cacciari[3], and M. Melato[4]

[1] Central Institute for Applied Mathematics,
Forschungszentrum Jülich, D-52425 Jülich, Germany
[2] Fujitsu Laboratories of Europe Ltd, Hayes Park Central,
Hayes End Road, Hayes, Middlesex, UB4 8FE, UK
[3] CINECA, via Magnanelli 6/3, 40033 Casalecchio di Reno, Bologna, Italy
[4] NICE, via Marchesi di Roero 1, 14020 Cortanze, Italy
r.menday@fz-juelich.de

Abstract. In order to foster uptake by scientific and business users we need an easy way to access Grid resources. This is the motivation for the A-WARE project. We build upon a fabric layer of Grid and other resources, by providing a higher-layer service for managing the interaction with these resources - A One-Stop, Fire-and-(almost)Forget, Dropping-off and Rendezvous Point. Work assignments can be formulated using domain specific dialects, allowing users to express themselves in their domain of expertise. Both Web service and REST bindings are provided, as well as allowing the component to be embedded into other presentation technologies (such as portals). In addition common desktop notification mechanisms such as Email, RSS/Atom feeds and instant messaging keep users informed and in control. We propose using the Java Business Integration specification as the framework for building such a higher-level component, delivering unprecedented opportunities for the integration of Grid technologies with the enterprise computing infrastructures commonly found in businesses.

1 Introduction

UNICORE[23],[18] has gained a reputation as a vertically integrated architecture. Sometimes referred to as a 'stovepipe' architecture, it offers a complete 'ready to run' solution. From a user and administrative perspective this is clearly attractive.

Recently, in the Grip[11] and then the UniGrids[9] projects, UNICORE has been prominent in promoting interoperable Grid middleware. Indeed, UNICORE emerged as an early adopter of Service Oriented (SOA)[20] approaches to building distributed systems[21]. The consequence of a good SOA design is that there is a loose-coupling between the components, thus loosening the links in the UNICORE stovepipe. Emerging from the current activity in the UNICORE

⋆ This work is partially funded through the European A-WARE project under grant FP6-2005-IST-034545.

W. Lehner et al. (Eds.): Euro-Par 2006 Workshops, LNCS 4375, pp. 225–234, 2007.

community will be a best-of-breed packaging of select components. In essence, the next-generation of UNICORE will become a stovepipe construction toolkit. Furthermore, there now exists the possibility for others to take individual components and use them for something else.

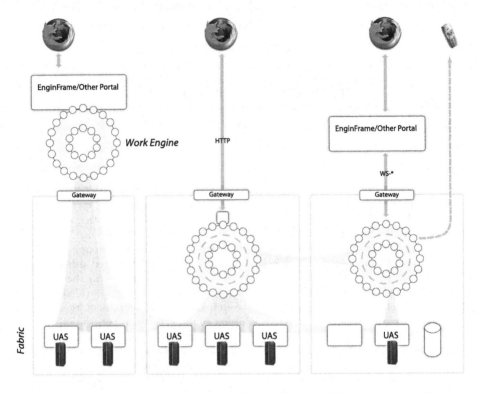

Fig. 1. Deployment possibilities of the work engine

Referring to Figure 1 we assume a cloud of services consisting at the lowest level of fabric services. These are services associated with particular computational or data resources. The UNICORE Atomic Services (UAS) developed in the UniGrids project provide us with a Web services based interface to such atomic grid functionality. Services which are not fabric services - i.e. not coupled to a particular computational or data resource - are termed higher-level services. Individual fabric services are not normally used in isolation. A set of resources and services are orchestrated into a complex workflow, business process, or service chain. This paper deals with such a 'work engine', acting on behalf of multifarious users that manages the multiple invocations of contributing services. This work will be carried out within the A-WARE project[1]. Examples of such functionality include atomic Grid jobs, other higher-level services, databases, legacy applications, etc. In short our work engine component can be described as a

one-stop, fire-and-(almost)forget, dropping-off and rendezvous point.

One-stop, because it presents a façade of the Grid to the user[1]. Fire-and-forget implies that the work engine manages the orchestration of the users' tasks over the Grid infrastructure. In many cases, after assigning work to the engine, the next contact the client makes is when the assignment is complete - the rendezvous. The 'almost' proviso implies that the client may wish to be notified during the execution of the assignment, either for purely informational purposes, or to participate in its execution, approving resource selection choices, or adding additional information not available at the time of submission.

Whilst the use-cases driving the development of the Grid lead to some special requirements, it has also become increasingly evident that businesses face similar issues related to internet-scale communication, cross-organisational interactions, and the accessing of services over the internet. This has resulted in the blurring of the lines between Grid computing in the scientific community, and the kind of the enterprise computing seen within business. Through the seamless integration of Grid resources and local (non-Grid) resources, using a very powerful and flexible orchestration environment, leading to the *disappearing grid*.

The further integration of a Grid portal component, such as EnginFrame[8], would allow organizations to provide application oriented computing and data services to users in a simplified Web browsing experience. Grid portal technology hides the complexity of the underlying Grid infrastructure and provides an additional user-oriented abstraction layer on the Grid.

This paper proposes using the Java Business Integration (JBI)[14] specification for building higher-level services for the Grid. We suggest that this will increase the ease of integration of Grid technologies into standard procedures and systems, bringing considerable advantage through extending the reach of Grid technologies. JBI provides the environment for the orchestration of resources. The JBI based work engine supports a wide variety of work description documents, external bindings, swappable and co-existing orchestration strategies

Domain Specific Languages (DSL) are used throughout the architecture. These can be used to expose a legacy application or process, or to provide core support for pre-designed 'canned' workflows which can be used as a top-level work description, or as embedded fragment in a larger orchestration language. As such, the use of DSLs can be seen as a logical progression of the software resource concept of UNICORE.

The design encourages an ecosystem of multiple clients all using the services offered by the work engine. These are supported through multiple external bindings. So, for example, it will offer a Web services interface as well as HTTP interface following the REST [19] architectural guidelines. Alternatively, the work engine can be deployed as an embedded component in other publishing frameworks, supporting the established portal technology EnginFrame[8], as well as opensource portlet[17] containers such as GridSphere[12]. Finally, there are further opportunities for building domain specific workbench applications leveraging the DSL support.

---

[1] For performance reasons data transfers occur in a point-to-point nature (bypassing the work engine).

This paper begins in section 2 by reviewing the status of UNICORE development highlighting the UNICORE Atomic Services (UAS) developed in the Uni-Grids project. We follow with high-level view of functionality targets in section 3. The JBI-based framework is introduced in section 4. In section 5 we outline some future strategies for workflow execution, including the use of rule technologies in section 5.3. Furthermore, we outline in section 5.1 how Domain Specific Languages are a core concept in the architecture. Finally, we conclude with a summary.

2   Atomic Services and Interoperation

Our primary interface for Grid tasks is the UNICORE Atomic Services (UAS) as developed in the UniGrids [9] project. The UAS covers the basic use-cases for 'atomic' Grid usage, e.g. submit and manage a job, elementary data management, at a single target system (a VSite in UNICORE terminology). This is done by defining a contract for Target System and related services.

At the time of writing, nothing exists as a standard - from the GGF (or elsewhere) - with the same level of usability and maturity as the UAS, although the Global Grid Forum has a number of initiatives in this area. Thus, for now, we support the UAS interface as the 'native' interface to atomic Grid functionality, until a concrete standard emerges. Indeed the UAS has provided an excellent input to the standardisation process

We introduce the term *willingness* to categorise levels of support. We see gradients of willingness. For example, a fabric service may use JSDL[10] for describing jobs, although alternative mechanisms for conveying this message are possible. Often a partial willingness to comply is due to the very nature of the standard. For example, JSDL has an extensible nature whereby open-content can appear at some points within the document.

What emerges is that some form of mediation strategy is necessary in almost every case. Sometimes this involves some simple protocol translation steps, but in other cases it may mean using ontological techniques to cope with different information models.

3   Functionality

This section contains an incomplete presentation of possible fields where the higher-level service discussed here may prove useful.

- **Workflowing**
  High-level, abstract workflows described by DSLs broken down into low-level, concrete workflows for execution by fabric services. Basic orchestration of concrete workflows.
- **Scheduling**
  Different approaches to scheduling can be enumerated as static, dynamic and hybrid scheduling. Static tasks are completely predefined or directly

authorised by the client. Dynamic tasks consist of a description of work to be done with no particular resources assigned to them. Dynamic scheduling involves lookup of appropriate resources with respect to an associated requirements description. We can combine both approaches in hybrid scheduling strategies.

– **Brokering**
Grids are subject to constant change. A dynamic broker supports the selection of resources according to the user's policies and currently offered resources. Reaction on changes of resources during execution of workflows closely links brokers and schedulers.

– **Negotiating**
Most real-life Grid applications involving multiple resources require scheduling and reservation steps. Coordinated time-dependant synchronised starting provides co-scheduling support.

– **Integrating**
With the Grid mainstream clearly moving towards web services based technologies, solutions supporting clean integration of 'legacy' business systems or processes are necessary.

– **Mediating**
In an environment dominated by open, extensible messaging formats, often collaboration between services entails some form of mediation. For example, various dialects of WS-Addressing or JSDL might coexist in a Grid.

– **Transforming**
In a similar vein, data might need transformation steps between services, for example in a multi-step workflow with data transfers.

– **Managing**
Specific services might expose administrative interfaces, for example security services might offer the possibility to add users or modify user permissions.

– **Informing**
The massive amount of both static and dynamic information available in next-generation Grids needs to be filtered for various needs - both end user and software agents. Web users are accustomed to using a wide variety of communication tools, such as e-mail, RSS feeds, SMS or instant messaging. These can be profitably leveraged for Grid users. A common use case is notifying users about an interesting status change of some resource, for example when jobs have finished, or results are available.

– **Interacting**
A particular work assignment may require input from the user during the course of its execution. Such interaction could be used to approve a dynamic resource selection, or could be used to adjust the rules governing the execution.

– **Securing**
In heterogeneous, truly service oriented Grids, the ability to use and mediate between various trust and security approaches may well become vital. Our work engine will use appropriate security services to achieve this.

4 Java Business Integration

In order to cover the wide-range of possibilities covered by the functional require-ments, we selected the Java Business Integration[14] specification as a framework technology. JBI promotes the idea of a loosely-coupled collection of components interacting with each other via the bus. It is an event-driven, component ar-chitecture. The specification defines a standard means to assemble integration components which are plugged into a JBI environment and can provide or con-sume services through it, in a loosely-coupled way. The JBI environment routes the exchanges between these components and offers a set of technical services.

JBI distils SOA concepts into the design of the internal interfaces collected around the bus. As such, JBI encourages the programmer to design and code in a loosely-coupled manner - e.g. between each module of code contributing to the system, there is a cleanly defined contract for the interactions.

JBI offers a lot of potential integration possibilities into existing enterprise Java deployments. Many businesses will find this a particularly compelling as-pect of JBI. Furthermore, as a standard Java specification there exists a number of JBI implementations, and lots of opportunity to re-use existing components. A deployment is declaratively configurable and manageable using standard mech-anisms. It is easy to 'customise' a particular JBI deployment, for example to support local processes using a DSL (see section 5.1).

Furthermore, JBI is an excellent framework for supporting multiple protocols and transports, through various binding components, such as

- **REST**
  A well designed HTTP interface following the guidelines of the REST archi-tectural style, offers an extremely attractive interface with an extremely low barrier to entry. Through interaction with a REST interface, browsers can construct Web applications using client-side scripting and using AJAX[2] approaches. It is clear that a number of other interesting Web techniques can also be applied here too.
- **WS-***
  Tool support for Web services is excellent. A good toolkit automates a sub-stantial amount of the process of building client tooling for web services. Businesses with commitment and expertise with Web services will find this channel for interaction appealing.
- **Embedded**
  This allows the work engine to be embedded into portals, and other presen-tation layer technologies.

The goal of each of the binding components above is to ultimately deliver a work assignment to the JBI bus for execution. The user submitting this can configure their work engine with notification preferences, such that they are contactable during and after the execution of their work. We propose using ServiceMix[3] as the implementation of JBI, and this comes with a number of notification mechanisms (such as Email, Jabber[13] messaging, RSS/Atom feeds) 'out of the box'.

5   Orchestration

The JBI-based core is a suitably powerful and generic framework to host the execution of scientific and business processes. Based on interacting with, and learning from, the computing world surrounding it, semi-autonomous agents form the conduit between the bus and the external world, through monitoring of the outside world, and reacting by sending events onto the bus. For example, agents could check and arrange QoS guarantees. Further agents could be responsible for negotiating trust relationships including security token exchange via a security token service. Co-allocated, time-dependent, synchronised start is also possible given an appropriate co-allocation agent.

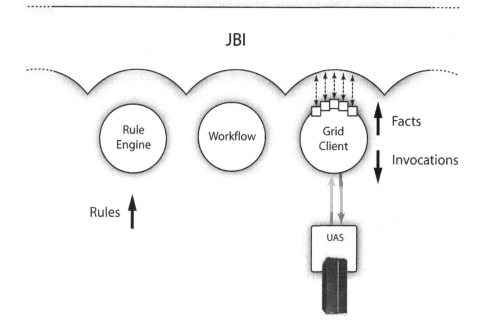

Fig. 2. The bus and some sample components

5.1   Domain Specific Languages

When dealing with workflows on the Grid, one inevitably has to deal with bridging the gap between the low-level, technical workflow (execution of small, "atomic" activities) and the high-level view taken by the end users, where maximising the impact application developers can make in their domain of expertise is important. One approach towards bridging the above-mentioned gap is the combination of orchestration engines and high-level DSLs. The idea is that higher level 'work assignments' use the underlying orchestration services to execute. Work assignments are, if necessary, mapped to a description understood by the core orchestration components. The work engine can advertise which DSLs it supports.

DSLs allow work assignments to be expressed in the level of abstraction of the problem domain. Consequently, domain experts themselves can understand, validate, modify, and often even develop DSL descriptions. There is some debate whether XML documents are in fact DSLs[22]. We argue that they are, and that it is the level of abstraction which allows classification as a DSL, although clearly it is nothing more than some XML conforming to a particular schema! We propose supporting both XML and non-XML DSLs as work assignments to be fed into the work engine. However, there is some merit to XML based DSLs[2] as it is then easier to embed inside other XML documents. It is also simple to use XML schema to validate XML-based DSLs.

Regardless of the means by which it was conveyed, at some point a work assignment exists as a message on the JBI bus. Some messages are said to be in a 'native' format, i.e. can be directly executed by one of the orchestration engines which are responsible for managing the workflow and persisting the state of the orchestration. Therefore, a process of transformation and mediation interprets the incoming work assignment. The JBI bus is the controlled environment responsible for managing this break-down. Intermediate steps may themselves be expressed in some DSL formulation. Eventually this process produces an execution plan that can be directly executed in the native format of the underlying orchestration engine.

5.2   BPEL

The Grid community seems to be somewhat split regarding the use of BPEL in conjunction with WSRF. Part of the problem is that the WSRF interactions between a service consumer and provider are quite verbose and fine-grained. While it is possible to describe this conversation as a BPEL workflow the result is somewhat long-winded.

The key is to use the bus for the invocation of services from BPEL. Each invocation breaks down to a series of WSRF-based invocations, but crucially the contract to the BPEL consumer on the bus is coarse-grained and service-oriented, and avoids the verboseness. Thus the usage of JBI as a mediating technology between BPEL and WSRF-based Grid services looks very promising to successfully use BPEL to orchestrate Grid (WSRF-based) services.

Furthermore through JBI multiple orchestration strategies can co-exist. Indeed, runtime selection of orchestration strategy may be based on the type of assignment passed to it. Other orchestration technologies which also look interesting include Business Process Management (BPM) workflow solutions (OSWorkflow[16], jBPM[15], etc), petri-net based solutions (Bossa [4]), continuation-based approaches (bpmscript[5], dalma[6]). As emphasised previously, these orchestration strategies can be swapped in and out much easier using the JBI infrastructure.

---

[2] Even if its just a trivial wrapping.

5.3 Rules

A rule engine is an example of another useful component that can be hosted by JBI. Prototype work to date has concentrated on the Java rules engine, drools [7]. This is based on the facts supplied from the computing environments, and a dynamically evolving rule base.

This provides an alternative approach to routing messages on the JBI bus, or initiating the delivery of new messages. This can be leveraged to reason on the state of a executing work assignment, using the rule base to make decisions, for example to assist with brokering decisions.

Potentially, a rule engine could be used to orchestrate an entire work assignment. This enables a declarative approach to workflow description. The rules can be changed during runtime opening up some very interesting runtime possibilities such as 'workflow rewriting'. Alternatively, the rule engine could be used at particular points within the course of a workflow execution, such as evaluations at decision points. This hybrid approach using multiple strategies is likely to be the most commonly used.

6 Summary

This paper has reported on some architectural approaches under consideration at the start of the A-WARE project. Clearly, the new breed of grid infrastructure is based on the SOA paradigm. Functional requirements pose a strong need for dynamic message exchanges between all components, which can be added to and removed from the infrastructure in dynamic ways.

A flexible architecture supporting the stated functional requirements is JBI, offering normalized message exchange between components plugged into a message bus. JBI offers general purpose components which will be useful in implementation of A-WARE infrastructure. Very importantly, we envisage re-using many existing opensource libraries for the implementation, writing code to integrate these using JBI. Finally, as a integration framework, JBI offers excellent support for the integration of existing systems and processes.

Work assignments may be described in terms of DSLs, allowing specialists to work in their domain. DSL work descriptions are abstract and will be broken down to concrete submissions of contributing resources. DSLs can be nested and provide a notion of 'canned' workflow. Furthermore, JBI allows for the integration of several orchestration strategies. They can be selected on the basis of particular work assignment. JBI comes with a component supporting BPEL, which can be used as a start, and support for other orchestration engines will be added. A rule engine hosted by JBI declaratively describes consequences of certain states of workflows or events in the environment. Rule engines can potentially be used to orchestrate entire work assignments.

An early prototype of the JBI based framework looks very promising. The great advantage of this approach is the possibility of rapid and flexible development. Development is incremental and highly modular, such that extensions can be added without interfering with the existing components.

References

1. A-WARE Project. http://www.a-ware.org/.
2. AJAX. http://adaptivepath.com/publications/essays/archives/000385.php.
3. Apache ServiceMix. http://incubator.apache.org/servicemix/.
4. Bossa. http://www.bigbross.com/bossa/.
5. Bpmscript. http://www.bpmscript.org/.
6. Dalma. https://dalma.dev.java.net/.
7. Drools. http://drools.codehaus.org/.
8. EnginFrame. http://www.enginframe.com/.
9. European UniGrids Project. http://www.unigrids.org.
10. GGF JSDL. https://forge.gridforum.org/projects/jsdl-wg/.
11. Grid Interoperability Project. http://www.grid-interoperability.org.
12. GridSphere. http://www.gridsphere.org/.
13. Jabber. http://www.jabber.org.
14. Java Business Integration. http://www.jcp.org/en/jsr/detail?id=208.
15. jBPM. http://www.jboss.com/products/jbpm.
16. OSWorkflow. http://www.opensymphony.com/osworkflow/.
17. Portlet Specification. http://www.jcp.org/en/jsr/detail?id=168.
18. D. Erwin, editor. *UNICORE Plus Final Report – Uniform Interface to Computing Resources*. UNICORE Forum e.V., 2003. ISBN 3-00-011592-7.
19. Roy Thomas Fielding. *Architectural styles and the design of network-based software architectures*. PhD thesis, 2000. Chair-Richard N. Taylor.
20. Ian Foster. Service-oriented science. *Science*, 308(5723):814–817, May 2005.
21. R. Menday and Ph. Wieder. GRIP: The Evolution of UNICORE towards a Service-Oriented Grid. In *Proc. of the 3rd Cracow Grid Workshop (CGW'03)*, Oct. 27–29 2003.
22. M.Fowler. Language Workbenches: The Killer-App for Domain Specific Languages? 2005. http://www.martinfowler.com/articles/languageWorkbench.html.
23. A. Streit, D. Erwin, Th. Lippert, D. Mallmann, R. Menday, M. Rambadt, M. Riedel, M. Romberg, B. Schuller, and Ph. Wieder. Unicore - From Project Results to Production Grids, 2005. Elsevier, L. Grandinetti (Edt.), Grid Computing and New Frontiers of High Performance Processing.

Grid-Based Processing of High-Volume Meteorological Data Sets

Guido Scherp[1], Jan Ploski[1], and Wilhelm Hasselbring[1,2]

[1] OFFIS, Escherweg 2, 26121 Oldenburg, Germany
{guido.scherp,jan.ploski}@offis.de
[2] University of Oldenburg, Software Engineering Group, 26111 Oldenburg, Germany
Hasselbring@informatik.uni-oldenburg.de

Abstract. Our energy production increasingly depends on regenerative energy sources, which impose new challenges. One problem is the availability of regenerative energy sources like wind and solar radiation that is influenced by fluctuating meteorological conditions. Thus the development of forecast methods capable of determining the level of power generation (e.g., through wind or solar power) in near real-time is needed. Another scenario is the determination of optimal locations for power plants. These aspects are considered in the domain of energy meteorology. For that purpose large data repositories from many heterogeneous sources (e.g., satellites, earth stations, and data archives) are the base for complex computations. The idea is to parallelize these computations in order to obtain significant speed-ups. This paper reports on employing Grid technologies within an ongoing project, which aims to set up a Grid infrastructure among several geographically distributed project partners. An approach to transfer large data sets from many heterogenous data sources and a means of utilizing parallelization are presented. For this purpose we are evaluating various Grid middleware platforms. In this paper we report on our experience with Globus Toolkit 4, Condor, and our first experiments with UNICORE.

1 Introduction

Regenerative energy sources are becoming more and more important for our energy supply. It is assumed that in the future the dependency on these sources will increase. However, the availability of these sources is highly influenced by meteorological factors which impose new challenges. Forecast models for simulations are needed to provide a near real-time estimation of the power generation (e.g., through wind power). Another scenario is the determination of appropriate locations for building power plants. For example, an analysis based on (archived) solar irradiation data combined with further geographical and commercial information (lakes, rivers, costs, etc.) can be used to find optimal spots for solar power plants. Each simulation or analysis is based on large heterogeneous data sets, which come from satellites, earth stations, data archives, or other sources. Due to the large amount of data, the computational power of a single computer

W. Lehner et al. (Eds.): Euro-Par 2006 Workshops, LNCS 4375, pp. 235–244, 2007.

is insufficient. Next generation satellites with higher resolution will further increase that amount of data. Today, approximately one terabyte of new data per month is received and continuously archived that are relevant for our project WISENT, which is introduced in the following.

A solution for speeding up the simulations and analyses lies in utilizing parallel computation. Therefore, the challenges of transferring large amounts of data as well as the parallelization of simulations and analyses have to be addressed. In the context of parallel execution and transfer of large data amounts, the term Grid [1,2] has become more and more familiar in the last years. In the project WISENT (wisent.d-grid.de) Grid technologies are employed to focus on these challenges in the domain of energy meteorology. The University of Oldenburg (ehf.uni-oldenburg.de), three departments of the German Aerospace Center (DLR DFD, DLR TT, DLR IPA, www.dlr.de) and the company meteocontrol (www.meteocontrol.de) are collaborating on WISENT with the institute OFFIS (www.offis.de) as project coordinator. The project started in October 2005 and is funded by the German Federal Ministry of Education and Research (BMBF, www.bmbf.de) over three years. One aim of the project is to set up a Grid infrastructure to support large data transfers and distributed processing.

Each project partner has different resources that are to be shared within the Grid infrastructure. The different types of resources and their utilization are shown in Figure 1. Firstly, data are received from many heterogeneous sources such as satellites and earth stations. Based on this data, various complex computations are performed utilizing resources such as desktop PCs, multi-processor servers or a dedicated cluster. Because of the planned integration into the German Grid initiative D-Grid (www.d-grid.de), we also intend to utilize D-Grid resources, for example high-performance computing (HPC) centers. The results of these computations are used for multiple purposes, for example to monitor the energy output of photovoltaic modules. To gather experience, we examined several Grid middleware platforms such as Globus Toolkit 4, Condor, and we are currently evaluating UNICORE.

The challenges of transferring large data sets and parallelization in energy meteorology are described in Section 2. In Section 3, we report on possible solution paths we have examined or intend to examine, and experiments with Globus Toolkit, Condor, and UNICORE we have already performed or planned. We conclude with an overview of future work in Section 4.

2   Challenges to Be Addressed

In WISENT, we are faced with various challenges. In the present paper we focus on the transfer of large data sets and the utilization of parallelism to run complex computations on these data sets. To fulfill the requirements associated with it, we intend to examine appropriate Grid technologies. The challenges are described in more detail in the following subsections.

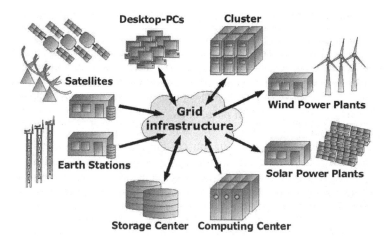

Fig. 1. Overview of the application scenario in WISENT

2.1 Transfer of Large Data Sets

One characteristic of the collaboration of our project partners is the exchange of data. Those data mostly consist of data products that are the results of computations on raw satellite data or previously processed data products. Data transfers are executed periodically (weekly, daily, hourly) or on demand. The number of daily data transfers at DLR-DFD concerning WISENT, for example, is up to 100, some of which are performed between the project partners and others which involve external facilities. The size of one transfer ranges from a few kilobytes up to several hundred megabytes. In the future, the number and size of data transfers will increase.

Today, a data transfer often relies on simple transfer protocols such as FTP or HTTP. Manual execution and monitoring are not uncommon. One goal is to execute these transfers reliably and automatically, which should result in less manual work and thus in a cost reduction. Each project partner intends to be able to easily share own data as well as access the data offered by any project partner within the Grid infrastructure. Furthermore, some transfers are used to deliver commercial data products. Thus, security for authentication and authorization as well as monitoring for accounting and billing are also required.

2.2 Parallelization

Our project partners use applications in the context of energy meteorology that can benefit from parallelization. The potential for parallel execution can be exploited in different ways, for example at data or program level. At data level, it means that the input data is first separated into several parts. The computation is next executed on distributed computing nodes, whereat the same program is running on each node without any network communication among the nodes.

Finally, after the parallel execution is completed, the corresponding output data parts are merged into the final result. In contrast, the parallelization at program level is used if a clear subdivision of input data for independent computation is not possible. Thus, parts of the programs are distributed and a frequent exchange of intermediate data is needed, for example via an implementation of the Message Passing Interface (MPI) [3]. The implementation of parallelization at data level is often much easier than parallelization at program level. Fortunately, we found out that most applications considered within our project can use parallelization at data level. No program sources have to be modified, which is an advantage because the source code of some programs concerned and libraries is not available. Our project partners already use parallelization at program level in a few applications. Thus, this parallelization technique will become relevant and has to be considered in the future.

3   Solution Approaches Employing Grid Technologies

Today, several Grid technologies and respective Grid middleware platforms are available. We gained initial experience with Globus Toolkit 4 and Condor and we are currently evaluating UNICORE, which we describe in the following sections. Our evaluation of each software considers the challenges mentioned in Section 2.

A typical classification of Grid infrastructures distinguishes Intra-, Extra-, and Inter-Grids (Figure 2). The term Intra-Grid refers to a Grid set up within a single organization. Several Intra-Grids of different organizations are connected to an Extra-Grid, which requires stronger security policies, such as virtual private networks (VPN). An Extra-Grid often provides access to a Grid infrastructure for a certain user group with established working relationships, thus it is typical for community grids. The final extension is Inter-Grid, a global Grid infrastructure for a wide range of independent users. The planned Grid infrastructure for our project conforms to the Extra-Grid definition. Grid technologies can be employed at Intra-Grid level, Extra-Grid level, or both levels, which is taken into account by our evaluation.

3.1   Globus Toolkit 4

As a start, we have chosen to evaluate the Grid middleware platform Globus Toolkit 4 [4]. This decision was made because of the comprehensive available documentation [5,6], the long development history since Globus Toolkit 1 (released in 1996), and the implementation of the Grid standard Web Services Resource Framework (WSRF) [7]. Globus Toolkit 4 offers comprehensive services for data transfers which motivates, due to the requirements of data transfers described in Section 2.1, the examination of these services. In this section we report on our first experiences.

Application Scenario. Our project partner DLR DFD is involved in most data transfers exchanging data products, which may be performed in two ways.

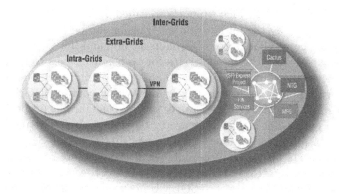

Fig. 2. Intra-, Extra-, and Inter-Grid (Source: IBM)

Either DLR DFD offers the desired data product for download on their "public" FTP server or the data product is transferred directly to an FTP server of a project partner or another external facility. This FTP-based approach has some disadvantages because errors are handled manually. For example, a transfer initiated by a cronjob might fail due to a temporary network timeout.

The manual error handling of failed FTP transfers gives rise to potential problems. The required data products are stored together with raw satellite data in an archive called *Digital Information and Management System* (DIMS), which is secured behind a firewall. Thus, a data product is periodically transferred to the "public" FTP server, which is located in a demilitarized zone (DMZ). The recipient is also periodically looking for changes on this FTP server and downloading the desired data products. This procedure does not include checks whether the required data products are completely available. If the series of uploads to the "public" FTP server was interrupted, an incomplete product might be downloaded, which might be discovered too late.

We have chosen five data transfers and their corresponding data products as an initial test scenario. Our approach is an incremental migration of each transfer toward Grid technologies in multiple steps. First, test data transfers are to be executed internally between DLR DFD and the OFFIS institute. Based on this experience, these transfers will be extended step-wise to cover the real-world scenarios.

Data Transfers with Globus Toolkit 4. Globus Toolkit 4 offers a layered architecture of services for data transfers. The GridFTP service belongs to the lowest layer. It is an enhanced FTP protocol with features such as the support of encryption and authentication based on X.509-certificates and several options to increase the data troughput, e.g. using several parallel data channels. The use of certificates allows Singe Sign-On (SSO), which is an important advantage. By default, GridFTP does not encrypt the data channel, but activation of encryption reduces the data throughput significantly. Globus Toolkit 4 comes with a standalone GridFTP server, which supports so-called *Third Party Transfers* between two hosts, whereby the initiator of the transfer can be located anywhere. With `globus-url-copy`, Globus Toolkit 4 offers a command-line GridFTP client.

GridFTP provides only the basic service for transferring files from a source to a destination. Higher-level services are supposed to build upon this capability. One practical application of this idea is the WSRF-compliant Grid service named Reliable File Transfer (RFT). Besides several configurable properties, the RFT service accepts a list of files that are to be transported. The data transfer is processed in a reliable manner. That means that the RFT-Service retries interrupted transports up to a limited number of times before reporting an error. For this purpose, the state of each transport is recorded in a database with transactional control (such as PostgreSQL). Unfortunately, the RFT service in the actual version of GT4 is not utilizing the support of data channel encryption in GridFTP, but this feature is proposed in the coming versions.

In our test scenario, we installed Globus Toolkit 4 on a host at DLR DFD and configured it to provide the RFT service. For testing purposes our institute assumed the role of a Certificate Authority (CA) to support authentication with signed X.509-certificates. For the future we plan to use D-Grid certificates. The first data transfers from DLR to Offis failed due to the strong firewall policies at DLR. Based on the FTP-based protocol GridFTP uses a new dynamic port for each data channel it creates. Typically, these ports are in a higher range and blocked by the firewall. However, the port range used by both client and server can be restricted by environmental variables. Because of the strong firewall policies DLR develops an Application Level Gateway (ALG), a proxy placed in the DMZ to support transparent communication for using (Grid) software such as GridFTP. These experiences are the basis for the incremental extension of data transfers based on GridFTP and RFT. Also planned is the full integration of the DIMS archive into the Grid infrastructure. Finally, a detour via the "public" FTP server should be avoided.

Currently, we are investigating the Monitoring and Discovery System (MDS) of Globus Toolkit 4. Each Grid service has properties which can be exploited with MDS, for example the RFT service reports the number of active transfers and transferred bytes as properties. This information can be used for live monitoring, accounting and billing.

Globus Toolkit 4 also supports distributed computing. However, in the context of connecting desktop PCs and PCs within clusters, the features of Globus Toolkit 4 do not appeal to us. The main reason is the apparent lack of a good scheduling mechanism available out-of-the-box. However, we have not fully evaluated the Community Scheduler Framework (CSF) it offers yet. Moreover, Globus Toolkit 4 does not specifically support pooling resources of desktop PCs. For example, some useful features such as the discovery of mouse and keyboard activities and job migration are not available. Consequently, we are currently considering using Globus Toolkit 4 only at the Extra-Grid level to improve the execution of data transfers within the Grid infrastructure.

3.2 Condor

Several work packages in WISENT focus on improving computational performance through data parallelization. One promising approach in that context is provided by Condor [8], a freely available software package developed at the University of Wisconsin-Madison. In this section, we briefly describe a parallelization scenario for which Condor appears to be one likely solution based on the insights gathered so far.

Application Scenario. Two- and three-dimensional computational models of radiative transfers are currently employed by our project partners to determine the amount of irradiation reaching the Earth's surface based on satellite observation. Apart from images, the algorithms rely on input data consisting of various atmospheric parameters, such as cloud profiles and water vapor distribution.

The three-dimensional radiative transport solver MYSTIC [9] developed by DLR utilizes a compute-intensive Monte Carlo simulation to track individual photons through multiple layers of the atmosphere. The simulation time varies depending on the covered geographic area and the number of simulated photons, which affects the precision of the results. For example, a modestly sized simulation for a 100x100 km area with 10,000 photons consumes more than 5 hours of CPU time on a 1 GHz PC.

From the parallel programming point of view, an attractive feature of both the 2D and 3D radiative transfer models is the ability to compute results for multiple distinct columns of the atmosphere separately, with a certain amount of input data overlap. The results can be combined into one large output image.

Our project partner DLR IPA intends to reduce the computation time of radiative transfer models by harnessing the power of a Linux cluster consisting of sixteen 2,4 GHz Pentium Xeon nodes, several servers and 10 to 15 Linux desktop workstations. Based on experiences gathered in the project, the network will be expanded to support other resource-intensive scientific computations. The initial experiments will process only several gigabytes of locally available input data and produce a fraction of that as output, meaning that large-scale data transports will not be of primary concern.

While parallel 3D computation models are being implemented by our colleagues with background in physics and meteorology, our task is to assist them with the technology to improve their existing parallel computing solution for 2D radiative transfer. The currently deployed solution, used to drive the Linux cluster mentioned above, distributes libRadtran [10] solver processes to multiple cluster nodes through a combination of shell scripts implemented in-house, the Parallel Virtual Machine (PVM) package [11] and a third-party PVM-based extension of the Unix `make` tool called `ppmake`. This solution does not scale to utilizing desktop resources nor does it permit unsupervised execution. Problems such as difficult-to-explain non-termination of PVM processes are common and call for manual interventions. Finally, the present solution does not stand software engineering scrutiny. For example, the `ppmake` tool, which is suspected to cause the encountered problems, is no longer actively maintained.

The Condor Approach. Condor is a software package that has been designed with the aim to support "cycle scavenging", that is, taking advantage of computational resources during idle periods without interfering with their primary users. A typical Condor installation consists of a central coordinator machine and one or more machines acting as compute and/or job submission nodes. Both classes of nodes communicate with the coordinator to announce the type and availability of resources or the availability of jobs to be computed. On each compute node, a Condor process monitors and regularly reports to a coordinator static attributes such as the operating system and processor architecture, as well as dynamic information such as the current CPU load, amount of disk space and keyboard and mouse activity.

A job submission node announces resource requirements for user-submitted batch jobs, which consist of binary executables along with a specification of input/output files. When an available resource has been matched with a job request by the coordinator, it arranges direct communication between the submitting and computing node. A shadow process is started on the submitting node in order to monitor and report the execution status of the job process on the computing node; both processes communicate using a proprietary protocol.

A job may be interrupted if its current computing node becomes unavailable or at its owner's discretion. If the executable program was linked with the Condor library and adheres to some restrictions, Condor is able to migrate a snapshot of the job process to another compatible compute node and continues its execution there. Furthermore, I/O system calls made by executable programs can be automatically re-routed to the original submission machine, which may be used to eliminate the need for explicit data transfer operations before and after job execution. Alternatively, Condor offers built-in file transport mechanisms.

Our tests with Condor consisted of migrating the current PVM-based scripts to utilize the Condor submission client instead of ppmake. They performed to our satisfaction on a network of several different Linux PCs. The small amounts of data we utilized for the tests helped us understand Condor's opportunistic approach to scheduling. In fact, because individual fragments of computations were very short, the measured clock time for the Condor run exceeded that of a sequential execution. We were able to influence the loss factor by modifying the Condor configuration, particularly the communication intervals between individual processes. Further improvements are expected by adjusting the job granularity, as we have already discovered in experiments on behalf of another project partner.

Based on our experience so far, the main weaknesses of Condor appears to be the platform dependence, caused in part by the restricted availability of its source code and in part by its operating-system dependencies, the complexity reflected in a daunting number of (well-documented) configuration settings, and the demands it places on the network connectivity. Condor's main advantages are the flexibility, long product history, solid documentation, wide user base and built-in features which facilitate the construction of "desktop Grids", which we intend to utilize at the Intra-Grid level. Those "desktop Grids" are very suitable for running our applications using parallelization at data level. Condor also supports parallelization at program level through MPI, but in this scenario the network bandwidth could quickly become a major bottleneck. Thus, the access to HPC centers is planned as well in order to support future demands for executing those applications within the Grid infrastructure.

3.3 UNICORE

Our evaluation of UNICORE has just begun. So far, we have successfully connected two UNICORE hosts and performed first tests using the UNICORE client. Thus, an assessment of whether we will use UNICORE at the Intra-Grid or Extra-Grid level is not yet possible. As UNICORE was originally used to connect distributed computing centers, its classification into the Extra-Grid category seems reasonable. One great advantage of UNICORE is that the communication between the client and the gateway is encrypted and uses only one fixed port for both job submissions and data transfers. This is very compliant with stronger firewall policies. Each connection of a GridFTP client in contrast needs at least one exclusive port for the data channel (c.f. Section 3.1). But a data transfer through the UNICORE Protocol Layer (UPL) has a noticeably less data throughput than GridFTP and thus it seems only suitable for small data transfers.

Another benefit of UNICORE is the possibility to model jobs and subjobs in a workflow-style. As mentioned in Section 3.1, the DIMS archive of DLR DFD stores both raw and post-processed satellite data. Some data products are computed regularly and stored in the archive for further purposes while others are computed on demand without persistent storage. The computation of each data product consists of several steps that form so-called process chains. However, as of today, each process chain is implemented individually and no consistent description format exists. We intend to assess the UNICORE client's utility for modeling and executing such process chains in WISENT. With the growing complexity of the process chains and their automation requirements, more sophisticated workflow elements supported by UNICORE's workflow description language may become useful. One disadvantage of the workflow model is that each workflow element must be bound to one specific resource (Virtual Site). For our scenario a dynamic resource selection with an optional limitation to specific resources ought to be possible as well.

4   Conclusions and Future Work

This paper has shown how the use of Grid technologies at different levels can address the challenges in the context of energy meteorology, including transfers of large data sets and parallelization of programs at data level. We are still in an evaluation phase, thus beside further tests with Globus Toolkit 4 and Condor we have to gain more experience with UNICORE for a more comprehensive comparison. Currently, we are also considering evaluating gLite [12] and the commercial platform Sun N1 Grid Engine [13]. Another task is to investigate the interoperability between these Grid middleware platforms. It is very likely that we will use different Grid middleware platforms for the Intra-Grid and Extra-Grid levels, thus interoperability is important.

In the view of the considerable data heterogeneity, one of our next steps will be to support a uniform access method. To this end, we plan to examine the OGSA-DAI (Data Access and Integration) [14], which provides a Grid-related standardized access to different data resources using Web Services technologies.

Moreover, we plan to further examine security issues. A critical success factor for a Grid infrastructure lies in achieving trustworthiness [15]. The project partners need to be sure that the Grid infrastructure is secure, and they intend to control their own participating systems. Therefore, services for monitoring, access control and logging are required. Finally, easy access to the Grid infrastructure as well as fast installation and deployment of new Grid nodes are also important concerns.

References

1. Foster, I.: What is the Grid? A Three Point Checklist. Grid Today **VOL. 1 NO. 6** (2002)
2. Foster, I., Kesselman, C., Tuecke, S.: The anatomy of the Grid: Enabling scalable virtual organizations. Lecture Notes in Computer Science **2150** (2001) 2–13
3. MPI: Message Passing Interface. (http://www-unix.mcs.anl.gov/mpi/standard.html) Retrieved: 2006-06-11.
4. Globus: The Globus Alliance. (http://www.globus.org/) Retrieved: 2006-06-11.
5. Globus: Globus Toolkit 4.0 Release Manuals. (http://www.globus.org/toolkit/docs/4.0/) Retrieved: 2006-06-11.
6. Jacob, B., Brown, M., Fukui, K., Trivedi, N.: Introduction to Grid Computing. (http://www.redbooks.ibm.com/redbooks/pdfs/sg246895.pdf) Retrieved: 2006-06-11.
7. WSRF: Web Services Resource Framework. (http://www.globus.org/wsrf/) Retrieved: 2006-06-11.
8. University of Wisconsin: Condor High Throughput Computing. (http://www.cs.wisc.edu/condor/) Retrieved: 2006-06-11.
9. Mayer, B.: I3RC phase 1 results from the MYSTIC Monte Carlo model. (I3RC (Intercomparison of 3D Radiation Codes) workshop, Tucson, Arizona, 1999)
10. Mayer, B., Kylling, A., Hamann, U.: libRadtran – library for radiative transfer. (http://www.libradtran.org) Retrieved: 2006-06-11.
11. Geist, A.: PVM: Parallel Virtual Machine: A Users' Guide and Tutorial for Network Parallel Computing. MIT Press (Scientific and Engineering Computation) (1994)
12. EGEE: gLite Lightweight Middleware for Grid Computing. (http://glite.web.cern.ch/glite/) Retrieved: 2006-06-11.
13. Sun: Sun N1 Grid Engine. (http://www.sun.com/software/gridware/) Retrieved: 2006-06-11.
14. OGSA-DAI: Open Grid Services Architecture - Data Access and integration. (http://www.ogsadai.org.uk/) Retrieved: 2006-06-11.
15. Hasselbring, W., Reussner, R.: Toward trustworthy software systems. IEEE Computer **39**(4) (2006) 91–92

BLAST Application on the GPE/UnicoreGS Grid

Marcelina Borcz, Rafał Kluszczyński, and Piotr Bała

Nicolaus Copernicus University,
Faculty of Mathematics and Computer Science,
ul. Chopina 12/18, 87-100 Toruń, Poland
{marbor,klusi,bala}@mat.uni.torun.pl

Abstract. Sequence analysis is one of the most fundamental tasks in molecular biology. Because of the increasing number of sequences we still need more computing power. One of the solutions are grid environments, which make use of computing centers. In this paper we present plug-in which enables the use of BLAST software for sequence analysis within Grid environments such as UNICORE (Uniform Interface to Computing Resources) and GPE (Grid Programming Environment).

1 Introduction

Grid technology is becoming very popular nowadays. Watching weather forecast we usually do not even realize how much computing power has been used to prepare it. Scientific research and technology development implies an increasing demand for distributed resources. We still need more precise weather forecast, longer molecular simulations and we want to determine genome structure and its functions. In other words, we need computing power which could easily be made available to the user by means of Grid middleware like UNICORE [16] or Globus [9].

The concept of Grid computing was first introduced in 1999 [7]. Its main idea is the use of multiple distributed resources combined together on a single application to work cooperatively. Definition of Grid computing have evolved during the time [8]. Ian Foster in [6] suggested simple checklist definition of the Grid technology composed of 3 points:

- The resources it uses should be distributed without any centralized control.
- All protocols used for authorization, resource discovery etc. should be open and standard.
- The use of distributed resources in combined form should be much more worthwhile than using them separately.

Grid middlewares find the use in much of scientific research. They have been successfully used in 3D graphics, quantum chemistry and molecular modeling. Since it became possible to determine the structure and sequence of DNA, many scientists have been trying to indicate the role of specific DNA motifs in human

W. Lehner et al. (Eds.): Euro-Par 2006 Workshops, LNCS 4375, pp. 245–253, 2007.

organism. This gives the opportunity to model many biological processes in our body. It may also help us to cure many diseases, for example by blocking expressions of specific genes. Since *Basic Local Alignment Search Tool* (BLAST) appeared [1], it has become one of the fundamental and widely used tools in molecular biology for sequence analysis. Prediction of human metabolic pathways can serve as an example of using BLAST to find human enzymes taking part in metabolic process (see [15]). Since 1990 we can observe an increasing need for the use of the BLAST application by molecular biology scientists. That is why there have been many attempts to optimize and speed up this biological tool. For example in 1998 in [5] there were presented modifications which produce a better suited code for comparing large numbers of sequence to several different databases. Grid middleware brings another way of comparing a large number of sequences. In a simple way comparisons of different sequences can be distributed within different computing centers where original BLAST code can be used. This eliminates errors which could occur during applying proposed modifications to the code. The purpose of this paper is to show Grid technology solution by presenting plug-in for BLAST application designed for UNICORE and GPE middleware.

2   BLAST Software

The Central Dogma of Molecular Biology states, that two-stranded helix of DNA composed of four types of nucleotides, is duplicated during replication process. Next, DNA is transcribed into one-stranded structure of RNA. Finally protein, i.e. polymers built from 20 different types of amino-acids, is synthesized from RNA by translation. Sequences can change over time due to mutation, natural selection and genetic drift. By means of BLAST biologists are able to compare DNA or protein sequences from the same or different organisms. In this way evolutionary relationship between organisms can be explored and biological functions of new sequences can be predicted.

BLAST (Basic Local Alignment Search Tool) became an important and the most widely used tool in the field of bioinformatics. It finds statistically significant local similarities between pairs: user-defined (protein or DNA) sequence and sequences from databases. As an output it gives gapped or gap-free alignments. Each alignment is a high-scoring segment pair with final score and estimation of statistical significance, called E-value (see [2] for details). During DNA sequences comparison, results contain also information about DNA strand direction where the similarity has been found.

The popularity of the program, apart from its functionality, is based on the programs following qualities: reliability, speed and flexibility. BLAST is the name of a package of software containing, among other things, *blastp* and *blastn* programs (see [3] for detailed description). Each of these components realizes algorithms for different types of sequences (*blastp* for proteins, *blastn* for nucleotides). BLAST uses a heuristic algorithm. First it takes subsequences of query sentence of fixed length and all words with the same length which alignment score is at least the pre-defined threshold value. Next, database is scanned for these words called "hot spots", which are extended in both directions until some threshold or cut-off value is reached. In this way BLAST finds similarities without exploring the entire search space which could take a lot of time.

There are also several variants of BLAST, such as PSI-BLAST (Position-Specific Iterated BLAST) or PHI-BLAST (Pattern-Hit Initiated BLAST), which run specialized version of BLAST programs. First of them uses PSSMs (Position-Specific Scoring Matrices) to improve sensitivity of searching. In the second program, there can be put a motif next to the query sequence, which is the most significant and is used to scan the database for, instead for the "hot spots".

BLAST software is being developed by many institutes. The most popular versions of the programs are NCBI-BLAST and WU-BLAST. The first of them originated from the National Center for Biotechnology Information, while the other is being developed at Washington University in St. Louis. Each version has plenty of options which makes BLAST a very flexible tool used by biologists all over the world.

3   Grid Programming Environment (GPE)

Grid environments are rapidly changing because of the new standards and application areas which are still emerging and are being discovered. That is why Grid Programming Environment (GPE) is being currently developed. The goal is to implement new grid middleware based on the experience gained during the implementation of UNICORE. Because the success of the Grid technology depends among others on interoperability between different Grid implementations, there is an idea to establish a stable interface between them. It is also important to build a more flexible and user-friendly client framework for heterogeneous Grid infrastructures with easy and secure access.

With the concept of GPE project (see [14] for more details) comes different client applications designed for three category of users:

- *Expert Users* – with the ability to construct complex workflows of their jobs (it is the successor of the UNICORE Client). This type of client is dedicated to users with some knowledge about the functionality of the Grid. Besides the workflow editor, the user has the ability to use different identities on different target systems.
- *Application Users* – it is sufficient for them to use only one application, so they do not need all of the tools available in order to build workflows. This client interface doesn't have all the functionality offered by the Grid. Instead of that it is very easy and intuitive to use. Application Client depicted in Fig. 1 is dedicated to scientists who usually run only one application at the time.
- *Unaware Users* who manage their tasks and Grid resources through portals in web browsers. Such users do not even have to install client application.

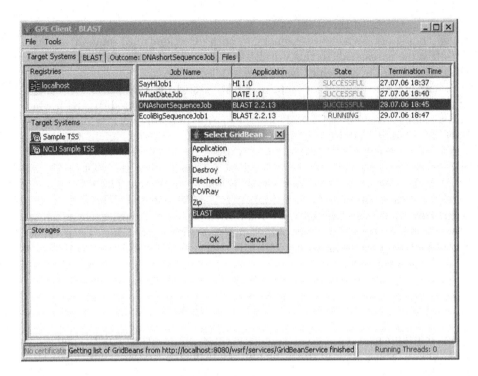

Fig. 1. GPE client interface dedicated for *application users*. It can be seen response from *GridBean service* presenting list of available plug-ins on specified target system.

Everything they need on the computer is just an Internet browser. Such a solution is also very attractive for mobile users who could manage their tasks from any Internet cafe.

GridBean approach is one of the major advantages of the GPE framework. Usually, when user wants to use specific application on the grid, he has to download corresponding plug-in and integrate it with the client program. Sometimes specific knowledge about the structure of the software is needed, like directory name where the plug-in files should be placed. However, not every user has some knowledge about computer science, even the basic one. That is why the concept of GridBeans is very promising. Instead of downloading and manually integrating the needed plug-in, GPE client software has the ability to check what applications are available on the remote system. In the case the list contains the application the user wants to work with, it is enough to select it by clicking resulting in the corresponding plug-in being automatically downloaded into the client's job managing interface and made ready to use.

In other words, GridBeans are plug-ins which can be loaded dynamically into any of the three different client applications. There is no need to have any detailed knowledge about the physical structure of the software in order to use available application. Ratering in [14] described four major advantages standing for presented GridBean approach:

- Easy distribution and update of plug-ins – updating plug-ins takes place in the same way like downloading. User has to ask *GridBean service* for available gridbeans and then downloads new version of plug-in he is interested in.
- Overview of supported applications – once the user asks the *GridBean service* for plug-ins available on the target system, he gets clearly presented list of actually supported applications.
- Flexibility of GridBeans – once the plug-in has been implemented, there is no need to create different copies for different types of client applications discussed above. Special care may be only needed for graphical layout depends on the type of client.
- Easy implementation of GridBeans – this is especially important for developers bringing applications on the Grid. The main effort to design plug-in is implementation of the graphical interface and construction of the job description. Thanks to recipe guide [11], even non-professional programmer will be able to design plug-in in reasonable time.

As it was mentioned, these points make the GridBean approach very promising. Thanks to inter-operating *GridBean service* and client programs, manipulating with plug-ins for different applications has become very easy for the user, even one without any computer science background.

Grid Programming Environment was designed based on the experience of UNICORE implementation. To keep track of all resources available on all target systems *Target System Registry* has been created. Every target system during start-up registers its resources properties as installed software, workload or running jobs. This information is being dynamically updated during runtime by *Target System Service*. Presented solution assures that when client asks for available resources he gets always up-to-date information. *Target System Registry*, besides managing the list of available target systems with their resources, plays another crucial role. In similar way to the classical UNICORE Gateway, it performs authentication to prevent any access to the services from unsafe Internet.

The GPE Clients can contact to the Target Systems Services available in the different hosting environments such as Globus or UnicoreGS. In result, it is possible to develop interface which can be used with the different middleware without any modifications.

4   BLAST GridBean for GPE

In this section we present GridBean designed for BLAST application. User interface is organized in a similar way to the one on NCBI-BLAST website [12]. The main reason of such design is to keep it easy to use, specially by scientists who are used to NCBI website and could have some troubles to migrate on the GPE

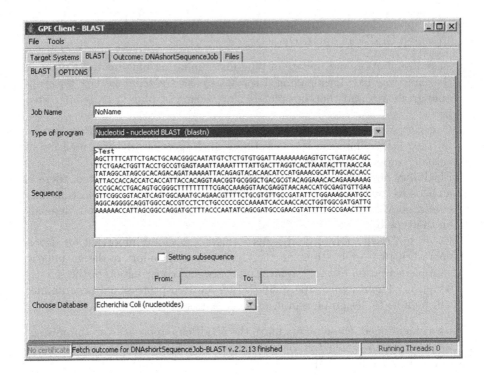

Fig. 2. Main graphical user interface panel. It grants a choice of the type of BLAST program and a possibility to enter the sequence and to specify a database which is to be browsed.

framework. In the main panel, besides the job name, a particular sequence can be entered so that it may be compared with the sequences from a chosen database. There can be typed either a whole sequence or the definition line only (Fig. 2). There also should be chosen the type of BLAST program and the database that the sequence will be search in.

The second panel provides optional choices which affect the presentation of the search results. In the case of using *blastp* program user should determine which scoring matrix he wants to use. Accuracy of matching can be controlled by changing appropriately expect and word size values (Fig. 3). There is also a possibility to determine gap penalties for protein comparison. Flexibility of searching similarities can be increased by using filters while readability of output by choosing alignment view and convenient number of descriptions and alignments to be displayed. In the last field of the panel advanced user can enter more sophisticated and more rarely used options, like "-i T" which show GenInfo Identifier number in definition lines.

Presented panels (*Input Panels*) are extensions of the GridBean panel designed for inputing the data by the user. Input Panel provides a visual area to display needed parameter controls, and also a technique enabling the correlation of values gained from those controls with the job description. Depending

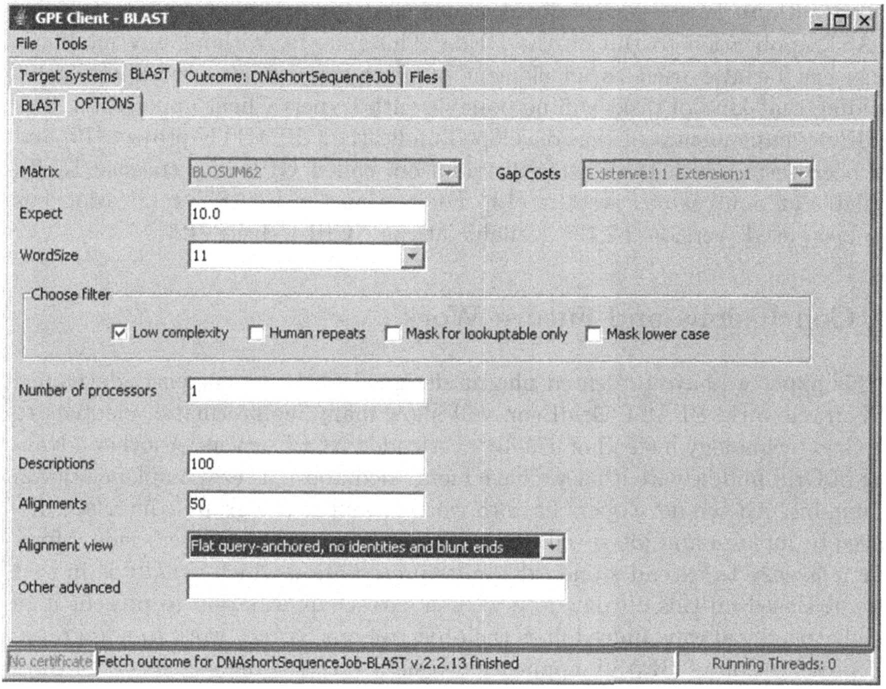

Fig. 3. Second panel of client's GUI. Here the user specifies more detailed data including the number of similar sequences or alignments to be show. It is also possible to enter some specific options for BLAST package expert users.

on the application and its complexity, which is usually determined by a large number of parameters, there is a possibility of using more than one input panel for grouping the varied options respectively. Another advantage determining the flexibility of the Gridbean approach is, that the plug-in can also have *Output Panels*. Their main purpose is to display the job results downloaded from the Grid, but the way of presenting the data can be made much more attractive and more intuitive than the raw text. It could for instance show some precomputed graphics, charts or tables with statistics. Of course, designing such panels would strictly be connected with the application the GridBean is meant to be used with. It is therefore recommended for the developers to have a good knowledge of the software or a good specialist nearby to address any questions. It is also important to mention that thanks to Grid Services the configuration of available databases for BLAST can be made on the Target System. Such solution ensures that there is no need to upgrade the GridBean after changing the database list on the server. BLAST plug-in simply checks the list before the user starts preparing the task. This means that the user can see only those databases which are currently available in the system.

The authors have successfully designed and implemented an interface for BLAST application to run on the Grids. This plug-in, without any modifications, can be also used as an element of more sophisticated workflow tasks. Building that kind of tasks will be possible with Expert Client application. The GridBean implementation was developed under the GPE4GTK project [10] and has been tested with the binary distribution, called GPE-Lite (release 1.0.0), available at SourceForge website [13]. For testing the BLAST application has also been used (version 2.2.13) available at the NCBI website [12].

5   Conclusions and Future Work

In this paper we have presented plug-in designed for being currently developed GPE framework. BLAST GridBean will allow many bioinformatic scientists to use Grid technology instead of usually overloaded NCBI servers. Another advantage of Grid middleware, that we have mentioned above, is easy implementation of plug-ins. All the developers have to worry about is the graphic interface displayed to the user and job specification to run on computing center's side. Moreover it is easy to extend standard results presentation which usually is in text format. Based on this output data we can extract information to present it in specific graphical way. Indeed, it is the object of our current work to add graphical presentation of BLAST application results (i.e. alignments) using BioJava package [4].

References

1. Altschul, S., Gish, W., Miller, W., Myers, E.W., Lipman, D.: A Basic Local Alignment Search Tool. Journal of Molecular Biology 215:403–410 (1990).
2. Altschul, S., Karlin, S.: Applications and Statistics for Multiple High-Scoring Segments in Molecular Sequences. Proceedings of the National Academy of Sciences 90:5873–5877 (1993).
3. Bedell, J., Korf, I., Yandell, M.: BLAST. O'Reilly & Associates (2003).
4. BioJava project website: http://www.biojava.org/wiki/Main_Page.
5. Camp, N., Cofer, H., Gomperts, R.: High-Throughput BLAST. SGI White Paper (1998).
6. Foster, I.: What is the Grid? A Three Point Checklist. Grid Today, Vol. 1(6), Argonne National Laboratory & University of Chicago (2002). Available at: http://www.gridtoday.com/02/0722/100136.html.
7. Foster, I., Kesselman, C. (Eds.): The Grid: Blueprint for a New Computing Infrastructure. Morgan Kaufmann Publishers (1999).
8. Foster, I., Kesselman, C., Tuecke, S.: The Anatomy of the Grid, Enabling Scalable Virtual Organizations. International Journal of Supercomputer Applications, 15(3):200–222 (2001).
9. Globus project website: http://www.globus.org.
10. GPE4GTK project website: http://gpe4gtk.sourceforge.net.

11. GridBean Cookbook. UniGrids project documentation (2005).
12. NCBI BLAST website: `http://www.ncbi.nih.gov/BLAST`.
13. Open Source projects website: `http://sourceforge.net`.
14. Ratering, R.: Grid Programming Environment (GPE) Concepts. Intel Corporation, GPE documentation (2005).
15. Romero, P., Wagg, J., Green, M.L., Kaiser, D., Krummenacker M., Karp, P.D.: Computational prediction of human metabolic pathways from the complete human genome. Genome Biology 6:R2 (2004).
16. UNICORE project website: `http://unicore.sourceforge.net`.

Job Management Enterprise Application

Thomas Soddemann

Rechenzentrum der MPG (RZG), Institut für Plasmaphysik, Boltzmann-Str. 2,
85748 Garching, Germany
soddemann@rzg.mpg.de

Abstract. This paper describes the development of a Job Management
Enterprise Application (JMEA) which was developed by the DEISA material science and plasma physics joint research activities. It is capable
of submitting jobs to a UNICORE server infrastructure and managing
them. Since it is a Java EE application, it can be used by multiple users
concurrently. Furthermore, it prefetches and caches request results in
order to able of responding as quick as possible to client requests. In
addition to normal user credentials it also supports the use of proxy
credentials and explicit trust delegation.

1 Introduction

The Distributed European Infrastructure for Supercomputing Applications
(DEISA) [1] is a consortium of leading national supercomputing centers that currently deploys and operates a persistent, production quality, distributed supercomputing environment with continental scope. The purpose of this FP6 funded
research infrastructure is to enable scientific discovery across a broad spectrum
of science and technology, by enhancing and reinforcing European capabilities
in the area of high performance computing. This becomes possible through a
deep integration of existing national high-end platforms, tightly coupled by a
dedicated network and supported by innovative system and grid software.

The DEISA supercomputing grid is a European research infrastructure resulting from the integration of national High Performance Computing (HPC) infrastructures. This integration of national resources – using modern grid technologies
such as UNICORE [6] – is expected to contribute to a significant enhancement
of HPC capability and capacity in Europe.

DEISA is structured as a layer on top of the national supercomputing services, and coexists with them. This infrastructure addresses the computational
challenges that require the coordinated action of the different national supercomputing environments and services for both efficiency and performance. DEISA
provides scientific users with transparent access to a European pool of computing
resources. The coordinated operation of this environment is tailored to enable
new, ground breaking applications in computational sciences.

Eleven partners contribute currently to the DEISA infrastructure with their
top level supercomputers: BSC, Spain; CINECA, Italy; CSC, Finland; ECMWF,
UK; EPCC, UK; HLRS, Germany, IDRIS, France; FZJ, Germany, LRZ, Germany;

W. Lehner et al. (Eds.): Euro-Par 2006 Workshops, LNCS 4375, pp. 254–263, 2007.

RZG ,Germany; SARA, Netherlands This heterogeneous grid of super-computers includes of the most recent systems from leading vendors (IBM – PowerPC970, Power 4, 4+, 5, SGI – ALTIX, NEC – SX8).

Science Gateways, Portals and Web Service interfaces, are crucial for enhancing the user's adoption of sophisticated supercomputing infrastructures, by hiding from them the complexities of the computational environment. This extends up to the point that users make in their view direct use of an application. The choice of resource utilization is completely left to the infrastructure providing access to this application. So the portal solutions play the role of an application service provider (ASP).

The DEISA joint research activities in material sciences and plasma physics were faced with the development of comfortable means of access to the DEISA resources for standard applications in their fields like CPMD [2] and CP2K [3] for material sciences, and TORB [4] for plasma physics, a so-called science gateway.

Within DEISA several options for job submission across Cluster boundaries exist, e.g., the Multi Cluster Load Leveler (MC-LL) allows submitting jobs from the command line to any of the connected Load Leveler clusters (but naturally not to non MC-LL sites). The middle ware service activity within the DEISA project chose to employ the UNICORE suite as the default job submission interface to all heterogeneous compute resources within DEISA. Hence any DEISA job submission portal solution should be able to interface the UNICORE infrastructure deployed in DEISA in order to submit jobs in behalf of its users.

The UNICORE suite consists essentially of three components and one local batch scheduling system interface implementation establishing the connection to resource management systems like MC-LL and others. On the server side the central component is the Network Job Scheduler (NJS). It takes care of job submission, job management as well as file transfers and work flow execution. A gateway is the central entry point from the client perspective and several NJSs can be connected to it. A grid infrastructure relying on UNICORE can have more than one gateway e.g. for an increased fault tolerance. A user typically employs a rich client application (the UNICORE Client) to connect to several NJSs via a single gateway (see Fig. 1).

In the case of a portal application the job submission and management part needs to be implemented by an appropriate interface component. This components has to be able to carry out all tasks usually performed by the UNICORE client. Ideally, this interface is as general as possible in order to be able to deal with resource management systems besides UNICORE. The next section describes the requirements for such general job submission and management interface component.

The UNICORE suite offers a client library, the Arcon library, implementing an API which offers most of the desired functionality in dealing with the UNICORE server side. Unfortunately, this client library suffers from some minor deficiencies which mainly affect its use in a multi user multi threaded environment. This will be discussed in section 3.1.

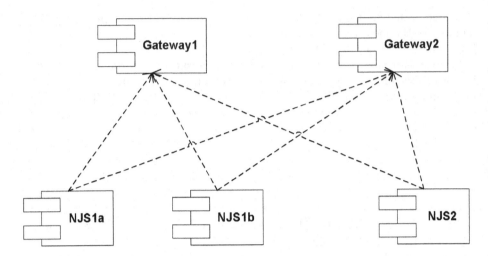

Fig. 1. Multiple gateways and NJSes deployed in a more sophisticated UNICORE infrastructure such as DEISA. Each NJS connects to all gateways. A user sees all NJSes regardless of the choice of the gateway.

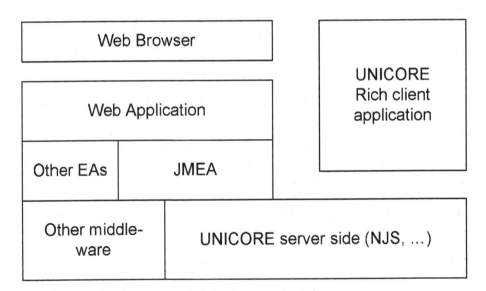

Fig. 2. Access to UNICORE within DEISA (without DESHL command line tooling). The Job Management Enterprise Application sits on top of the UNICORE server side infrastructure. A user accessing DEISA resources via the material science and plasma physics portal application makes direct use if JMEA for submission and handling. Furthermore, he is able to make use of other non-UNICORE related features like file system access. The UNICORE rich client application offers essentially the same functionality with respect to UNICORE. It speaks directly with the UNICORE server side. Pros and Cons of these approaches are discussed in detail in [5].

Section 3.2 describes a implementation of the Job Submission and Management interface described in section 2. It is implemented in form of a J2EE [10] version 1.4 compliant Job Management enterprise application developed within the DEISA project. It circumvents most of the problems described in section 3.1 and provides in addition functionality well beyond those of an ordinary client library. It is being used in the DEISA material science and plasma physics portal application.

The last sections provides conclusion and outlook of future developments.

2   Requirements for a Job Management Application

Ideally, a job management component for a portal application is able to connect to all grid job managers and local batch scheduling system. Adhering to this design principle, an API has to be used which is independent of any underlying job management component specifics. E.g. it should be able to work with a Globus [11] GRAM as well as with UNICORE NJS.

Hence, we identified a set of operation which we think are applicable to all kinds of resource management systems. This particular choice was motivated by the APIs of several different resource management systems [7,8] and the job management API of GGF's SAGA research group specified in their strawman API document [9]. It contains: submit (submitting the job request, cancel (canceling a job request which is not being executed), delete (delete a finished job request), kill (kill a running job request), halt (halt a job which is being executed), resume (resume a previously halted job). In addition it should be possible to retrieve information about the available resources, information about the status of jobs owned by a particular user, and results from a finished job such as console output. It is evident that the implementation of the job manager has to be able to deal with the identity management of the underlying resource management system. Fig. 3 shows the JobManager interface definition. All previously mentioned methods have been integrated into the definition. Additionally the client can select the gateway machine (if necessary as in the case of UNICORE) and retrieve a list of all known virtual site, which are part of the resource infrastructure.

Furthermore, the application should be able to cope with different instances of the same or different job management implementation at the same time, e.g. UNICORE and Globus. Hence, in addition to the JobManager Interface, a Factory for the JobManager is needed.

In the following we motivate two further requirements based on our experiences with UNICORE. But they should also hold for other Grid and cluster middle ware such as the Globus Tool Kit. The the requirements are derived from the design principle of giving best user satisfaction by having shortest request to response times and letting the user/client select the level of sophistication.

The UNICORE NJS requires a considerable amount of time to process many of its requests, e.g. request for a job status. Since it is undesirable to have users wait for anything, the application should answer right away. It should even answer if the targeted NJS should be unavailable for whatever reason.

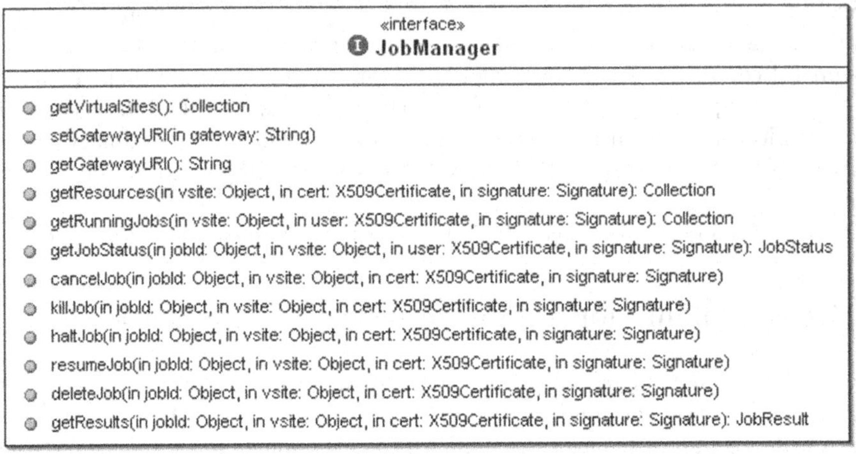

«interface»
🟦 **JobManager**

- getVirtualSites(): Collection
- setGatewayURI(in gateway: String)
- getGatewayURI(): String
- getResources(in vsite: Object, in cert: X509Certificate, in signature: Signature): Collection
- getRunningJobs(in vsite: Object, in user: X509Certificate, in signature: Signature): Collection
- getJobStatus(in jobId: Object, in vsite: Object, in user: X509Certificate, in signature: Signature): JobStatus
- cancelJob(in jobId: Object, in vsite: Object, in cert: X509Certificate, in signature: Signature)
- killJob(in jobId: Object, in vsite: Object, in cert: X509Certificate, in signature: Signature)
- haltJob(in jobId: Object, in vsite: Object, in cert: X509Certificate, in signature: Signature)
- resumeJob(in jobId: Object, in vsite: Object, in cert: X509Certificate, in signature: Signature)
- deleteJob(in jobId: Object, in vsite: Object, in cert: X509Certificate, in signature: Signature)
- getResults(in jobId: Object, in vsite: Object, in cert: X509Certificate, in signature: Signature): JobResult

Fig. 3. JobManager Interface definition

Ideally, the application should find out about available resources and cache those information without user interaction.

3   Architecture and Implementation of the Job Management Enterprise Application (JMEA)

The Job Management Enterprise Application (JMEA) is as its designed as a multi-component application. It offers interface components such as the Job-Manager for synchronous interaction and a similar component for asynchronous interaction. Furthermore, persistent classes store jobs, job results, and job stati. Service components take care of the things like the periodic querying of job stati and automatic retrieval of results.

In the following we discuss why we choose to develop a new UNICORE client library as a replacement for the current Arcon release. The second subsection deals with the details of the implementation of JMEA.

3.1   The Arcon Client Library

In the case of UNICORE a client library exists which could be used to implement the UNICORE job manager: the Arcon library.

The Arcon library is client library which allows application to interact with the UNICORE server side (gateway and NJS). Although it works perfectly well for a single user application with a single thread interacting with the UNICORE server side, it has some limitations when employed in a multi user – multi thread environment.

In order to avoid race conditions in multi threaded applications, one should omit static variables, unless they are used for communication between the threads and their access is synchronized. In the latter case, one has to make sure that synchronized access does not lead to a performance bottleneck.

In the case of the JobManager class of the Arcon library three static variables can be identified which can have an impact on the use in a multi user environment:

- outcome_dir which specifies the directory, streamed files will be stored
- buffer_size which reflects the buffer size for connections
- always_poll which tells if request are always asynchronous or not.

The static nature of those variables make it difficult to have per user settings which are desirable at least for the outcome_dir.

The abstract class *Connection* implements three static variables:

- keep_open which defines if the next connection is kept open after use.
- compression which tells if the transmission should be compressed or not.
- encrypt which defines whether the next retrieved connection should be encrypted communication or not.

While encryption may never been turned off, it is trivial to see that compression and keep open will definitely affect a multi user multi thread environment. Setting keep_open to true in one thread in order to retrieve a connection which stays open may result in an error, if that connection has been closed after first use, if a concurrent thread acquired a keep_open=false connection.

There are other classes which implement static variables such as *VsiteManager*. In our opinion, those do not limit the Arcon libraries use in a multi user environment.

The Arcon library implements its own proprietary logging mechanism which is not compatible with any of the existing logging mechanisms as log4j and the java.logging API. This makes it difficult to route logging messages to the applications or container logging files and impossible to influence the logging in the standard way.

Furthermore, in the implementation of some classes, e.g. in *Connection*, exceptions are used for flow control (bad style).

The main disadvantage at the time of the decision not to use the the original Arcon library was the missing support for Explicit Trust Delegation (ETD) [12] which allows a special user (agent) to formulate job request on behalf of a user by using his own set of credentials. In the mean time an EDT supporting Arcon library can be found at [13], which has not officially been released yet.

Furthermore, the Arcon library does not directly support Proxy Certificates which are favored in some environments over the use of ETD.

Based on these facts the decision was made to re-implement a client library, the new JobManager Library. Parts of the Arcon library's code base have been reused and those parts which were identified to be problematic have been replaced.

The new JobManager library now supports ETD, Proxy Certificates and is usable in a multi-user/multi threaded environment without limitations.

3.2 The Components

Fig. 4 sketches the main components of the job management enterprise application.

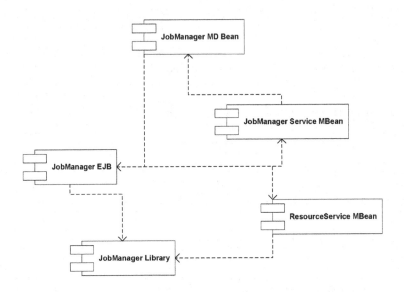

Fig. 4. Main components of the Job Management Enterprise Application

The JobManager Enterprise Java Bean (EJB) is the workhorse of the application. Its main purpose is to wrap calls to the JobManager library and handle persistent objects (see below). This EJB is implemented as a stateless EJB which allows the Application container to pool instances. The size of the pool can adjusted to optimize the balance the number of concurrent requests and responsiveness of the application.

The JobManager Message Driven Bean wraps the job update methods JobManager EJB and allows each of them to be executed asynchronously. These beans receive messages containing the AJO id of the job whose status is to be updated or performs a bulk update on all jobs for a particular user.

The JobManager Service Management Bean performs service tasks such as the periodic initiation of status update requests for running jobs. It employs the JobManager Message driven beans and the RunningJobs persistent object in order to achieve its goal.

Resources such as available software etc. are published in UNICORE on a per site basis. Hence the enterprise application queries the resources in configurable intervals and makes the result available to all of its users. The JobManager Resource Service Management Bean is responsible for querying the various NJS server instances and caching the query result. Furthermore, it provides information about the status of an NJS (available/not available). Since it has cached information, a temporarly unavailability does not mean that a client does not get information from it.

Several persistent classes have been used in order to store job requests, job stati, job results, and running jobs. Hibernate [14] is used for object relational (O/R) mapping purposes. Among the persistent classes are the JobStatus class, JobResult class and the RunningJobs class. JobStatus contains all status information available from UNICORE. In a similar way the results of finished jobs including stdout and stderr are stored in objects of the JobResult class. Objects of the RunningJobs class contains information about all running jobs for a particular user who is known to the enterprise application.

As mentioned in sec. 2, especially the UNICORE NJS requires some time to process certain requests. To circumvent this behavior and give an answer to the client side as quick as possible, information such as the job status, which obviously can change (e.g. from queued to running to finished), is polled periodically and cached by the application. The JobManagerService MBean operates a timer service which periodically initiates the polling by sending messages to the Job-Manager Message Driven Beans. These utilize instances of the JobManager EJB in order to retrieve and cache the required information from the NJS. When a client asks the JobManager EJB about such information it answers by querying the cache. A similar conecept is applied for the caching of a site's resource information.

Generally, requests should be grouped, if possible. Refreshes should be performed at suitable intervals. This shortens the time to a response on client requests for information using the application. Only information which need a special user authorization have to be retrieved separately, e.g. asynchronously after the user/client has authenticated himself to the application and provided the necessary credentials so that the application is able to act on the user's behalf. If the concept of explicit trust delegation (ETD) is enabled in the targeted UNICORE deployment, this task is trivial, since the NJS checks the authorization.

3.3   Security

In the current version of the enterprise application different security models are in place. The enterprise application itself can be protected by employing the container's security mechanisms. Once authenticated and authorized, a client can access all methods of all EJBs deployed in this application. There are currently no priviledged methods which need role based access restrictions. Access to the NJS via the enterprise application can be achieved in three ways of authentication by either employing

- the concept of Explicit Trust Delegation [12],
- Proxy credentials [15], or
- using the user credentials,

UNICORE is able to deal with one level of proxy certificates if the NJS check_signers property has been switched on. In that case the client needs to provide a *Signature* object for most of the operations which result in an interaction with a NJS. The JobManager library act in this case as a mediating NJS.

In the case of Explicit Trust delegation, it is not necessary for the client to provide a *Signature* object. The JobManger library here is used in the Agent mode, and signs all request with as a agent while setting the user attribute fields in the AJOs accordingly. The only thing needed in both cases is the X509 certificate.

In the case of ETD it should in principle suffice to provide a X500 DN of the user. This would simplify things, if the certificate is not available to the JobManager application. In discussions with the NJS authors it became clear that the certificate is used to identify users in cases where different users have the same DN. Nevertheless, in our opinion, only certificate authorities should be used which do not reuse their DNs or have overlapping name spaces. E.g. CAs, which are members of the EUGridPMA, qualify for that.

4 Conclusion and Outlook

The Job Management is currently in production use for the materials science and plasma physics portal of the according DEISA activities. In the mean time extension plans towards an integration of WS-GRAM are made. Furthermore, JMEA will be used in the implementation of the UNICORE connector for Grid-SAM [16]. This will allow users of the OMII [16] and hence users of the AHE [17] to submit jobs to the DEISA infrastructure. The reader might have missed a treatment of file transfer. Currently, file transfers are not implemented in JMEA itself (apart from the fact, that files can be embedded in job objects). An additional file management application is used to transfer file with appropriate mechanisms. Since JMEA will primarily be used in connection with web applications and web services, transfers of large volume data using HTTP/HTTPS as the transport layer in form of a multi-part HTTP request or embedded in an XML document is for performance reasons not advisable.

The Arcon library could easily be extended to integrate seamlessly into a multi threaded – multi user environment and a part of this work has already been incorporated in the unreleased version [13]. This would certainly help to build third party applications on top of UNICORE which do not rely on the UNICORE rich client.

A few ideas came into the authors' minds when working on JMEA. These could also enhance the integrability of UNICORE into third party applications. In some applications connectors to UNICORE have easy access to the *Principal* class of the user, but often accessing the user/client certificate or even the whole certificate chain is everything else than trivial. Hence, *X500Principal* object should be used rather than the whole certificate in the class *UserAttributes*.

Furthermore, there are requests which do not necessarily need to be performed by a user. E.g. requests for resource information could be performed by an ETD agent as well. Result could then be handled by the agent itself and e.g. delivered to all users. This would reduce the number of request to make to an NJS. But currently an agent is not allowed to make such a request. This should be changed in our opinion.

References

1. DEISA, Distributed European Infrastructure for Supercomputing Applications, http://www.deisa.eu
2. CPMD, http://www.cpmd.org
3. CP2K, http://cp2k.berlios.de
4. Hatzky R, Tran TM, Knies A, Kleiber R, Allfrey SJ. Phys.Plasmas 2002; 9: 898.
5. Thomas Soddemann, Concurrency Computat.: Pract. Exper., DOI 10.1002/cpe
6. UNICORE, http://unicore.sf.net/
7. http://www-03.ibm.com/servers/eserver/clusters/software/loadleveler.html
8. http://gridengine.sunsource.net
9. A. Merzky, et al., https://forge.gridforum.org/sf/docman/do/downloadDocument/projects.saga-rg/docman.root/doc12183
10. Java 2 Enterprise Edition
11. The Globus Toolkit, http://www.globus.org/
12. D. Snelling, et al., http://www.fujitsu.com/downloads/MAG/vol40-2/paper12.pdf
13. Unreleased EDT version of the arcon library, http://fisheye1.cenqua.com/browse/unicore/unicore/optional/arconclient
14. Hibernate http://www.hibernate.org/
15. Proxy Certificate RCF3820
16. Open Middleware Infrastructure Institute, http://www.omii.ac.uk/
17. Application Hosting Environment, http://www.realitygrid.org/AHE/

UNICORE Deployment Within the DEISA Supercomputing Grid Infrastructure

Luca Clementi[1], Michael Rambadt[2], Roger Menday[2], and Johannes Reetz[3]

[1] CINECA
Via Magnanelli 6/3,
40033 Casalecchio di Reno, Italy
l.clementi@cineca.it
[2] Central Institute for Applied Mathematics
Forschungszentrum Jülich GmbH
D-52425 Jülich, Germany
{m.rambadt,r.menday}@fz-juelich.de
[3] Garching Computing Centre of the Max Planck Society
Max-Planck-Institute for Plasma Physics
D-85748 Garching, Germany
johannes.reetz@rzg.mpg.de

Abstract. DEISA is a consortium of leading national supercomputing centers that is building and operating a persistent distributed supercomputing environment with continental scope in Europe. To integrate their resources, the DEISA partners have adopted the most advanced middleware and applications currently available. The consortium decided to embrace UNICORE as a job submission interface for the DEISA grid infrastructure. UNICORE is the foremost European grid technology able to hide the complexity of the underlying resources providing a user-friendly graphical user interface for job submission. This paper presents the deployment solution and strategies implemented by DEISA in order to adapt UNICORE for their infrastructure.

Keywords: UNICORE, computational grid, middleware deployment, interoperability.

1 Introduction

The DEISA project [1] started in May 2004 with the goal of providing a persistent and production quality, distributed supercomputing environment. The members of the consortium want to improve the level of exploitation of their systems and, at the same time, to provide a higher Quality of Service to the users, being able to offer a larger joint resource pool [9]. When building such an infrastructure, the DEISA partners considered several applications and middleware technologies that are providing the functionalities necessary to integrate their high-performance computing systems.

The DEISA consortium decided to use UNICORE to establish a grid infrastructure. UNICORE is one of the leading grid middleware used in production in several supercomputing centers. It hides the complexity of the underlying systems and

W. Lehner et al. (Eds.): Euro-Par 2006 Workshops, LNCS 4375, pp. 264–273, 2007.

architectures and it provides a single sign-on mechanism based on X.509 certificates from a Public Key Infrastructure (PKI).

The DEISA partners adapted and customized the UNICORE architecture according to the specific DEISA requirements. This paper presents the deployment strategies adopted to incorporate UNICORE into the DEISA grid. The second and the third chapters describe the DEISA infrastructure and the UNICORE architecture. The fourth chapter presents a detailed analysis of the solutions and the adaptations put in place by the partners in order to deploy UNICORE without interference with their local policies. To conclude, the fifth chapter presents a summary and considerations regarding our experience with UNICORE.

2 DEISA

DEISA (Distributed European Infrastructure for Supercomputing Applications) is a consortium of leading national supercomputing centers that is funded by the 6th European Framework Program. It deploys a production-quality supercomputing environment by exploiting the grid paradigm.

The DEISA infrastructure has two levels of integration. The inner level comprises strongly coupled clusters running IBM AIX on IBM POWER systems located at CINECA, CSC, FZJ, IDRIS, and RZG. Due to the same operating system and batch scheduler in common, these coupled clusters establish a *homogenous* super-cluster. An outer level of *heterogeneous* supercomputer clusters comprises by the SGI ALTIX Linux cluster located at SARA, the IBM PowerPC Linux system at BSC, the SGI ALTIX Linux at LRZ [1, 2], plus other resources provided by EPCC and ECMWF.

All the DEISA sites are linked together via a dedicated network provided by GEANT and National Research and Education Network providers (NREN). Beside the provision of high-performance compute facilities, the DEISA consortium offers also services such as help desk, documentation, technical and scientific workshops.

2.1 Infrastructure Overview

To achieve a higher level of interoperability between the different resources, the partners decided to harmonize their user management systems and to establish a DEISA user administration system by deploying a distributed set of LDAP servers [3] used to propagate information about DEISA users from the user's home site to all the partner sites. A standardization of the naming schema for DEISA users, and the assignment of site-specific ranges of UIDs and GIDs, ensures that DEISA user accounts are replicable on every system belonging to the DEISA infrastructure.

A user who wants to use DEISA resources needs to apply for an account only at his home site. The user record information (user name, UID, GID, the subject of his certificate, etc.) propagates via LDAP from his home site to all the other DEISA sites.

It is possible to identify three types of users for every DEISA site:

1. Internal users: they have an account only on a single resource and their user records are not published via the DEISA LDAP servers.
2. Local DEISA users: as home site users, they belong to the DEISA site locally, and their user records are published via the DEISA LDAP server at their site. Hence, their user accounts are automatically replicated on all the other DEISA resources.

3. External DEISA users: they belong to other DEISA sites, but their user records have to be imported from the DEISA LDAP servers at their home site in order to replicate their accounts on the local systems.

All the resources belonging to the inner level of the DEISA infrastructure have been integrated also by means of a shared file system. The consortium has decided to use GPFS-MC (General Parallel File System-Multi Cluster) [4] to achieve a transparent high-performance data access over the Wide Area Network. All the shared instances of the GPFS-MC are mounted on specific paths beginning with /deisa/<site acronym>, and so they are accessible from all the DEISA sites in the same manner.

Finally, a grid-enabled version of the LoadLeveler Batch Scheduling System [5] from IBM has been adopted for the intra-cluster scheduling of jobs. This product allows that a job submitted, e.g., by a CINECA user to CINECA's IBM P5 cluster can be routed to another DEISA site, depending on the resource requirements of the job and the availability of appropriate resources at the other DEISA sites.

Thank to the shared file system and to the distributed user administration system, migrated jobs can be executed under the same UID used on the user's home site cluster; the ownership of files on the shared file system needs not to be translated.

Users can access DEISA resources via UNIX shell, UNICORE, or Web Portals. DEISA partners decided to adopt the shell access because it is still the most common user interfaces for UNIX systems, and because it is useful for debugging applications during the development phase. On the other side, UNICORE provides an abstract view of the underlying system with its powerful Graphical User Interface, and the single sign-on mechanism simplifies the access to the distributed DEISA resources.

In the next chapters, the integration of UNICORE with the other components of the DEISA infrastructure is explained in more detail.

3   UNICORE

UNICORE (UNiform Interface to Computing REsources) provides a seamless interface for preparing and submitting jobs to a wide variety of heterogeneous distributed computing resources and data storages. It supports users for running scientific and engineering applications in a heterogeneous Grid environment.

The UNICORE software has been developed in the UNICORE and UNICORE Plus [6, 12] projects funded by the German Ministry of Education and Science (BMBF) until the end of 2002. After that, its functionalities and its robustness were enhanced within the EU-funded projects EUROGRID [7], OpenmolGrid [8]. Since 2004, several supercomputing centers are employing UNICORE in production.

In UNICORE every job represented by Java based abstract job formulation, the so called Abstract Job Object (AJO). This gives the user the possibility to prepare jobs on an abstract level without having to know deep details of the target system. With the abstract formulation, the job can be submitted to different target architectures running different batch schedulers without significant changes.

3.1 UNICORE Components

UNICORE is designed as vertically integrated three-tier architecture. It provides client and server components. The server-side consists of the Gateway, Network Job Supervisor (NJS) including an Incarnation Database (IDB), a UNICORE User Database (UUDB), and the Target System Interface (TSI). All components (except the TSI) are written in Java allowing to install UNICORE on a large variety of operating systems.

3.1.1 UNICORE Client

The UNICORE client GUI is used for preparation, submission, monitoring, and administration of complex multi-site and multi-step jobs. It provides the user with an extensible application support, resource management of the target system and an integrated security mechanism.

Every submitted request (AJO) is signed using the personal X.509 certificate of the user. Thus, other UNICORE server components can perform authentication and authorization relying on the PKI in use. The client allows also performing data management and transfer provided by an intuitive GUI.

3.1.2 Gateway

The Gateway is the site's point of contact for all connections relative to a UNICORE site (Usite). It accepts SSL connections from clients and one or more NJSs, but only if the incoming certificate is signed by a trusted Certification Authority (CA). Moreover, it verifies if received AJOs have been signed with trusted and valid certificates. If the authentication is successful, the AJO is redirected to the corresponding NJS, otherwise it is rejected.

3.1.3 Network Job Supervisor (NJS)

The Network Job Supervisor (NJS) operates as a UNICORE scheduler and is responsible for the virtualization of the underlying resources. It receives/sends AJOs from/to the Gateway, translates them into concrete instances and sends them to the target system component, called Target System Interface (TSI) (see next paragraph). The NJS dispatches jobs to a dedicated target machine or cluster (Virtual site, Vsite), and handles dependencies and data transfers for complex workflows. It transfers the results of executed jobs from the target machine and forwards them via the Gateway to the UNICORE client.

The abstract definition of the Job is translated to a concrete job in the NJS with the help of the Incarnation Database (IDB). The IDB contains all target system specific information regarding computing resources typology and availability of applications. Therefore, each NJS has a dedicated IDB that describes its specific target system.

Finally, the NJS implements the UNICORE security model for user authorization. All public user certificates are stored in the UNICORE User Database (UUDB) and they are mapped with an account existing on the target system. Every time the NJS receives an AJO, it checks if the signer's certificate is present in the UUDB, and on success, the job is forwarded to the target system and assigned to the corresponding user account.

3.1.4 Target System Interface (TSI)

The TSI running on the target machine is the interface to the batch scheduler. It comprises a set of Perl libraries that implements the specific target system commands for job submission, status query, file handling, etc. A variety of TSI implementations are available for different batch schedulers and operating systems, e.g., LoadLeveler under AIX, LSF, PBS-Pro, and CCS.

4 UNICORE and DEISA Infrastructure Integration

The main design guidelines for the DEISA grid can be summarized as follows:

- The grid middleware has to present the different target architectures and resources with a seamless view hiding all the underlying complexity
- Users need to be able to address the various DEISA resources without perceiving the complexity behind them
- Grid middleware needs to provide reliability to software and hardware failures
- As far as possible, the deployment of the DEISA grid infrastructure should not interfere with the local site policies and security requirements. The middleware must be easily adaptable to the local site procedures and policies.

The following paragraphs demonstrate how UNICORE respects these principles.

4.1 UNICORE Deployment for DEISA

According to the conventional UNICORE deployment pattern, every site represents a separate Usite: every NJS, located at one site, connects only to its site Gateway. A virtual organization (VO) could provide several Gateways, whereas each represents the entry point to a separate site. The advantage of this pattern is a better scaling due to the decoupling of the UNICORE deployment at the VO member sites. On the other hand, the composition of the VO is not pervasive in this case, and has to be defined on the client-side using a list of references to all the member gateways.

Due to the limited amount of DEISA members and an efficient cooperation of all the DEISA sites, the consortium decided that all the UNICORE Gateways are to be connected to all NJSs and vice versa. Exploiting the UNICORE dynamic Vsite registrations feature a fully meshed UNICORE infrastructure was built up to allow for the job submission via all Gateways to all target systems. Figure 1 depicts the DEISA UNICORE infrastructure showing five DEISA sites as an example. As a result the DEISA UNICORE infrastructure provides a distributed set of DEISA access points. Each represents an entry point to the whole infrastructure. Preferentially, DEISA users shall use the Gateway at their home site for submitting jobs to any DEISA target system.

The benefits of the deployment pattern shown, where all the NJSs are *jointly connected* to one or more gateways are:

- Pervasive visibility of all the available NJSs in DEISA at every single Gateway
- As an expression of corporate identity, all the DEISA sites are committing to maintain the persistent availability and accessibility of their DEISA resources via any Gateway of the DEISA partner sites.

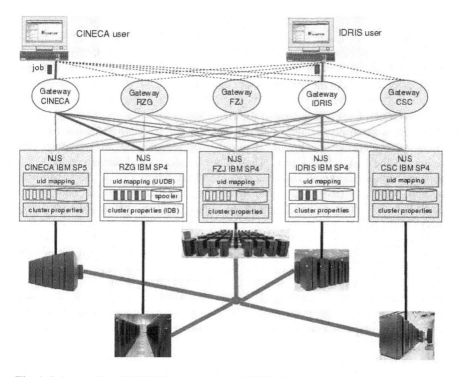

Fig. 1. Schema of the UNICORE deployment at DEISA for the homogeneous infrastructure

- On the client-side, DEISA users need only to validate the issuer of the server certificate of their home site Gateway. It is usually not necessary to import certificates of trusted CAs from other countries to authenticate other Gateways.
- The DEISA NJSs can be enabled to communicate with each other allowing to employ implementations of an alternative file transfers mechanism (AFT). This is useful particularly if the file system of a site does not share GPFS-MC.

The benefits of the *fully meshed deployment* pattern with multiple gateways are:

- Reliability of service: if the DEISA gateway of the user's home site is down or heavily loaded, other DEISA gateways can substitute it completely. In order to access another Gateway, the user must of course import the certificate of the trusted issuer of the Gateway's certificate into his client's truststore.
- Load balancing on the multiple Gateways: provided an appropriate information system is available, it is possible to select a gateway according to the actual load or number of the available DEISA gateways.

Additional UNICORE gateways can be added to the infrastructure in cooperation with the DEISA partners. As a proof of concept, the DEISA UNICORE infrastructure is now composed of 10 sites, and every site is allowing access to its resources through a UNICORE Gateway and NJS, configured as explained in this paragraph.

This fully meshed infrastructure remains firewall friendly and flexible in terms of deployment. Each site has to open only two ports in the site firewall (DMZ or intranet firewall); one is the Gateway port that has to be opened for the whole outside world for inbound connections. While the NJS port needs to be opened only for inbound connections coming from the other DEISA Gateways.

4.1.1 UNICORE User Management for DEISA

In order to get the user and server certificates needed for UNICORE a set of root Certification Authorities (CA) are required. DEISA decided to use only CAs accredited at the EUGridPMA [10]. These CAs issue user and server certificates in compliance with the minimal requirements of the EUGridPMA [11]. Users who want to submit jobs via UNICORE need to obtain a DEISA user account and a EUGridPMA compliant personal certificate signed by their national EUGridPMA member CA.

As an improvement of the UNICORE authorization system the DEISA consortium requested to implement a modification in the UUDB internal management. The standard UUDB implementation maps the complete public part of the user's certificate to a user's account, while the modified DEISA UUDB checks only whether the Distinguished Name (DN) of the certificate used to sign the AJO is present in the UUDB. With support of the UNICORE developers, the implementation of the UUDB authorization mechanism has be adapted accordingly.

4.1.1.1 Security implications. For the DEISA infrastructure, checking only for the DN it is not a security risk since DEISA relies on the joint PKIs of the EUGridPMA member CAs that ensure the uniqueness of the DN within their joint domain. Therefore, a DEISA UNICORE Gateway will never authenticate a certificate issued by a suspected CA.

The benefits of the modified implementation of the UUDB are that

- the DEISA user management system needs to store and exchange only the DN
- there is no need to update the content of the UUDB when a user certificate has been reissued, because the DN of the reissued certificate remains unchanged.

Furthermore the DEISA consortium is considering the adoption of some Globus components (GRAM and GridFTP) [15] for its infrastructure which are based on the Grid Security Infrastructure (GSI) security protocol [16]. GSI authentication is based on PKI and the credentials needed for the authentication are the DN of the user's certificate with the corresponding user name. Hence, the integration of the GSI authentication system with the DEISA infrastructure will be straightforward due to the similarities with the current UNICORE UUDB.

Users can also access DEISA resources using an interactive shell where the authentication is performed by means of username and password. In order to avoid the distribution of users password hash, the DEISA partner decided not to publish this information on the DEISA LDAP servers. Therefore, DEISA users are allowed for interactive access only on the home site system. By this approach, no security sensible user information is published on the DEISA LDAP servers.

4.2 Integration of UNICORE with DEISA Batch Scheduling Systems and GPFS Multi Cluster

A key feature of the DEISA infrastructure is the capability of migrating submitted job to other clusters using LoadLeveler (LL). Migrated jobs can use the DEISA wide shared file system to access data, or write output and debugging information to the user directory allowing a user to monitor the execution status of his application. In order to keep this functionality several modifications have been performed to its standard configuration.

4.2.1 Adaptations Regarding LoadLeveler

To accomplish this goal all the LL instances running on the homogeneous POWER/AIX clusters needed to be configured in a coherent way. This was achieved defining a set of mandatory parameters that have to be specified by a user to describe the resource requirements of his job to be executable on DEISA resources. These parameters are *total tasks*, *threads per task*, *wall clock limit*, *data memory limit*, *stack memory limit*, and a keyword notifying that the job has to be handled by the underlying queuing system as an explicit DEISA job.

The DEISA partners agreed that this set of parameters is adequate for all the cluster configurations and Batch Scheduling Systems currently employed at all the DEISA sites (LL, PBSPro, LSF, and Torque).

The standard version of UNICORE allows for specifying mainly up to six, partly different parameters of an abstract job. Since these parameters can not be mapped directly to those identified to be relevant for DEISA, it was decided to use only four of the original parameters (*total tasks*, *threads per task*, *wall clock limit*, *data memory limit*) and to specify the remaining two parameters as environment variables in an adequate way. UNICORE allows specifying additional environment variables for each submitted job. To take the additional environment variables into account, the TSI, in particular the script that creates the submission command for the LL, has been easily adapted.

4.2.2 Adaptations Regarding GPFS-MC

When submitting a job, UNICORE creates a temporary working directory (called USPACE) at the target system where, among others, batch scripts, input, output and error files are placed. At first, DEISA partners did not require a common path for USPACE. The USPACE was simply located on a site local file system. However, when submitted UNICORE jobs are to be migrated to other clusters by the Multi-Cluster LoadLeveler, the USPACE at the originating cluster needs to be transparent.

As a solution, the different partners have decided to configure UNICORE in order to use a common USPACE path on the GPFS-MC. In this way, UNICORE jobs submitted to the homogeneous super-cluster have always a consistent reference to the USPACE and thus to the files needed by the NJS at the originating site for monitoring the job status and fetching the output. The implementation of this solution required the modification of some TSI scripts.

4.3 The Final Picture

The figure 2 shows how the various components of the DEISA Grid interact. GPFS-MC provides a common shared storage resource available for the different DEISA resources. The Multi-Cluster LoadLeveler enables to address local and remote computational resources. The interactive usage of a Shell provides access to the local instance of LL while UNICORE allows additionally to access remote instances of LL.

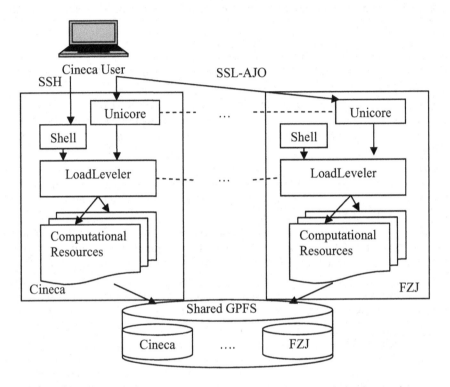

Fig. 2. A general schema of the DEISA Infrastructure representing only two sites

5 Conclusions

UNICORE has proven to be a suitable solution for a complex environment like the DEISA infrastructure. Its high customization level has been shown to be essential for its deployment in accordance with DEISA requirements. Its open source characteristic allows the DEISA partners to modify the default behaviors. Its wide adoption and its long-standing production usage guarantee the Quality of Service that is required for the DEISA infrastructure.

The functionalities of UNICORE within the DEISA consortium have been demonstrated at the IST conference of 2004 where a demonstration showed the seamless usage of the homogeneous resources by means of UNICORE. The number of DEISA users is rising constantly and the commitment to serve their aims with

UNICORE will remain part of the consortium goal. Moreover, UNICORE will play a central role for the integration of the heterogeneous clusters considering its capability to support various system platforms and batch schedulers.

Finally, GRIP [13] and UniGrids [14], two EU funded research projects have developed UNICORE extensions for interoperability with Globus [15]. These enhancements allow DEISA to integrate Globus components, such as GridFTP and GRAM, into the DEISA UNICORE infrastructure.

Acknowledgments. A special thank to all the DEISA partners that contributed their solutions and ideas to the current deployments and configurations.

References

1. http://www.deisa.org - Distributed European Infrastructure for Supercomputing Appications
2. DEISA Primer. DEISA Consortium, Version 1.2 (02/2006)
3. K. Zeilenga, and OpenLDAP foundation: Lightweight Directory Access Protocol (LDAP): Technical Specification Road Map. RFC 4510, (06/2006)
4. Frank Schmuck, Roger Haskin: GPFS: A Shared-Disk File System for Large Computing Clusters. Proceedings of the FAST, Monterey, (01/2002)
5. LoadLeveler for AIX 5L and Linux V3.3.1 Using and Administering. IBM (11/2005)
6. D. Erwin (Ed.): UNICORE Plus Final Report - Uniform Interface to Computing Resources. Forschungszentrum, Julich, (2003)
7. K. Nowinski, B. Lesyng, M. Niezgódka, P. Bala: Project EUROGRID. Proceeding of the PIONIER 2001, Poznan (2001)
8. S. Sild, U. Maran, M. Romberg, B. Schuller, E. Benfenati: OpenMolGRID: Using Automated Workflows in GRID Computing Environment. Proceedings of the European Grid Conference 2005, Amsterdam, (02/2005)
9. I. Foster, C. Kesselman (Eds.). The Grid 2: Blueprint for a New Computing Infrastructure. Morgan Kaufmann Publishers Inc. San Fransisco (2004)
10. www.eugridpma.org
11. Profile for Traditional X.509 Public Key Certification Authorities with secured infrastructure. EUGridPMAa, Version 4.0 (10/2005)
12. D. W. Erwin, D. F. Snelling: UNICORE: A grid computing environment. Proceedings of Euro-Par 2001, Springer, Machester (08/2001)
13. Michael Rambadt, Philipp Wieder. UNICORE - Globus: Interoperability of Grid Infrastructures. Proceedings of the Cray User Group 2002, Manchester (05/2002)
14. http://www.unigrids.org/ - Uniform Interface to Grid Services
15. I. Foster: Globus Toolkit Version 4: Software for Service-Oriented Systems. International Conference on Network and Parallel Computing, Springer LNCS, (2005)
16. R. Butler, D. Engert, I. Foster, C. Kesselman, S. Tuecke, J. Volmer, V. Welch: A National-Scale Authentication Infrastructure. IEEE Computer, 33(12):60-66, (2000)

Petascale Computational
Biology and Bioinformatics

Introduction

Craig A. Stewart

Workshop Chair

Multiple plans to create petascale computing environments have been announced. This workshop addressed what bioinformatics or computational biology applications can or should accomplish with such facilities, and what obstacles must be overcome in order to implement and use effective and important problems in the life sciences (biology, biochemistry, environmental sciences, etc.).

Five papers were presented at this workshop:

Progress Towards Petascale Applications in Biology: Status in 2006 by Craig A. Stewart, Matthias Mueller, and Malinda Lingwall examines current trends in computing power and explains the need for petascale computing in the life sciences.

Progress in Scaling Biomolecular Simulations to Petaflop Scale Platforms by Blake G. Fitch, Aleksandr Rayshubskiy, Maria Eleftheriou, T.J. Christopher Ward, Mark Giampapa, Michael C. Pitman, and Robert S. Germain describes some issues involved with scaling biomolecular simulations onto massively parallel machines, and examines what it will take to overcome the challenges of petascale computing.

Toward a Solution of the Reverse Engineering Problem Using FPGAs by Edgar Ferrer, Dorothy Bollman, and Oscar Moreno looks at the reverse engineering problem for genetic networks and proposes an efficient approach to finding a solution.

Two Challenges in Genomics That Can Benefit from Petascale Platforms by Catherine Putonti, Meizhuo Zhang, Lennart Johnsson, and Yuriy Fofanov addresses the computational challenges necessary to successfully examine the ever-increasing amount of biological data available, including the number of genomic sequences made publicly available.

High-Throughput Image Analysis on Petaflop Systems by Robert Henschel, Yannis Kalaizidis, and Matthias Mueller describes software developed to assist biologists with image analysis work, integrating high-performance computing systems into their workflow.

W. Lehner et al. (Eds.): Euro-Par 2006 Workshops, LNCS 4375, p. 277, 2007.
© Springer-Verlag Berlin Heidelberg 2007

Progress in Scaling Biomolecular Simulations to Petaflop Scale Platforms

Blake G. Fitch[1], Aleksandr Rayshubskiy[1], Maria Eleftheriou[1], T.J. Christopher Ward[2],
Mark Giampapa[1], Michael C. Pitman[1], and Robert S. Germain[1]

[1] IBM Thomas J. Watson Research Center, 1101 Kitchawan Road/Route 134, Yorktown
Heights, NY 10598, USA
[2] IBM Hursley Park, Hursley, Hursley SO212JN, United Kingdom

Abstract. This paper describes some of the issues involved with scaling biomolecular simulations onto massively parallel machines drawing on the Blue Matter application team's experiences with Blue Gene/L. Our experiences in scaling biomolecular simulation to one atom/node on BG/L should be relevant to scaling biomolecular simulations onto larger peta-scale platforms because the path to increased performance is through the exploitation of increased concurrency so that even larger systems will have to operate in the extreme strong scaling regime. Petascale platforms also present challenges with regard to the correctness of biomolecular simulations since longer time-scale simulations are more likely to encounter significant energy drift. Total energy drift data for a microsecond-scale simulation is presented along with the measured scalability of various components of a molecular dynamics time-step.

1 Introduction

IBM's Blue Gene project was announced in December 1999 with the twin goals of advancing the state of the art in all aspects of computer systems while building a petaflop-scale machine and of using the computational power enabled by this work to explore important issues in the life sciences. This paper describes some of the challenges and issues encountered by the Blue Gene application and science team in the course of creating a molecular simulation environment to both support our scientific goals and to facilitate the exploration of parallel algorithms and programming models suitable for massively parallel machines. The largest installation of the first member of the Blue Gene family, Blue Gene/L[13], is a 65,536 node system at Lawrence Livermore National Laboratory with a theoretical peak performance of 360 TFlop/second. Our application development efforts and simulation science within the Blue Gene project target the 20,480 node, 112 TFlop/s peak performance Blue Gene/L installation at the IBM Thomas J. Watson Research Center (BGW) which is currently the largest unclassified supercomputing facility in the world.

Of all the subfields of computational biology, molecular simulation is almost certainly the most mature in its ability to exploit high performance computing. Most of the biology-related work on the Blue Gene/L facility at Watson (BGW) has thus far been in that area, with projects targeting studies of protein folding mechanisms[5] and structural and dynamical studies of membrane proteins[18,16]. All of these projects share

W. Lehner et al. (Eds.): Euro-Par 2006 Workshops, LNCS 4375, pp. 279–288, 2007.

a requirement for very long time-scale simulations (microseconds) of modestly sized molecular systems (10,000-100,000 atoms). The need for long time-scale simulations drives requirements for both (strong) scalability and correctness that will be discussed below.

The original target architecture for the Blue Gene project[1] had characteristics (very small amount of memory per node, millions of processing elements) that drove a specific set of design goals for the Blue Matter application framework that we have developed as part of the Blue Gene project[8]. The design goals included:

- Running only the computationally intensive molecular dynamics core on the massively parallel Blue Gene platform to reduce the memory footprint of the code.
- Leveraging existing applications as much as possible for problem set-up and other non-performance-critical functionality.
- Separating the complexity of domain-specific aspects of molecular dynamics from the complexity of the parallel communications required. The goal was to allow exploration of parallel decompositions without requiring the involvement of the domain experts.

2   Experiences with Blue Gene/L

We believe that our experiences in developing the Blue Matter simulation code and in running simulations on BGW are relevant to discussions about biomolecular simulations on future peta-scale systems since the BGW facility already has a peak capability of over 0.1 PFlop/s. BGW is typically operated in partitioned fashion where most of the partitions comprise 4096 or 8192 nodes. The allocation and usage patterns of the BGW facility reflect the usual tradeoffs between supporting a range of projects, carrying out the ensembles of simulations required for scientific validity, maximizing overall throughput, and the drive to reduce the total time to solution for a single researcher or simulation. Using this resource, we have been able to run a number of large scale simulation experiments including

- 26 separate 100 nanosecond simulations of Rhodopsin in a membrane environment (44K atoms)[16].
- several microsecond-scale simulations of the same membrane protein system including a pair of simulations totaling 3.5 microseconds.
- several 700 nanosecond simulations of Lysozyme (41K atoms)[5].

and additional long time-scale simulations of a fast folding Lambda Repressor protein are currently underway. Although Blue Matter continues to speed up through 16,384 nodes on the systems being studied[11,10,9], these microsecond scale simulations typically use 4096 node partitions since this currently represents the best tradeoff between throughput and total time to solution. Large Replica Exchange[22] simulations running as a single MPI job on up to 8192 nodes[6] have also been run to obtain temperature dependent thermodynamic information about protein systems.

While the I/O bandwidth requirements of molecular dynamics are quite modest since the entire state of the system is represented by the positions and velocities of the particles in the system, the aggregate storage requirement is potentially quite large. Archival storage for the molecular simulation work is provided by a 500 TB capacity tape library which backs approximately 8 TB of disk storage being managed with Tivoli^{TM} Space Manager (hierarchical storage management).

3 Peta-scale Challenges

3.1 Molecular Simulation Validity

As the target time-scale for typical molecular simulations increases from tens of nanoseconds to microseconds or more, the stringency of the requirements on simulations will also increase. In particular, the permissible rate of total energy drift in constant energy, volume, and particle number (NVE) simulations will have to decrease as the length of the NVE simulations increases. An increase in total energy of the system will cause a rise in the instantaneous temperature of the system (defined by the kinetic energy) of the same order. It is useful to measure the energy drift relative to the average kinetic energy in the system (and actually to do so in units of temperature) to make the scale of the effect clear. For example, an energy drift of 7×10^{-2} K/nanosecond results in a an increase in total energy equivalent to 0.7 K over a 10 nanosecond simulation. This is quite small in comparison with biological temperatures on the order of 310 K, but the same energy drift in a 1 microsecond simulation would result in an increase of 70 K in the total energy which is no longer negligible.

One of the principal rationales for believing in the relevance of long term simulations is that for a symplectic integrator such as velocity Verlet[23], used to numerically integrate Hamiltonian systems, there exists a "modified" Hamiltonian whose exact (continuous time) dynamics at integer multiples of the numerical integration time-step coincides with the discrete dynamics generated by the symplectic integrator[17,2,24,19]. This modified Hamiltonian may be "close" to the original in the sense that it can be expressed as a formal expansion in powers of the time-step size about the original Hamiltonian. The existence of this modified Hamiltonian means that the trajectory computed by the numerical integrator should exactly conserve the total energy as computed by the modified Hamiltonian (up to numerical roundoff) and hence should approximately conserve the energy as computed by the original Hamiltonian. The popularity of various forms of Verlet integrators for molecular dynamics simulation is largely due to their simplicity and long term energy stability which stems from the symplectic property that these integrators possess[12].

In general, a computational scientist will want to use the largest time-step size possible consistent with "correctness" in order to maximize throughput. Other performance-critical simulation parameters affecting simulation accuracy and stability include the FFT mesh spacing for Particle-Particle-Particle-Mesh (P3ME) methods[3] and the force-splitting scheme and time-step ratios chosen for symplectic multiple time-stepping methods[21,25,26]. Determining the optimal parameters for simulations enabled by multi-teraflop and larger machines that involve billions or tens of billions of time-steps provides a considerable challenge. Figure 1 shows a plot of the change in the total energy in a simulation of a 43,222 atom system containing Rhodopsin running with a velocity Verlet integrator using a 2 femtosecond time-step where all heavy-atom to hydrogen bonds are constrained (eliminating the highest frequency vibrations from the system).

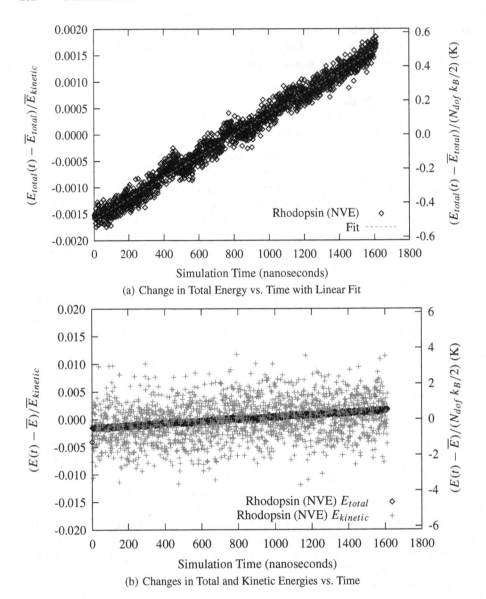

Fig. 1. Energy drift of NVE molecular dynamics simulation of Rhodopsin in a solvated membrane environment over a 1.6 microsecond run using a 2 femtosecond time-step

The energy drift measured by a linear fit to the data is about 6×10^{-4} K/ns where the left-hand axis shows the energy change as a fraction of the average kinetic energy and the right-hand axis expresses the change in energy as the change in instantaneous temperature that would result if all of the change were in the kinetic energy. For the parameters used in this production simulation, the total change in energy over 1.6 microseconds was slightly larger than 1 K. This is smaller than the fluctuations observed in the kinetic energy during the simulation as shown in Figure 1b and uncertainty of the temperature in the experimental data that we compare with. This time-step size was chosen based on experiences with shorter (10-100ns) simulations, but it is entirely possible that those estimates could have been too low. It should also be noted that the execution time required for a time-step is essentially independent of the choice of integration time-step size while the energy drift is a very strong function of time-step size. Therefore, longer simulations could be carried out with acceptably small energy drift simply by reducing the integration time-step size somewhat and our benchmarking data for the amount of wall clock time required per time-step would still be valid.

Using the normal system Hamiltonian makes it difficult to estimate the long term energy drift without very long simulation runs because of the short term energy fluctuations observed when using a discrete time integrator. Because of the computational expense involved, it has been impractical to carry out a systematic exploration of the tradeoffs between parameter choices such as time-step size and magnitude of the energy drift. In principle, such a study might have to be carried out for each new molecular system. In practice, a choice of parameters is made based on experience with shorter simulations, the drift is monitored as the simulation progresses, and the simulation would have to be rerun with a less aggressive choice of parameters if excessive energy drift were observed. Recently there have been results reported on the numerical estimation of the modified Hamiltonian from the simulation data[7] and this may allow more extensive explorations of the parameter space affecting tradeoffs between simulation quality and computational throughput without prohibitively large expenditures of computational time.

3.2   Performance and Scalability

It is likely that future peta-scale architectures will achieve their performance through massive concurrency (large numbers of CPUs per chip, massive parallelism). Given that this is the case, the application challenge for biomolecular simulations that require strong scaling will be considerable. Within Blue Matter, we have had to be very careful to root out any non-scalable operations from our implementation. As the scale of hardware available to us grew from a single 512-node prototype to the current 20 rack system we repeatedly went through cycles of identifying previously insignificant non-scalable operations that had to be eliminated.

Our current algorithms as implemented on Blue Gene/L can execute a time-step in fewer than 600,000 processor clock cycles (at 700MHz), including the processing associated with the global data dependency necessitated by the FFTs in the P3ME module. We have found that the velocity Verlet integrator which requires the P3ME operation to be carried out on every time-step enables us to run with very small amounts of energy drift in NVE simulations. If no significant increases in processor clock speeds are anticipated, then each order of magnitude decrease in time to solution will require each time-step to execute in a correspondingly smaller number of cycles. Since our scalability is now limited by the execution speed of the FFTs required for the P3ME method as shown in Figure 2, it is likely that investigations of alternative methods for treatment of the long range electrostatics and/or coarse-graining methods will be required to realize additional improvements to the current strong scaling results.

While the BG/L architecture is a relatively "pure" message-passing machine with two identical processing elements per node, each of which can participate in either communication-related activities or computation, there are other ways to deploy additional processing elements. For example, the use of additional specialized processors for DMA operations or communications could enable more overlap of communications and computation, but it isn't clear that this would increase the limits of scalability where data dependencies prevent further computation until communication operations complete.

4   Algorithmic Explorations

One way to place bounds on the potential scalability of an algorithm is to determine the amount of concurrency available in principle for that algorithm. As a concrete example, Table 1 enumerates the concurrency available in various portions of the molecular dynamics time-step iteration for a 43,222 atom simulation of Rhodopsin using the Particle-Particle-Particle-Mesh Ewald (P3ME) technique. The P3ME technique requires the evaluation of at least two three-dimensional FFTs on each time-step and development of a highly scalable distributed memory 3D-FFT[4] has been one of the key enablers of Blue Matter's current scalability. As shown in Figure 2, it appears that the three-dimensional FFT is the limiting factor of performance in the extreme strong scaling limit.

The other major component of the computational load in a typical molecular dynamics simulation comes from the finite-ranged pair interactions between particles. We have explored several different algorithms for parallelizing these operations, from simple replicated data decompositions[8,15], to a geometrically-based interaction decomposition with a minimal communication radius[11,14], and most recently, a set-based

Table 1. Degree of concurrency in computational modules within a single molecular dynamics time-step for a 43,222 atom membrane/protein system (Rhodopsin) using a $128 \times 128 \times 128$ mesh for P3ME. The system parameters were those used in production simulations[18,16]. The last column is the concurrency possible for that computational module based on the number of independent calculations required (assuming a "reasonable" level of granularity). The number of real-space pair interactions to be computed will actually fluctuate somewhat during the course of the simulation because of particle diffusion.

Stage	Major Computational Kernel	Independent Computation Count
Real-space Non-bond (9/1 Å cutoff/switch)	Pairwise forces (L-J and Ewald real-space)	9,113,514
Bonded	bond stretches, angle bends, torsions, Urey-Bradley	126,730
P3ME Meshing/Un-meshing	$4 \times 4 \times 4$ stencil	43,222
P3ME Convolution	3D Fast Fourier Transform (FFT)	16,384
Propagation of Dynamics	Verlet integration	43,222

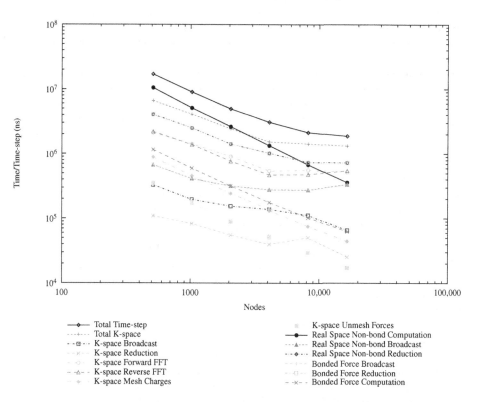

Fig. 2. This is a plot of the scalability of various components of the molecular dynamics time-step as a function of node count. The system is the same Rhodopsin membrane/protein system described in Table 1. There are data dependencies between some of these components and since we schedule modules on both CPUs of the BG/L node, some of the components are executing concurrently.

optimization technique that uses a geometrically derived heuristic as a starting point[9]. The most recent performance results demonstrate time-step execution times below one millisecond for a β-hairpin system and continued speedups through approximately one atom per node[9].

5   Conclusions

Experiences with scaling the Blue Matter biomolecular simulation application to run effectively on the 112 TFlop/s BGW system should be relevant to any efforts to run such codes on future petaflop-scale platforms because the design philosophy of Blue Gene/L required the kind of massive parallelism that is likely to be needed for such platforms. As the development of novel algorithmic techniques was required to realize improved time-to-solution for biomolecular simulations on Blue Gene/L, it is likely that significant additional innovation will be needed in order to continue to increase the time scales accessible via simulation. These innovations will almost certainly be related to the parallelization of the long range electrostatic interactions and may involve the adoption of alternative algorithms for the computation of those interactions such as multi-grid[20].

Even without additional algorithmic improvements, it is likely that increasing the system size studied (weak scaling) will enable effective use of peta-scale platforms to extend the accessible time-scales for those systems into the microsecond regime that Blue Gene/L has opened up for smaller systems ($< 100,000$ atoms). Also, the availability of peta-scale platforms will enable studies involving larger ensembles of long trajectories that can give improved sampling and allow the generation of statistical error estimates[16].

Acknowledgements. We would like to thank the members of the Blue Gene systems team including P. Heidelberger, A. Gara, M. Blumrich, J. Sexton as well as others who have made use of and contributions to the Blue Matter code over the course of its development, particularly Y. Zhestkov, Y. Sham, F. Suits, W. Swope, J. Pitera, A. Grossfield, and R. Zhou.

References

1. F. Allen, G. Almasi, W. Andreoni, D. Beece, B. J. Berne, A. Bright, J. Brunheroto, C. Cascaval, J. Castanos, P. Coteus, P. Crumley, A. Curioni, M. Denneau, W. Donath, M. Eleftheriou, B. Fitch, B. Fleischer, C. J. Georgiou, R. Germain, M. Giampapa, D. Gresh, M. Gupta, R. Haring, H. Ho, P. Hochschild, S. Hummel, T. Jonas, D. Lieber, G. Martyna, K. Maturu, J. Moreira, D. Newns, M. Newton, R. Philhower, T. Picunko, J. Pitera, M. Pitman, R. Rand, A. Royyuru, V. Salapura, A. Sanomiya, R. Shah, Y. Sham, S. Singh, M. Snir, F. Suits, R. Swetz, W. C. Swope, N. Vishnumurthy, T. J. C. Ward, H. Warren, and R. Zhou. Blue Gene: a vision for protein science using a petaflop supercomputer. *IBM Journal of Research and Development*, 40(2):310–327, 2001.
2. Giancarlo Benettin and Antonio Giorgilli. On the hamiltonian interpolation of near-to-the-identity symplectic mappings with application to symplectic integration algorithms. *J. Statist. Phys.*, 74:1117–43, 1994.
3. Markus Deserno and Christian Holm. How to mesh up Ewald sums. ii. an accurate error estimate for the particle-particle-particle-mesh algorithm. *J. Chem. Phys.*, 109(18): 7694–7701, 1998.
4. M. Eleftheriou, B. Fitch, A. Rayshubskiy, T.J.C. Ward, and R.S. Germain. Performance measurements of the 3d FFT on the Blue Gene/L supercomputer. In J.C. Cunha and P.D. Medeiros, editors, *Euro-Par 2005 Parallel Processing: 11th International Euro-Par Conference, Lisbon, Portugal, August 30-September2, 2005*, volume 3648 of *Lecture Notes in Computer Science*, pages 795–803. Springer-Verlag, 2005.
5. M Eleftheriou, R Germain, A Royyuru, and R Zhou. Thermal denaturing of mutant lysozyme with both oplsaa and charmm force fields. to appear in J. Am. Chem. Soc., 2006.
6. M. Eleftheriou, A. Rayshubskiy, J. W. Pitera, B. G. Fitch, R. Zhou, and R. S. Germain. Parallel implementation of the replica exchange molecular dynamics algorithm on Blue Gene/L. In *Fifth IEEE International Workshop on High Performance Computational Biology*, April 2006.
7. Robert D. Engle, Robert D. Skeel, and Matthew Drees. Monitoring energy drift with shadow hamiltonians. *Journal of Computational Physics*, 206(2):432–452, 2005.

8. B.G. Fitch, R.S. Germain, M. Mendell, J. Pitera, M. Pitman, A. Rayshubskiy, Y. Sham, F. Suits, W. Swope, T.J.C. Ward, Y. Zhestkov, and R. Zhou. Blue Matter, an application framework for molecular simulation on Blue Gene. *Journal of Parallel and Distributed Computing*, 63:759–773, 2003.

9. Blake G. Fitch, Aleksandr Rayshubskiy, Maria Eleftheriou, T.J. Christopher Ward, Mark Giampapa, Michael C. Pitman, and Robert S. Germain. Blue matter: Approaching the limits of concurrency for molecular dynamics. Research Report RC23956, IBM Research Division, April 2006. To appear in the Proceedings of the 2006 ACM/IEEE conference on Supercomputing.

10. Blake G. Fitch, Aleksandr Rayshubskiy, Maria Eleftheriou, T.J. Christopher Ward, Mark Giampapa, Yuri Zhestkov, Michael C. Pitman, Frank Suits, Alan Grossfield, Jed Pitera, William Swope, Ruhong Zhou, Scott Feller, and Robert S. Germain. Blue Matter: Strong scaling of molecular dynamics on Blue Gene/L. In V. Alexandrov, D. van Albada, P. Sloot, and J. Dongarra, editors, *International Conference on Computational Science (ICCS 2006)*, volume 3992 of *LNCS*, pages 846–854. Springer-Verlag, 2006.

11. Blake G. Fitch, Aleksandr Rayshubskiy, Maria Eleftheriou, T.J. Christopher Ward, Mark Giampapa, Yuri Zhestkov, Michael C. Pitman, Frank Suits, Alan Grossfield, Jed Pitera, William Swope, Ruhong Zhou, Robert S. Germain, and Scott Feller. Blue matter: Strong scaling of molecular dynamics on Blue Gene/L. Research Report RC23688, IBM Research Division, August 2005.

12. D. Frenkel and B. Smit. *Understanding Molecular Simulation*. Academic Press, San Diego, CA, 1996.

13. A. Gara et al. Overview of the Blue Gene/L system architecture. *IBM Journal of Research and Development*, 49(2/3):195–212, 2005.

14. Robert S. Germain, Blake Fitch, Aleksandr Rayshubskiy, Maria Eleftheriou, Michael C. Pitman, Frank Suits, Mark Giampapa, and T.J. Christopher T.J. Christopher Ward. Blue Matter on Blue Gene/L: massively parallel computation for biomolecular simulation. In *CODES+ISSS '05: Proceedings of the 3rd IEEE/ACM/IFIP international conference on Hardware/software codesign and system synthesis*, pages 207–212, New York, NY, USA, 2005. ACM Press.

15. R.S. Germain, Y. Zhestkov, M. Eleftheriou, A. Rayshubskiy, F. Suits, T.J.C. Ward, and B.G. Fitch. Early performance data on the Blue Matter molecular simulation framework. *IBM Journal of Research and Development*, 49(2/3):447–456, 2005.

16. Alan Grossfield, Scott E. Feller, and Michael C. Pitman. A role for direct interactions in the modulation of rhodopsin by omega-3 polyunsaturated lipids. *PNAS*, 103(13):4888–4893, 2006.

17. Benedict Leimkuhler and Sebastian Reich. *Simulating Hamiltonian Dynamics*, volume 14 of *Cambridge Monographs in Applied and Computational Mathematics*. Cambridge University Press, 2004.

18. Michael C. Pitman, Alan Grossfield, Frank Suits, and Scott E. Feller. Role of cholesterol and polyunsaturated chains in lipid-protein interactions: Molecular dynamics simulation of rhodopsin in a realistic membrane environment. *Journal of the American Chemical Society*, 127(13):4576–4577, 2005.

19. Sebastian Reich. Backward error analysis for numerical integrators. *SIAM Journal on Numerical Analysis*, 36(5):1549–1570, 1999.

20. C. Sagui and T. Darden. Multigrid methods for classical molecular dynamics simulations of biomolecules. *Journal of Chemical Physics*, 114(15):6578–6591, April 2001.

21. J.C. Sexton and D.H. Weingarten. Hamiltonian evolution for the hybrid Monte Carlo algorithm. *Nuclear Physics B*, 380:665–677, 1992.

22. Y. Sugita and Y. Okamoto. Replica-exchange molecular dynamics method for protein fold-ing. *Chem. Phys. Lett.*, 314:141–151, 1999.
23. W.C Swope, H.C. Andersen, P.H. Berens, and K.R. Wilson. A computer simulation method for the calculation of equilibrium constants for the formation of physical clusters of molecules: Application to small water clusters. *Journal of Chemical Physics*, 76:637–649, 1982.
24. Søren Toxvaerd. Hamiltonians for discrete dynamics. *Phys. Rev. E*, 50(3):2271–2274, Sep 1994.
25. M. Tuckerman, B.J. Berne, and G.J. Martyna. Reversible multiple time scale molecular dynamics. *J. Chem. Phys.*, 97(3):1990–2001, August 1992.
26. R. Zhou, E. Harder, H. Xu, and B.J. Berne. Efficient multiple time step method for use with Ewald and particle mesh Ewald for large biomolecular systems. *Journal of Chemical Physics*, 115(5):2348–2358, August 2001.

Progress Towards Petascale Applications in Biology: Status in 2006

Craig A. Stewart[1], Matthias Müller[2], and Malinda Lingwall[3]

[1] Office of the Vice President for Information Technology, Indiana University, Bloomington, IN
[2] Center for Information Services and High Performance Computing, Technische Universitaet Dresden
[3] University Information Technology Services, Indiana University, Bloomington, IN
stewart@iu.edu, matthias.mueller@tu-dresden.de, mlingwal@indiana.edu

Abstract. Petascale computing is currently a common topic of discussion in the high performance computing community. Biological applications, particularly protein folding, are often given as examples of the need for petascale computing. There are at present biological applications that scale to execution rates of approximately 55 teraflops on a special-purpose supercomputer and 2.2 teraflops on a general-purpose supercomputer. In comparison, Qbox, a molecular dynamics code used to model metals, has an achieved performance of 207.3 teraflops. It may be useful to increase the extent to which operation rates and total calculations are reported in discussion of biological applications, and use total operations (integer and floating point combined) rather than (or in addition to) floating point operations as the unit of measure. Increased reporting of such metrics will enable better tracking of progress as the research community strives for the insights that will be enabled by petascale computing.

Keywords: Computational biology, grand challenge problem, high performance computing, life sciences, peak theoretical capacity, petabytes, petaflops, petascale computing.

1 Introduction

The worldwide high performance computing (HPC) community is at present highly focused on petascale computing – a common topic of discussion in press releases, grant solicitations, conferences, and technical papers. Biology in general and protein structure in particular are often important themes in discussion of petascale computer applications. The government of Japan and the Institute of Physical and Chemical Research (RIKEN) announced in 2003 plans to create a high performance computing system with 1 petaflops peak theoretical capability to model protein folding [1]. In the United States, the National Science Foundation (NSF) and the US Department of Energy (DOE) have each announced programs designed to develop and implement petaflops supercomputers, in both cases with biology among the driving applications. The DOE has announced

W. Lehner et al. (Eds.): Euro-Par 2006 Workshops, LNCS 4375, pp. 289–303, 2007.

plans to install a supercomputer with 1 petaflops peak theoretical capability in 2008 [2], while the NSF's target is 1 petaflops sustained performance achieved by 2010–2011 [3]. Most recently, the RIKEN Institute announced that their Protein Explorer system has been clocked at a peak theoretical capability of 1 petaflops [4]. The era of petascale computing in biology is here – at least by one measure.

The purpose of this paper is to assess the current state of progress toward petascale computing in biology. Petascale is used here to indicate applications that use petaflops of computing power, petabytes of data, or both. We present data combed from the literature on execution rates of applications in biology and other sciences, as well as information on the size of publicly available data sets. Based on examination of the currently available data, we make recommendations about ways in which performance of applications and size of databases could be reported so that the research community could better track progress in capabilities of biological applications.

2   Methods and Materials

There are several ways to measure computational speed: peak theoretical capability (the maximum number of operations that could possibly be completed by a computer given the number of instructions per clock cycle and number of clock cycles per second); peak achieved performance on benchmark applications (especially the Linpack benchmark program, which is used in rankings for the Top500 List of the fastest supercomputers in the world [5]); and peak achieved performance on a "real" applications that solve some current scientific problem.

To assess progress in scale of applications in biology and other disciplines, we combed the literature and the World Wide Web for examples of particularly large computations in biology and, for purposes of comparison, other scientific disciplines. Because there is little consistency in how the performance of large biological applications is reported, we also solicited information directly from leading supercomputing centers. The progress of application performance can be understood only in the context of the progress in the capabilities of hardware systems. For comparisons of hardware capabilities we compiled information on the peak theoretical capability of general and special-purpose supercomputers. Key sources of information included papers about Gordon Bell prizes from the ACM/IEEE SCxy supercomputing conferences [6, 7, 8, 9] and the Top500 List [5]. To assess progress toward petascale data used in biology, we examined the current sizes of major public biological data sources.

3   Results

Figure 1 demonstrates the well-understood progress of the peak theoretical capability of the top-ranked system on the Top500 List. In terms of systems that run the Linpack benchmark, statistical extrapolation from all previous Top500 Lists suggests that the top system on that list will reach a peak theoretical

capability of 1 petaflops in November 2009 and achieved Linpack performance
of 1 petaflops in June 2012.

Figure 1 also shows peak theoretical capability of several special-purpose systems of note. The MD-GRAPE and GRAPE systems are not included on the
Top500 list since they perform molecular dynamics and astrophysical N-body calculations, respectively, and cannot run the Linpack benchmark suite. Figure 1
also shows current aggregate TFLOPS for the combined BOINC project [10],
and two subcomponents of that system – SETI@Home [11], the largest BOINC
project overall, and ROSETTA@Home [12], the largest biological application
within the BOINC system for which aggregate performance data are available.
Table 1 details the systems shown in Figure 1.

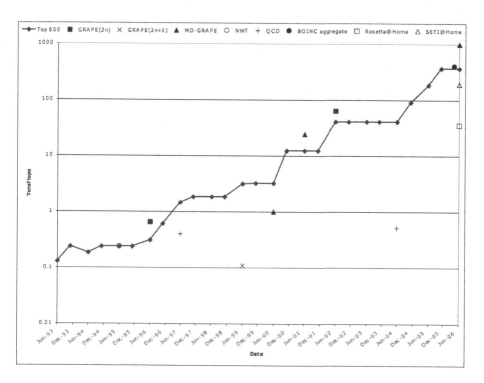

Fig. 1. Peak theoretical capacity of high performance computing systems over time.
Shown are the peak theoretical capacity of the #1 ranked system on the Top500 List since
its inception, along with the peak theoretical capability of selected special-purpose computing systems. Special-purpose systems represented include the Numerical Wind Tun-
nel, GRAPE family, MD-GRAPEs, specialized QCD systems, and distributed BOINC
applications [4], [5], [8], [10, 11, 12, 13, 14, 15, 16, 17, 18, 19].

Figure 2 shows progress in sustained performance on several applications since
the inception of the Top500 List. Included are the top achieved Linpack performance from the Top500 List and the top performance achieved on several heroic
applications. Table 2 details the applications shown in Figure 2.

Table 1. Data about systems in Figure 1

System	Classification	Peak theoretical capacity	Year	Reference
MDGRAPE-3	MD-GRAPE	1 PF	2006	[4]
BOINC combined statistics	BOINC aggregate	400.85 TF	2006	[10]
SETI@Home	SETI@Home	191.233 TF	2006	[11]
GRAPE-6	GRAPE(2n)	63.4 TF	2002	[13]
Rosetta@Home	Rosetta@Home	35.654 TF	2006	[12]
MDGRAPE-2	MD-GRAPE	24.6 TF	2001	[14]
MDGRAPE-2	MD-GRAPE	1 TF	2000	[15]
GRAPE-4	GRAPE(2n)	0.66 TF	1996	[8]
QCDOC	QCD	0.512 TF	2004	[16]
QCDSP	QCD	0.4 TF	1997	[17]
Numerical Wind Tunnel	NWT	0.2 TF	1995	[18]
GRAPE-5	GRAPE(2n+1)	0.11 TF	1999	[19]

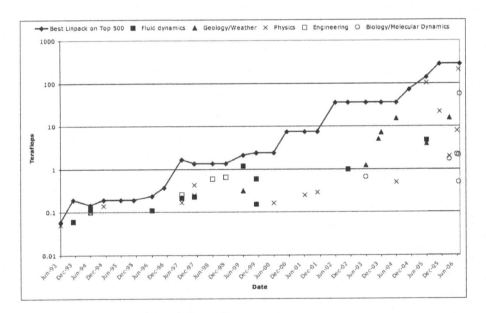

Fig. 2. Achieved floating point computation rates for applications in several disciplines. Included are the Linpack performance data of the #1 system on the Top500 List since its inception, and other applications that have reported high floating point operation rates. [5,6,7,8,9], [20,21,22,23,24,25,26,27,28,29,30,31,32,33,34,35,36,37,38,39,40, 41,42,43,44].

We collected information about the size of data sets used in several fields of research in order to study progress in data-centric life sciences research as compared to other disciplines. Table 3 shows the sizes of several important data sets. In many cases these databases tend to report their size in terms of numbers

Table 2. Data about applications in Figure 2

Application	Discipline	Peak achieved rate	Year	Reference
Qbox	Physics	207.3 TF	2006	[20]
Solidification simulations	Physics	103 TF	2005	[21]
Peptide simulation	Biology/Molecular dynamics	55 TF	2006	[22]
Qbox	Physics	22.02 TF	2005	[23]
Corona simulation	Geology/Weather	15.6 TF	2006	[24]
Earth Simulator	Geology/Weather	15.2 TF	2004	[25]
LSMS	Physics	8 TF	2006	[26]
Weather forecast (NWS)	Geology/Weather	7.3 TF	2003	[27]
Earth Simulator	Geology/Weather	5 TF	2003	[28]
Lattice Boltzmann model	Fluid dynamics	4.7 TF	2005	[29]
Weather forecast (NOAA)	Geology/Weather	4 TF	2005	[30]
Blue Matter	Biology/Molecular dynamics	2.2 TF	2006	[31]
NAMD	Biology/Molecular dynamics	2.08 TF	2006	[32]
VASP	Physics	2 TF	2006	[33]
CPMD	Biology/Molecular dynamics	1.7 TF	2006	[33]
Wave propagation solver	Geology/Weather	1.21 TF	2003	[34]
Turbulence simulation	Fluid dynamics	1.18 TF	1999	[35]
DOWSER	Fluid dynamics	1 TF	2002	[36]
First principles calculation	Engineering	0.657 TF	1998	[37]
NAMD	Biology/Molecular dynamics	0.65 TF	2003	[38]
Parallel Eigensolver	Engineering	0.605 TF	1998	[39]
Turbulence simulation	Fluid dynamics	0.6 TF	1999	[35]
NAMD	Biology/Molecular dynamics	0.5 TF	2006	[32]
Finite element analyses	Physics	0.5 TF	2004	[40]
Tree-code method	Physics	0.43 TF	1997	[9]
Hairpin vortices simulation	Geology/Weather	0.319 TF	1999	[41]
Cactus	Physics	0.292 TF	2001	[42]
MP-QUEST	Engineering	0.256 TF	1997	[9]
Cactus	Physics	0.249 TF	2001	[42]
Quark modeling	Physics	0.246 TF	1997	[9]
Pronto	Fluid dynamics	0.225 TF	1997	[9]
MPSalsa	Fluid dynamics	0.212 TF	1997	[9]
Tree-code method	Physics	0.17 TF	1997	[9]
Bunyip	Physics	0.163 TF	2000	[43]
Unstructured mesh CFD	Fluid dynamics	0.156 TF	1999	[44]
Sound wave computation	Physics	0.143 TF	1994	[7]
Numerical Wind Tunnel	Fluid dynamics	0.12 TF	1994	[7]
Numerical Wind Tunnel	Fluid dynamics	0.111 TF	1996	[8]
Composite modeling	Engineering	0.1 TF	1994	[7]
Radar scattering	Physics	0.1 TF	1994	[7]
Boltzmann equation	Fluid dynamics	0.06 TF	1993	[6]
Crack modeling on CM-5	Physics	0.05 TF	1993	[6]

of records (or in the case of sequence databases number of sequences). Indiana University maintains a repository of copies of many of these data sets, and we determined the size in petabytes of these data sets from those copies.

Table 3. Current size of some exemplars of databases used in the life sciences as of summer 2006, compared with key exemplars from other disciplines. The size of datasets marked with an * were determined from copies of data downloaded to Indiana University from the original resources.

Database name	Discipline	Current estimated size
BaBar	High-energy physics	2 PB [45]
National Virtual Observatory	Astronomy	~ 0.5 PB [46]
NCBI*	Biology	0.005 PB
Regenstrief Medical Records System	Medicine	0.004 PB [47]
Protein Data Bank	Biology	0.0007 PB [48]
EarthScope	Geology	0.0004 PB [49]
PubChem*	Chemistry	0.0001 PB
Swiss-Prot*	Biology	0.00000087 PB

4   Discussion

There are notable accomplishments in terms of peak performance of biological applications. The top performance in terms of floating point execution rate that we have been able to find for a biological application is 55 teraflops on a special-purpose MDGRAPE-3 system with a peak theoretical capability of 415 teraflops (an efficiency of 13.25%) [22]. This application simulated the formation of amyloid fibrils including 14 million atoms. The top performance in terms of floating point execution rates on a general-purpose supercomputer is approximately 2.2 teraflops with Blue Matter software on 80% of an 11.5 teraflops Blue Gene/L supercomputer (an efficiency of approximately 24%) [31], using the 92,000 atom ApoA1 benchmark. (The Blue Matter software is discussed in this volume in the paper by Fitch et al, "Progress in Scaling Biomolecular Simulations to Petaflop Scale Platforms.") Another application of note in terms of instruction rate is NAMD, which can operate at 2.08 teraflops in a 2.7 million atom simulation on a system with a peak theoretical capacity of 9.83 teraflops (an efficiency of approximately 21%) [32]. Based on the data we have been able to obtain, these seem to be the top biologically-oriented applications in terms of rates of floating point executions. There is a fairly strong contrast between the achieved rate of floating point operations on biological codes, the peak theoretical performance of systems available today, and the peak achieved performance on other scientific applications.

The progress of the peak theoretical capability of HPC systems, and of Linpack performance on these systems, is progressing steadily toward petascale computing. Special-purpose systems based on GRAPE and MD-GRAPE boards have on several occasions managed faster peak theoretical capability than the top system on the Top500 List. This trend is in evidence at present, as the MDGRAPE-3 is the basis for the RIKEN Institute's Protein Explorer, the first system with a reported peak theoretical capability of 1 petaflops. The fastest supercomputer in the world according to the June 2006 Top500 List (among those capable of running the Linpack Benchmark) is the 367 teraflops IBM BlueGene/L system at Lawrence Livermore National Laboratory, larger than but otherwise similar to the system used for the Blue Matter software calculations mentioned above. Plans announced by the US Department of Energy and National Science Foundation will thus result in implementation of systems of 1 petaflops peak theoretical capability (2008) and 1 petaflops achieved performance (2010–2011) more quickly than would be predicted on the basis of extrapolation from the existing Top500 list data.

In terms of performance of applications other than Linpack, the highest rate of floating point executions reported to date are from simulation of crack formation in 1,000 Molybdenum atoms with the Qbox application [20], [23]. Qbox on the 367 teraflops LLNL BlueGene/L system has achieved a peak execution rate of 207.3 teraflops – 56.5% of peak theoretical capability (as compared to 73.8% of peak achieved on the Linpack benchmark). Another notable physics application is LSMS [26], which ran on Pittsburgh Supercomputer Center's Cray XT3 at just over 8 teraflops – 82% of peak theoretical capability (as compared to 80.2% of peak achieved on the Linpack benchmark.) This LSMS run performed an ab initio quantum calculation of an iron nanoparticle of more than 4,400 atoms.

There seem to be fewer data available at present regarding high rates of floating point executions for heroic biological applications – and fewer than seem available for other disciplines. This is at least in part because HPC applications in biology have been in existence for less time (and are still less prevalent) than disciplines such as material science, physics, and computational fluid dynamics. In addition, performance results for biological codes are most often reported in ways that are directly meaningful to the time to solution of the particular problem at hand. Wall clock times, and decreases thereof, to solve a particular problem are perhaps the most common metric overall; total CPU hours used is also a common metric. In the case of protein folding, wall clock time per time step (or simulated time steps per unit wall clock time) is often used. In the case of genome sequence comparisons, number of sequences compared per unit time is a common metric. In the case of phylogenetic inference, the number of evolutionary trees analyzed per wall clock hour is commonly used. Researchers in the life sciences often do not collect and report the performance of their applications in terms of floating point operations. For example, two of the authors of this paper participated in an HPC Challenge project at SC2003, in which many collaborators created a global computational grid to run fastDNAml, a program for inferring evolutionary relationships [50]. We reported our results in terms of rate of analysis of trees, total number of processors used, etc. but did not instrument the code to measure actual floating point executions. Had we tried to do so, we would not have managed to get the application running during the time period of the HPC Challenge at SC2003. Similarly, high throughput applications such as Folding@home [51] and fightAIDS@home [52] involve thousands of computers working simultaneously on particular parts of a large-scale biological problem, but the rate at which work is done is not reported in terms of floating point calculation rates.

Floating point operation rates are mentioned specifically in major grant solicitations, and are thus of some practical import to the high performance computing community [53]. However, rates of floating point operations have two limitations as a measure of biological applications. One is that improving time to solution may involve decreasing execution rates. For example, the floating point rates for NAMD today are roughly 30% lower than in the code version in 2002 [38] because the underlying algorithms are more efficient [32].

A second limitation, perhaps more specific to biological operations, is the relative importance of integer operations in biological applications. The performance of the DOTTER program [54] was carefully analyzed in terms of total operations because of the predominance of integer mathematics in that application [55]. Understanding the application performance was possible only by including integer operations in the analysis. BLAST and other important bioinformatics applications also use integer operations extensively. Roughly two thirds of the mathematical operations in NAMD are integer operations [32]. To the extent that execution rates provide a means to compare the behavior of diverse biological applications, total operation rates (integer and floating point) would likely be a better basis for comparison than floating point operation rates alone. This poses the question of how to factor in the importance of operand length. Double precision reals are the basis for the standard Linpack benchmark, and there seems little reason at present to deviate from that approach in general (although there may be interesting exceptions [56]). As regards integer operations, when reporting rates it is probably best to specify the integer length – but it may make sense in the context of biological applications to count operations without regard to operand length. To do otherwise and somehow correct for length would likely penalize clever coding schemes that take advantage, for example, of the four letter alphabet of nucleotides (A,C,G,T).

In addition to measuring rates of operation execution, it will likely be useful to measure the total amount of computation that contributed to a particular analysis or simulation. For example, some of the largest biological computations performed to date in terms of total computer operations involve NAMD simulations of an entire ribosome in 2005 [57] and the tobacco mosaic satellite virus [58]. The former seems to be the largest simulation of a biological structure (in terms of CPU hours) ever published; the latter is the first ever molecular simulation of an entire life form. A useful measure of total calculation effort comparable across applications and systems might be simply total operations, or a measure analogous to the kilowatt-hour – that is, the PetaOPS-hour. Given the diversity of biologically oriented applications, it simply may not be possible to capture the performance of applications with a single metric. However, reporting total operation rates (integer and floating point) and total operation counts or PetaOPS-hours, in addition to other measures, will enable better comparisons among biological applications. Such comparisons are only a means, and the ends desired are biological insights rather than high operation rates. Still, tracking the progress of operation rates as a means will enable us to better determine if the oft-discussed ends (new insights and knowledge) are in evidence as the capabilities of our means progress.

The sizes of public biological data sets are growing rapidly, but life sciences data sets are still well away from the petabytes range and well smaller than the size of data sets found in other disciplines. Data sets in the range of 2 petabytes are available now in high energy physics research with 20 petabytes planned by 2008 [45]. The Terashake earthquake simulation run at the San Diego Supercomputer Center generated a data set of 45 TB [59]. In contrast, the largest publicly available biological data set is at present approximately 5 TB. Graphs of the amount of data contained in NCBI's Genbank data set show dramatic rates of growth [60], and that dramatic rate of growth creates an impression that may obscure the size of the actual data set: in 2006, the actual aggregate size of the data set is still well under a terabyte. Likewise, a recent demonstration at Indiana University included an analysis of some of the chemical properties of all of the compounds in PubChem [61] in less than 10 minutes – a significant accomplishment from the standpoint of obtaining information from a comparatively large data set (more than 19 million records), yet the input data amounted to less than 100GB. Other very large and notable data-centric initiatives in the life sciences include BIRN [62], eDiaMoND [63], and NEON [64]. Aggregated sets of data in clinical practice and held by pharmaceutical companies may be much larger. For example, the Regenstrief Institute [47] holds an aggregate of 4 TB of clinical data. While reporting of biological database size in number of records, or number of sequences, or number of compounds is common, more routine reporting of database size in terms of actual disk storage space would be useful in comparability across disciplines in discussing the size of data sets.

Sterling et al [65] produced the first careful analysis of the opportunities and challenges in achieving petascale computing. In their 1994 workshop, they identified several candidates for petaflops applications, including protein folding, modeling of circulation in the human body, and data-intensive applications using petabytes or exabytes of data. Stevens [66], CIBIO [67], and Atkins [68] provide more recent analyses of opportunities for petascale biological applications. Stevens [66] outlined eight categories of potential petascale applications; of these, five categories were related to molecular structure, function, and dynamics; other categories included sequence analysis, whole genome metabolic modeling, and population modeling. A recent NSF-sponsored workshop on petascale applications in biology reinforced many of these ideas, and added novel ideas such as ecological simulations linked to climate models and real-time patient profiling [69].

Based on data currently available, molecular dynamics codes clearly scale to the highest operation rates achieved on monolithic supercomputers and are likely candidates to be the first applications to achieve petaops calculation rates. One model of circulatory function in the human body – ATREE – creates large-scale models of biological function by employing computational physics codes (including turbulence) to solve biological problems. These codes have been implemented on the NSF-funded TeraGrid [70], linking simulation of many components of the human arterial system. By linking many HPC systems ATREE is a likely candidate to achieve extremely high mathematical operation rates in a grid environment. In terms of data-intensive applications, several examples given by Stevens [66] involve coarse-grained (and often very complex) parallel analyses of large data sets; such data-parallel applications are also good candidates for achieving very high operation rates. All in all, the current state of affairs is consistent with many of the predictions made by Sterling et al. more than a decade ago.

5   Conclusion

There are many ways to count what are petascale applications in biology; by one measure at least the era of petascale biology begins in 2006 with the successful operation of the Protein Explorer at a peak theoretical capability of 1 petaflops. Many obstacles remain between the state of the art in 2006 and biological applications that achieve petaops calculation rates and process petabytes of data. In tracking the progress toward petascale biological applications it will be helpful to report application characteristics in ways that will enable better comparisons across applications. For applications, routine reporting of calculation rates in terms of total petaoperations per second, and total computing power in petaoperations or PetaOPS-hours for particular simulations, would be helpful. For data-intensive applications, more routine reporting of data set size in tera- or petabytes would be helpful. Petascale applications are only a means to an end; the ends are new insights about the function of biological systems and better human health. Still, tracking progress of the means will enable some insight as to whether the ends anticipated are being achieved.

Acknowledgements. This research was supported in part by the Indiana Genomics Initiative and the Indiana Metabolomics and Cytomics Initiative. The Indiana Genomics Initiative of Indiana University and the Indiana Metabolomics and Cytomics Initiative of Indiana University are supported in part by Lilly Endowment, Inc. The authors also wish to thank IBM, Inc. for support via Shared University Research Grants and partnerships via IU's relationship as an IBM Life Sciences Institute of Innovation. Indiana University also thanks the TeraGrid partners; IU's participation in the TeraGrid is funded by National Science Foundation grant numbers 0338618, 0504075, and 0451237. The early development of this paper was supported by a Fulbright Senior Scholars award from the Council for International Exchange of Scholars (CIES) and the United States Department of State to Dr. Craig A. Stewart; Matthias Mueller and the Technische Universität Dresden were hosts. Many reviewers contributed to the improvement of the ideas expressed in this paper and are gratefully appreciated; Thom Dunning, Robert Germain, Chris Mueller, Jim Phillips, Richard Repasky, Ralph Roskies, and Allan Snavely are thanked particularly for their insights.

References

1. Taiji M., Narumi T., Ohno Y., Futatsugi N., Suenaga A., Takada N., Konagaya A. "Protein Explorer: A Petaflops Special-Purpose Computer System for Molecular Dynamics Simulations," *sc*, p. 15, ACM/IEEE SC 2003 Conference (SC'03), 2003. http://csdl.computer.org/dl/proceedings/sc/2003/2113/00/21130015.pdf
2. "Vendor Spotlight: Cray to Deliver Petaflop Supercomputer to ORNL in 2008." HPCwire. 16 June 2006. Accessed 30 August 2006. http://www.hpcwire.com/hpc/694425.html
3. NSF Cyberinfrastructure Council. "NSF's Cyberinfrastructure Vision for 21st Century Discovery." 20 July 2006. Accessed 31 August 2006. http://www.nsf.gov/od/oci/ci-v7.pdf
4. Taiji M., Yamashita Y., Nakanishi T. "Completion of a one-petaflops computer system for simulation of molecular dynamics." 2006. Press release. Accessed 17 August 2006. http://www.riken.go.jp/engn/r-world/info/release/press/2006/060619/index.html
5. Top500 Supercomputer Sites. Accessed 1 September 2006. http://top500.org/lists
6. Karp A.H., Heller D., Simon H. "1993 Gordon Bell Prize Winners." *IEEE Computer,* January 1994, pp. 69–75. http://csdl.computer.org/dl/mags/co/1994/01/r1069.pdf
7. Karp A.H., Heath M., Heller D., Simon H. "1994 Gordon Bell Prize Winners." *IEEE Computer*, January 1995, pp. 68–74. http://csdl.computer.org/dl/mags/co/1995/01/r1068.htm
8. Karp A.H., Geist A., Bailey D. "1996 Gordon Bell Prize Winners." *IEEE Computer*, January 1997, pp. 80–85. http://csdl.computer.org/dl/mags/co/1997/01/r1080.htm
9. Karp A.H., Lusk E., Bailey D.H. "1997 Gordon Bell Prize Winners." *IEEE Computer*, January 1998, pp. 86–92. http://csdl.computer.org/dl/mags/co/1998/01/r1086.htm
10. BOINC combined statistics (BOINCstats). Accessed 20 September 2006. http://www.boincstats.com/stats/project_graph.php?pr=bo
11. SETI@Home (BOINCstats). Accessed 20 September 2006. http://www.boincstats.com/stats/project_graph.php?pr=sah
12. Rosetta@Home (BOINCstats). Accessed 20 September 2006. http://www.boincstats.com/stats/project_graph.php?pr=rosetta
13. Makino J., Kokubo E., Fukushige T., Daisaka H. "A 29.5 Tflops simulation of planetesimals in Uranus-Neptune region on GRAPE-6." *sc*, p. 34, ACM/IEEE SC 2002 Conference (SC'02), 2002. http://csdl.computer.org/dl/proceedings/sc/2002/1524/00/15240034.pdf
14. Narumi T., Kawai A., Koishi T. "An 8.61 Tflop/s Molecular Dynamics Simulation for NaCl with a Special-Purpose Computer: MDM." *sc*, p. 11, ACM/IEEE SC 2001 Conference (SC'01), 2001. http://csdl.computer.org/dl/proceedings/sc/2001/1990/00/19900011.pdf
15. Narumi T., Susukita R., Koishi T., Yasuoka K., Furusawa H., Kawai A., Ebisuzaki T. "1.34 Tflops Molecular Dynamics Simulation for NaCl with a Special-Purpose Computer: MDM." *sc*, p. 54, ACM/IEEE SC 2000 Conference (SC'00), 2000. http://csdl.computer.org/dl/proceedings/sc/2000/9802/00/98020054.pdf

16. Boyle P.A., Chen D., Christ N.H., Clark M., Cohen S., Dong Z., Gara A., Joo B., Jung C., Levkova L., Liao X., Liu G., Mawhinney R.D., Ohta S., Petrov K., Wettig T., Yamaguchi A., Cristian C. "QCDOC: A 10 Teraflops Computer for Tightly-Coupled Calculations." *sc*, p. 40, ACM/IEEE SC 2004 Conference (SC'04), 2004. http://csdl.computer.org/dl/proceedings/sc/2004/2153/00/21530040.pdf

17. Chen D., Chen P., Christ N.H., Edwards R.G., Fleming G., Gara A., Hansen S., Jung C., Kahler A., Kasow S., Kennedy A.D., Kilcup G., Luo Y.B., Malureanu C., Mawhinney R.D., Parsons J., Sexton J., Sui C., Vranas P. "QCDSP: A Teraflop Scale Massively Parallel Supercomputer." *sc*, p. 52, ACM/IEEE SC 1997 Conference (SC'97), 1997. http://csdl.computer.org/dl/proceedings/sc/1997/1982/00/19820052.pdf

18. Yoshida M., Nakamura A., Fukuda M., Nakamura T., Hioki S. "Quantum Chromodynamics Simulation on NWT." *sc*, p. 65, ACM/IEEE SC 1995 Conference (SC'95), 1995. http://csdl.computer.org/dl/proceedings/sc/1995/2568/00/25680065.pdf

19. Kawai A., Fukushige T., Makino J. "$7.0/Mflops Astrophysical N-Body Simulation with Treecode on GRAPE-5." *sc*, p. 67, ACM/IEEE SC 1999 Conference (SC'99), 1999. http://csdl.computer.org/dl/proceedings/sc/1999/1966/00/19660067.pdf

20. Johnston D., Smith J., Acocella K. "NNSA announces new mark for world's fastest supercomputer." Press release. 22 June 2006. Accessed 12 September 2006. http://www.llnl.gov/pao/news/news_releases/2006/NR-06-06-07.html

21. Streitz F.H., Glosli J.N., Patel M.V., Chan B., Yates R.K., deSupinski B.R., Sexton J., Gunnels J.A. "100+ TFlop Solidification Simulations on BlueGene/L." November 2005. SC 2005. 14 August 2006. http://sc05.supercomputing.org/schedule/pdf/pap307.pdf

22. Narumi T., Ohno Y., Okimoto N., Koishi T., Suenaga A., Futatsugi N., Yanai R., Himeno R., Fujikara S., Taiji M., Ikei M. "A 55 TFLOPS Simulation of Amyloid-forming Peptides from Yeast Prion Sup35 with the Special-purpose Computer System MDGRAPE-3." To appear in *sc*, ACM/IEEE SC 2006 Conference (SC'06), 2006.

23. Gygi F., Yates R.K., Lorenz J., Draeger E.W., Franchetti F., Ueberhuber C.W., de Supinski B.R., Kral S., Gunnels J.A., Sexton J.C.. "Large-Scale First-Principles Molecular Dynamics simulations on the BlueGene/L Platform using the Qbox code," *sc*, p. 24, ACM/IEEE SC 2005 Conference (SC'05), 2005.

24. "SDSC Helps Scientists Accurately Simulate Sun's Corona." 3 August 2006. Accessed 11 October 2006. http://www.sdsc.edu/Press/2006/08/080306_corona.html

25. Kageyama A., Kameyama M., Fujihara S., Yoshida M., Hyodo M., Tsuda Y. "A 15.2 TFlops Simulation of Geodynamo on the Earth Simulator," *sc*, p. 35, ACM/IEEE SC 2004 Conference (SC'04), 2004. http://csdl.computer.org/dl/proceedings/sc/2004/2153/00/21530035.pdf

26. "Science, the XT3 and TeraGrid: An Interview with PSC Scientific Directors Michael Levine and Ralph Roskies." Pittsburgh Supercomputing Center. June 2006. Accessed 13 September 2006. http://www.psc.edu/publicinfo/news/2006/2006-06-09-xt3.php

27. Handwerk, B. "Faster Supercomputers Aiding Weather Forecasts." National Geographic News. 29 August 2005. Accessed 11 October 2006. http://news.nationalgeographic.com/news/2005/08/0829_050829_supercomputer.html

28. Komatitsch D., Tsuboi S., Ji C., Tromp J., "A 14.6 billion degrees of freedom, 5 teraflops, 2.5 terabyte earthquake simulation on the Earth Simulator," *sc*, p. 4, ACM/IEEE SC 2003 Conference (SC'03), 2003. http://csdl.computer.org/dl/proceedings/sc/2003/2113/00/21130004.pdf

29. Lammers P., Wellein G., Zeiser T., Hager G. "Have the Vectors the Continuing Ability to Parry the Attack of the Killer Micros." In: Resch M., Bnisch T., Benkert K., Furui T., Seo Y., Bez W. (eds) High Performance Computing on Vector Systems, Volume 1, 25–37. Springer: 2006.
30. Curns, T. "WEATHER FORECASTING ON A 'REMOTE' SUPER-COMPUTER?" HPCwire. 13 August 2004. Accessed 11 October 2006. http://www.hpcwire.com/hpcwire/hpcwireWWW/04/0813/108178.html
31. Germain, Robert. "Re: Two requests." Personal correspondence. 29 August 2006.
32. Phillips, Jim. "Re: Fwd: NAMD performance in FLOPS?" Personal correspondence. 29 August 2006.
33. Tiyyagura S.R. et al. "TERAFLOPS Sustained Performance with Real World Applications." Accepted for publication in: "Performance Characterization of the World's Most Powerful Supercomputers," special issue of the International Journal of High Performance Computing Applications (IJHPCA). Guest-edited by L. Oliker and R. Biswas. To appear 2007.
34. Akcelik V., Bielak J., Biros G., Epanomeritakis I., Fernandez A., Ghattas O., Kim E.J., Lopez J., O'Hallaron D., Tu G. Urbanic J. "High Resolution Forward And Inverse Earthquake Modeling on Terascale Computers," sc, p. 52, ACM/IEEE SC 2003 Conference (SC'03), 2003. http://csdl.computer.org/dl/proceedings/sc/2003/2113/00/21130052.pdf
35. Mirin A.A., Cohen R.H., Curtis B.C., Dannevik W.P., Dimits A.M., Duchauneau M.A., Eliason D.E., Schikore D.R., Anderson S.E., Porter D.H., Woodward P.R., Shieh L.J., White S.W. "Very High Resolution Simulation of Compressible Turbulence on the IBM-SP System." sc, p. 70, ACM/IEEE SC 1999 Conference (SC'99), 1999. http://csdl.computer.org/dl/proceedings/sc/1999/1966/00/19660070.pdf
36. Tajkhorshid, E., Nollert, P., Jensen, M. O., Miercke, L. J., O'Connell, J., Stroud, R. M., and Schulten, K. "Control of the Selectivity of the Aquaporin Water Channel Family by Global Orientational Tuning." (2002) Science 296, 525–530
37. "Gordon Bell Prize Winners." June 2000. SC2000. 14 August 2006. http://www.sc2000.org/bell/pastawrd.htm
38. Tajkhorshid E., Aksimentiev A., Balabin I., Gao M., Isralewitz B., Phillips J.C., Zhu F., Schulten K. "Large scale simulation of protein mechanics and function." In: Frederic M. Richards, David S. Eisenberg, and John Kuriyan, editors, Advances in Protein Chemistry, volume 66, pp. 195–247. Elsevier Academic Press, New York, 2003.
39. Sears M.P., Stanley K., Henry G. "Application of a High Performance Parallel Eigensolver to Electronic Structure Calculations." sc, p. 54, ACM/IEEE SC 1998 Conference (SC'98), 1998. http://csdl.computer.org/comp/proceedings/sc/1998/8707/00/87070054.pdf
40. Adams M.F., Bayraktar H.H., Keaveny T.M., Papadopoulos, P. "Ultrascalable Implicit Finite Element Analyses In Solid Mechanics With Over A Half a Billion Degrees of Freedom." sc, p. 34, ACM/IEEE SC 2004 Conference (SC'04), 2004. http://csdl.computer.org/dl/proceedings/sc/2004/2153/00/21530034.pdf
41. Tufo H.M., Fischer P.F., Papka M.E., Blom K. "Numerical Simulation and Immersive Visualization of Hairpin Vortices." sc, p. 62, ACM/IEEE SC 1999 Conference (SC'99), 1999. http://csdl.computer.org/dl/proceedings/sc/1999/1966/00/19660062.pdf

42. Allen G., Dramlitsch T., Foster I., Karonis N.T., Ripeanu M., Seidel E., Toonen B. "Supporting Efficient Execution in Heterogeneous Distributed Computing Environments with Cactus and Globus." *sc*, p. 52, ACM/IEEE SC 2001 Conference (SC'01), 2001. http://csdl.computer.org/dl/proceedings/sc/2001/1990/00/19900052.pdf

43. "ANU DCS Technical Services Group." 2000. Australian National University. Accessed 14 August 2006. http://tsg.anu.edu.au/Projects/Beowulf/

44. Anderson W.K., Gropp W.D., Kaushik D.K., Keyes D.E., Smith B.F., "Achieving High Sustained Performance in an Unstructured Mesh CFD Application," *sc*, p. 69, ACM/IEEE SC 1999 Conference (SC'99), 1999. http://csdl.computer.org/dl/proceedings/sc/1999/1966/00/19660069.pdf

45. Teige, Scott. "Re: Question about HEP databases." Personal correspondence. 31 August 2006.

46. Hanisch, Robert. "Re: [nvo-feedback] Amount of data currently accessible through NVO?" Personal correspondence. 31 August 2006.

47. Miller, Theda. "Re: [Fwd: Size of RMRS database(s)?]" Personal correspondence. 31 August 2006.

48. Research Collaboratory for Structural Bioinformatics. "Protein Data Bank Annual Report for July 2004 – June 2005." Accessed 15 September 2006. http://www.rcsb.org/pdbstatic/general_information/news_publications/annual_reports/annual_report_year_2005.pdf

49. "EarthScope Distribution Statistics from the IRIS DMC." Accessed 15 September 2006. http://www.iris.edu/earthscope/stats/

50. Stewart C.A., Hart D., Aumller M., Keller R., Mller M., Li H., Repasky R., Sheppard R., Berry D.K., Hess M., Wssner U., Colbourne J. "A Global Grid for Analysis of Arthropod Evolution." *grid*, pp. 328–337, Fifth IEEE/ACM International Workshop on Grid Computing (GRID'04), 2004. http://csdl.computer.org/dl/proceedings/grid/2004/2256/00/22560328.pdf

51. "Folding@Home Stats." Folding@home distributed computing. 2006. Accessed 14 August 2006. http://folding.stanford.edu/stats.html

52. fightAIDS@home. Accessed 31 August 2006. http://fightaidsathome.scripps.edu/

53. NSF Solicitation 06-573. "Leadership-Class System Acquisition - Creating a Petascale Computing Environment for Science and Engineering." 5 June 2006. Accessed 31 August 2006. http://nsf.gov/funding/pgm_summ.jsp?pims_id=13649

54. Dotter: A dot-matrix program with interactive greyscale rendering for genomic DNA and Protein sequence analysis. Accessed 1 September 2006. http://www.cgb.ki.se/cgb/groups/sonnhammer/Dotter.html

55. Mueller C., Dalkilic M., Lumsdaine A. "High-Performance Direct Pairwise Comparison of Large Genomic Sequences," *ipdps*, p. 199a, 19th IEEE International Parallel and Distributed Processing Symposium (IPDPS'05) - Workshop 7, 2005. http://csdl.computer.org/dl/proceedings/ipdps/2005/2312/08/23120199a.pdf

56. Feldman, M. "Less is More: Exploiting Single Precision Math in HPC." HPCwire. 16 June 2006. Accessed 21 September 2006. http://www.hpcwire.com/hpc/692906.html

57. Sanbonmatsu K.Y., Joseph S., Tung C.S. "Simulating movement of tRNA into the ribosome during decoding." Proc Natl Acad Sci U S A. 2005 Oct 25.

58. Freddolino, P.L. et al. "Molecular dynamics simulations of the complete satellite tobacco mosaic virus." Structure 14, 437–449 (2006).

59. Tooby P. "TeraShake: Simulating the BIG ONE on the San Andreas Fault." *EnVision* Volume 20, Number 1, 2004. pp 4–7. http://www.npaci.edu/envision/v20.1/Envision-2004.pdf

60. Genbank Statistics. Accessed 1 September 2006. http://www.ncbi.nlm.nih.gov/ Genbank/genbankstats.html
61. PubChem. Accessed 1 September 2006. http://pubchem.ncbi.nlm.nih.gov/
62. Biomedical Informatics Research Network (BIRN). Accessed 1 September 2006. http://www.nbirn.net/
63. eDiaMoND grid computing project. Accessed 1 September 2006. http://www.ediamond.ox.ac.uk/index.html
64. "NEON: National Ecological Observatory Network." Accessed 15 September 2006. http://www.neoninc.org/
65. Sterling T., Messina P., and Smith P.H. *Enabling Technologies for Petaflops Computing*. Cambridge, Massachusetts: MIT Press, 1995.
66. Stevens R. "Trends in Cyberinfrastructure for Bioinformatics and Computational Biology." *CTWatch QUARTERLY* Volume 2, Number 3, August 2006. http:// www.ctwatch.org/ quarterly/ articles/ 2006/ 08/ trends-in-cyberinfrastructure-for-bioinformatics-and-computational-biology/
67. "Building a Cyberinfrastructure for the Biological Sciences (CIBIO): A BIO Advisory Committee Workshop." July 2003. Accessed 31 August 2006. http://research.calit2.net/cibio/archived/CIBIO_Overview_Report.pdf
68. Atkins D.E., Droegemeier K.K., Feldman S.I., Garcia-Molina H., Klein M.L., Messerschmitt D.G., Messina P., Ostriker J.P., Wright M.H. "Revolutionizing Science and Engineering Through Cyberinfrastructure: Report of the National Science Foundation Blue-Ribbon Advisory Panel on Cyberinfrastructure." January 2003. Accessed 1 September 2006. http://www.nsf.gov/od/oci/reports/atkins.pdf
69. Petascale Computing in the Biological Sciences. NSF-funded workshop held August 29–30, 2006. Accessed 11 October 2006. http://www.sdsc.edu/PMaC/ BioScience_Workshop/biosciences.html
70. Dong S., Insley J., Karonis N.T., Papka M., Binns J. and Karniadakis G.E. 'Simulating and visualizing the human arterial system on the TeraGrid." Future Generation Computer Systems, The International Journal of Grid Computing: Theory, Methods and Applications, to appear, 2006.

Toward a Solution of the Reverse Engineering Problem Using FPGAs[*]

Edgar Ferrer[1], Dorothy Bollman[2], and Oscar Moreno[3]

[1] PhD. CISE Program, University of Puerto Rico, Mayagüez, PR 00681
eferrer@cs.uprm.edu
[2] Department of Mathematics, University of Puerto Rico, Mayagüez, PR 00681
bollman@cs.uprm.edu
[3] Department of Computer Science, University of Puerto Rico, Rio Piedras, PR 00931
moreno@uprr.pr

Abstract. An important issue in computational biology is the reverse engineering problem for genetic networks. In this ongoing work we consider reverse engineering in the context of univariate finite fields models. A solution to the reverse engineering problem using multipoint interpolation relies on intensive arithmetic computations over finite fields, where multiplication is the dominant operation. In this work, we develop an efficient multiplier for fields $GF(2^m)$ generated by irreducible trinomials of the form $\alpha^m + \alpha^n + 1$. We propose a design described by a parallel/serial architecture that computes a multiplication in m clock cycles. This approach exploits symmetries in Mastrovito matrices in order to improve time complexities of an FPGA (Field Programmable Gate Array) implementation. According to preliminary performance results, our approach performs efficiently for large fields and has potential for an efficient solution of the reverse engineering problem for large genetic networks, as well as other finite fields applications such as cryptography and Reed-Solomon decoders.

1 Introduction

An important problem in computational biology is modeling gene regulatory networks in order to determine gene behavior in biological systems and how they interact with each other. The reverse engineering problem for genetic networks is the problem of determining the network that describes functional relations between genes, given a set of experimental data.

In this ongoing work we consider the reverse engineering problem in the context of univariate finite fields models [1,12]. The reverse engineering problem can then be stated more precisely as follows:

Given a time series $s_0, s_1, \ldots, s_{k-1}$ of measurements of gene expression data representing the states of m genes at times $t_0, t_1, \ldots t_{k-2}$, and a set of conditions χ, the reverse engineering problem is the problem of finding a function f such

---

[*] This research is supported by grants NSF-CISE EIA-0080926 and NIH-MBRS (SCORE) S06-GM08102.

W. Lehner et al. (Eds.): Euro-Par 2006 Workshops, LNCS 4375, pp. 304–312, 2007.

that $f : GF(q) \rightarrow GF(q)$ has the property that $f(s_j) = s_{j+1}$, where $s_j = (a_0, a_1, \ldots, a_{m-1})$, and f satisfies the conditions in χ.

Our solution $f(x)$ to the reverse engineering problem then involves the determination of a polynomial $P(x)$, such that $f(x) = P(x) + g(x)$, and $P(s_i) = P(s_{i+1})$, and $g(x)$ is a polynomial such that $g(s_i) = 0$, for $i = 0, 1, \ldots, k - 2$. The polynomial $P(x)$ can be determined interpolating over the points s_i. Once having determined $P(x)$, the polynomial $g(x)$ can be used to adjust the model in order to satisfy the conditions in χ.

A classical method such as Lagrange interpolation formula can be used, but it has computational complexity $O(n^2)$, where n is the number of points to be interpolated. In contrast, Lipson's algorithm has complexity $O(n \log^2 n)$. Bollman et al [1] have shown that a parallel version of this algorithm is efficient for reverse engineering univariate genetic networks.

2 Dealing with Large Genetic Networks

Various approaches have been taken for modeling gene regulatory networks, including linear models, Bayesian networks, neural networks, nonlinear ordinary differential equations, stochastic models and Boolean models. One model that has received considerable attention is the Boolean model e.g., [6]. Recently, several researchers have pointed out advantages in generalizing the Boolean model to finite fields. Two types of finite field models have emerged, a multivariate model as described by Laubenbacher [8] et al and a univariate model as described by Moreno et al [12], [1]. The multivariate model gives local information at each gene, whereas the univariate model gives global information about the network.

In a finite field model for genetic networks we assume that gene expression is discretized so that there are a prime number p of levels. There are several ways to discretize the real-valued microarray data. One way is by thresholding. Another way is to normalize gene expressions and use the deviation from the mean to discretize the data. Inconsistencies due to either noise or biological variance can be resolved by using information theoretic error correction [13].

If there are m genes and p levels of expression, then there are p^m states and we model such a network by the elements of $GF(p^m)$. In this work we consider univariate finite field models with $p = 2$, so that each gene assumes just two states, either *on* or *off*. Thus we restrict interpolation to fields $GF(2^m)$.

In practice, m can be quite large. For example, [10] outlines a study of gene regulatory networks in yeast. Yeast has 6000+ genes. This study includes a subset of 106 transcription factors and 2343 genes for which strong empirical evidence of interaction was found using the experimental technique outlined in the paper. Advances in techniques should yield data on all 6270 genes in yeast, and eventually similar data will be available for all 20,000+ human genes. It is thus of vital interest to develop algorithms to reverse engineering very large networks.

A solution to the reverse engineering problem for large values of m using multipoint interpolation relies on intensive arithmetic computations over finite fields. Addition and multiplication are two basic operations. Addition is easily realized at very low computational cost, but multiplication is costly in terms of computation time and circuit complexity. Other arithmetic operations on finite fields used for reverse engineering such as inversion, squaring, exponentiation and divisions are performed by repeated multiplications. Thus, in order to solve the reverse engineering problem for the very large genetic networks that biologists would like to consider, it is essential to develop capacity for performing fast and efficient arithmetic over very large finite fields, especially multiplication.

One very fast method for performing arithmetic on $GF(2^m)$ involves the use of Zech logarithm tables. By using lookup tables we can perform arithmetic operations at "almost no cost", but the memory space becomes a great limitation. For instance, a 32-bit word length for storing the elements of $GF(2^{30})$ in a table, requires $2^2 \cdot 2^{30}$ bytes $= 4$ GB in main memory. This method is efficient for small finite fields, but it is not practical for the large fields that arise in real reverse engineering problems.

A natural approach for multiplication in $GF(2^m)$ is to multiply two elements in the field as polynomial multiplication modulo a m-degree irreducible polynomial over $GF(2)$. This operation is accomplished by simply using left-shifts and exclusive or's. In this simple procedure (also known as the direct or classical method) the memory space is not a limitation, but in a basic CPU based implementation the field size is limited by the architecture word-length.

One approach to addressing large finite fields is to use composite fields $GF((2^r)^s)$, combining lookup tables with direct multiplication. The acceleration of finite fields multiplication using FPGAs is determined mainly by access time between FPGAs and memory, since a composite field multiplication requires multiple accesses to lookup tables stored in memory.

In this work, we present an FPGA-based implementation of a multiplication algorithm over $GF(2^m)$ that exploits the symmetries in the Mastrovito matrix [11]. The proposed approach performs efficiently for large fields and has potential for efficient solutions of the reverse engineering problem for large genetic networks, as well as other finite fields applications such as cryptography and Reed-Solomon decoders.

3   Finite Field Multiplication

An element in the finite field $GF(2^m)$ can be represented as a sequence of m bits in $GF(2)$ describing the coefficients of a binary polynomial. This representation is useful for manipulating finite field elements via bitwise operations, so we can exploit the hardware architecture of computers by carrying out finite field arithmetic by means of bit-level operations. In essence the arithmetic computation over $GF(2^m)$ is suitable for FPGAs implementations. We take advantage of reconfigurable hardware resources with the aim of accelerating computations considerably.

Multiplication is the dominant operation in the interpolation phase of the solution of the reverse engineering problem for genetic networks. Many solutions have been proposed for efficient multiplication over finite fields. Solutions are based on purely software approaches, purely hardware approaches, and more recently, on hardware/software using reconfigurable computing.

The representation of the field elements distinguishes the particular features of a finite field multiplier. The most common representations are dual basis, normal basis, and standard basis. In this work we deal with finite field elements represented in standard (or polynomial or canonical) basis, such that the finite field $GF(2^m)$ consists of a finite set of all binary polynomials of degree less than m. For example $GF(2^2) = \{0, 1, \alpha, \alpha + 1\}$, where α is a root in $GF(2^2)$ of the irreducible polynomial $\alpha^2 + \alpha + 1$.

A very natural approach for standard basis multiplication in $GF(2^m)$ is to multiply two elements in the field as polynomial multiplication modulo an irreducible polynomial. This operation is typically accomplished in two stages: polynomial multiplication and modular reduction.

Let $A(\alpha)$, $B(\alpha)$, $C(\alpha)$ elements in $GF(2^m)$ and $f(\alpha)$ the irreducible polynomial generating $GF(2^m)$. Then the finite field multiplication $C(\alpha) = A(\alpha)B(\alpha)$ is accomplished by calculating

$$C(\alpha) = A(\alpha) * B(\alpha) \mod f(\alpha) \tag{1}$$

where $*$ denotes polynomial multiplication. In a first stage the product $A(\alpha) * B(\alpha)$ is calculated, resulting in a polynomial $Q(\alpha)$ of degree at most $2m - 2$.

$$Q(\alpha) = A(\alpha) * B(\alpha) = \left(\sum_{i=0}^{m-1} a_i \alpha^i \right) \left(\sum_{i=0}^{m-1} b_i \alpha^i \right) \tag{2}$$

In a second stage the modular reduction is performed on $Q(\alpha)$, that is, $C(\alpha) = Q(\alpha) \mod f(\alpha)$, resulting in the polynomial $C(\alpha)$ of degree at most $m - 1$.

It is easy to show that the expansion of equation (2) can be expressed as a matrix-vector product $Q = MB$, where Q is a vector of dimension $2m - 1$, which consists of the coefficients of $Q(\alpha)$. In the same way B is a m dimensional vector which consists of the coefficients of $B(\alpha)$, while the $(2m - 1) \times m$ matrix M involves coefficients of $A(\alpha)$ (see for example [14]).

Notice that the last $m - 1$ components of the vector Q (i.e. $[q_m, \ldots, q_{2m-2}]$) contain terms with degree greater than $m - 1$. These terms must be reduced modulo the irreducible polynomial $f(\alpha) = \alpha^m + g(\alpha)$ in order to express them as polynomials in the field $GF(2^m)$. This reduction is obtained by using the reducing identity $\alpha^m = g(\alpha)$, so all the terms with degree greater than $m - 1$ will be reduced to terms with degree in the proper range $[0, m-1]$. Each reduced term is added to the respective terms in $[q_0, \ldots, q_{m-1}]$, and so we get $C(\alpha)$. A particular term may need to be reduced several times. The maximum number of reductions is determined by:

$$N[m, n] = \left\lceil \frac{m - 1}{\Delta} \right\rceil$$

where $\Delta = m - n$ [5].

For example, let $m = 3$ and $f(\alpha) = \alpha^3 + \alpha^2 + 1$, thus $\alpha^3 = \alpha^2 + 1$ and $\alpha^4 = \alpha^3 + \alpha$. Using these identities the term $q_3\alpha^3$ is reduced only once: $q_3\alpha^3 = q_3\alpha^2 + q_3$, while $q_4\alpha^4$ is reduced twice: $q_4\alpha^4 = q_4\alpha^3 + q_4\alpha = q_4\alpha^2 + q_4\alpha + q_4$, and so we get $C(\alpha) = q_4\alpha^4 + q_3\alpha^3 + q_2\alpha^2 + q_1\alpha + q_0 = (q_4 + q_3 + q_2)\alpha^2 + (q_4 + q_1)\alpha + (q_4 + q_3 + q_0)$. Notice that the maximum number of reductions is $N[3, 2] = 2$.

An alternative to the two-stage method, described above, for computing C is to perform the reduction directly over the matrix M, obtaining an already reduced $m \times m$ dimensional matrix Z, such that $C = ZB$. Z is called the Mastrovito matrix [11].

4   A New FPGA-Based Approach

A common approach to the design of multipliers that is based on the Mastrovito matrix Z is to compute Z and then do the multiplication in $GF(2^m)$ by means of matrix-vector multiplication. In our approach, we exploit the symmetry of Z without actually computing Z.

A method for constructing the Mastrovito matrix is proposed in [5]. According to this method if $GF(2^m)$ is defined by the trinomial $\alpha^m + \alpha^n + 1$ then Z is given by

$$Z = \begin{bmatrix} U \\ L \end{bmatrix}$$

where U and L are Toeplitz matrices defined as follows:

Let $F = [0 \, a_{m-1} \, a_{m-2} \ldots a_1]$ and for each $i = 0, 1, \ldots, m-1$, let $F[i \rightarrow]$ be the result of shifting F i positions to the right (vacated positions on the left are filled with zeros). Also let $G = [a_n \, a_{n-1} \ldots a_1 \, a_0 \, a_m \ldots a_{n+1}]$

U is $n \times m$, its first column is $[a_0 \, a_1 \ldots a_{n-1}]^T$, and its first row is

$$[a_0] \| \sum_{i=0}^{N-1} F[i\Delta \rightarrow]$$

where $\Delta = m - n$, $\|$ represents concatenation, and N is a short notation for $N[m, n]$.

L is $\Delta \times m$, its first column is $[a_n \, a_{n+1} \ldots a_{m-1}]^T$, and its first row is

$$G + \sum_{i=0}^{N-1} F[i\Delta \rightarrow]$$

Although the previously described method is used for constructing the entire Mastrovito matrix Z, in this work we construct only one row of Z which is sufficient in our approach for carrying out multiplications in $GF(2^m)$. By constructing the n-th row Z_n (where rows are numbered $0, 1, \ldots$), the remaining rows of Z can be obtained by means of right-shifts and concatenations over Z_n.

Example: If $GF(2^7)$ is defined by $\alpha^7 + \alpha^4 + 1$, then $\Delta = 3$, $N = 2$, and

$$G = [a_4 \, a_3 \, a_2 \, a_1 \, a_0 \, a_6 \, a_5]$$

$$\sum_{i=0}^{N-1} F[i\Delta \rightarrow] = F + F[\Delta \rightarrow] = [\,0\ a_6\ a_5\ a_4\ a_3\ a_2\ a_1\,] + [\,0\ 0\ 0\ 0\ a_6\ a_5\ a_4\,]$$

and so L_0 is

$$Z_4 = [\,a_4\ a_3 + a_6\ a_2 + a_5\ a_1 + a_4\ a_0 + a_3 + a_6\ a_6 + a_2 + a_5\ a_5 + a_1 + a_4\,]$$

The proposed multiplier is implemented in a parallel/serial architecture which computes a multiplication in m clock cycles. One output bit of C is obtained in each cycle by multiplying (inner product) the current row Z_i by B, the current row is obtained by right-shifting the previous row and filling the vacated position on the left with a_i. Algorithm 1 shows this process.

Algorithm 1

Input: $A(\alpha), B(\alpha), Z_n;\ A(\alpha), B(\alpha) \in GF(2^m)$
Output: $C(\alpha) = A(\alpha)B(\alpha);\ C(\alpha) \in GF(2^m)$

$S \leftarrow Z_n$
for $i = 0$ to $m - 1$
    $c_{(i+n)\ \text{mod}\ m} \leftarrow S \cdot B$
    $S \leftarrow \text{right-shift}(S)$
    $s_0 \leftarrow a_{i+n}$
end for
return(C)

5   Experimental Results

This research is focused on accelerating finite field arithmetic using FPGAs for efficient solution to the reverse engineering problem in a hardware/software environment. Our target platform is a Cray XD1 system which includes six FPGAs units tightly integrated to 12 2.2 GHz Opteron AMD processors through a high bandwidth interconnection system. FPGA units are Xilinx Virtex II-Pro xc2vp50-7.

We have done an initial test to determine the acceleration gained by using FPGAs versus a high performance processor. A performance evaluation of FPGA and CPU implementation for the direct multiplier in $GF(2^{63})$ was made on the Cray XD1 platform; the field size was chosen to fit the 64-bit word-length of the target CPU architecture. A performance comparison between these multipliers and our approach is presented in Table 1. Times are measured for stand alone designs, in order to avoid the high overhead times arising from communication between the FPGA and the processor. Notice that the direct multiplier implementation on CPU is faster than the FPGA version of this method, a fact that is attributable to differences in clock speed: The clock rate for the Virtex-2P FPGA is about one-tenth that of on Opteron processor. However, when implemented on the same FPGA, our approach is about 65 % faster than the direct multiplier method.

Table 1. Multipliers comparison for the field $GF(2^{63})$ on Cray XD1: FPGA Virtex-2P xc2vp50-7, CPU 2.2 GHz AMD-Opteron

Multiplier	Time	Clock-period (Frequency)
Our approach on FPGA	0.62 μs	9.838 ns 101.6 MHz
Direct multiplier on FPGA	1.02 μs	16.168 ns 61.9 MHz
Direct multiplier on CPU	0.78 μs	

In Table 2 we compare our approach with other efficient multipliers reported in [3,4]. The field sizes used in this experiment are the same as those used in the cited references, the only suitable benchmarks for comparisons that are known to us. However, our approach can be implemented for larger finite fields. Finite fields elements are represented as bit-arrays in our implementation. These arrays are a part of the entire logic design, which uses a small number of slices. For instance, the multiplier for $GF(2^{239})$ uses 1.53 % of slices in the Cray XD1 FPGA (see Table 3). Therefore there are many slices available for implementing larger finite field multipliers.

The times in Table 2 have been measured using FPGA synthesis results reported by Xilinx tool XST (Xilinx Synthesize Technology) included in the package ISE Foundation 7.1. Our implementations are synthesized without area and timing constraints.

Table 2. Multipliers comparison

Field	Target FPGA	Implementation	Time (μs)	Space (slices)
$GF(2^{210})$	Xilinx Virtex xcv-300-6T	Reference [7] This work	12.30 2.21	343 334
$GF(2^{233})$	Xilinx xc2v-6000-4	Reference [4] This work	2.58 2.42	not reported 415
$GF(2^{239})$	Xilinx Virtex xcv-300-6	Reference [3] This work	3.10 2.47	359 385

According to the given results, our implementation exhibits the best time performance, whereas the area is not the most favorable for some cases. However our main goal is to achieve very fast computation using reasonably the physical devices.

Higher acceleration rates are obtained using the Cray XD1 FPGA (see Table 3). According to our results, there are significant opportunities for speeding up reverse engineering for large genetic networks on the Cray XD1 using reasonably the FPGA's physical space, however the communication time between CPU and

Table 3. Multipliers comparison on the Cray XD1 FPGA

Field	Time (μs)	Space (slices)	Space Utilization
$GF(2^{210})$	1.85	305	1.29%
$GF(2^{233})$	2.02	369	1.56%
$GF(2^{239})$	2.04	363	1.53%

FPGA becomes an obstacle. The communication model that we have used is a simple push-model in which the CPU pushes the input data to the FPGA's registers, and reads the output data from a destination register on the FPGA. Our experimental results indicate that this is a costly communication model, for example the direct multiplier for $GF(2^{63})$ spent 2.77 μs for communications and 0.62 μs for computations. Other works such as [2] have reported similar comunication problems with the Cray XD1.

6   Conclusions and Future Work

Finite field multiplication is the dominant operation in the interpolation phase of reverse engineering genetic networks. Traditionally CPU-based implementations of large finite field multiplication have implied challenging efforts in order to deal with common limitations such as architecture word-size, and storage space. Our approach overcomes these traditional obstacles and at the same time contributes to improved performance, achieving better times than other efficient FPGAs finite field multipliers.

Although our approach has shown to be efficient for the finite fields reported in the previous section, it promises more efficient results for multipliers on larger fields. In order to efficiently solve the reverse engineering problem for large genetic networks, our FPGA implementation could be used as a co-processor for accelerating a CPU-based interpolation algorithm, provided that we can resolve the problem of high communication costs. An alternative would be to shift more of the computational burden to FPGAs by embedding our multiplier in an FPGA-based interpolation algorithm.

Future work includes extending our implementation of multiplication to finite fields generated by irreducible pentanomials. We also would like to extend the field size in order to deal with genetic networks with $m \geq 500$ genes. The multipoint interpolation for solving reverse engineering problem for very large genetic networks using high performance reconfigurable computing requires a judicious partitioning of the problem between high performance CPUs and FPGAs. Here the communication overhead is an important issue and we have to consider alternative ways of communication in order to improve the overall performance. A potential solution is to take advantage of FPGA Transfer Region of Memory using the I/O subsystem developed by OSC [2].

Finally, we could extend our ideas for doing multiplication over finite fields $GF(2^m)$ to doing the algebra of polynomials over $GF(2^m)$. This would enable us to carry out the whole interpolation algorithm needed for the reverse engineering problem in FPGAs, thus allowing better optimization of the ration of computation time to communication time.

References

1. D. Bollman, E. Orozco, O. Moreno, "A Parallel Solution to Reverse Engineering Genetic Networks", in Gavrilova et al (eds), Lecture Notes in Computer Science, Springer-Verlag, Part III, 3045, pp. 490-497, 2004.
2. J.Fernando, D. Dalessandro, A. Devulapalli, and A. Krishnamurthy, "Enhancing FPGA Based Encryption", Ninth Workshop on High Performance Embedded Computing (HPEC). Sept. 2005.
3. M.A. Garcia-Martinez, R. Posada-Gomez, G. Morales-Luna, F. Rodriguez-Henriquez. "FPGA implementation of an efficient multiplier over finite fields $GF(2^m)$", Proceedings of International Conference on Reconfigurable Computing and FPGAs, 2005 (ReConFig'05), September 2005.
4. C. Grabbe, M. Bednara, J. Shokrollahi, J. Teich and J. Von Zur Gathen, "FPGA Designs of parallel high performance Multipliers", Proceedings of the IEEE International Symposium on Circuits and Systems (ISCAS-03), volume II, 268-271. Bangkok, Thailand.
5. A. Halbutogullari, Ç. Koç, "Mastrovito Multiplier for General Irreducible Polynomials", IEEE Transactions on Computers, volume 49, number 5, pp. 503-518, 2000
6. T.E. Ideker, V. Thorsson, and R.M. Karp, "Discovery of regulatory interactions through perturbation: Inference and experimental design," Pacific Symposium on Biocomputing, No. 5 pp 302-313, 2000.
7. P. Kitsos, G. Theodoridis, and O. Koufopavlou, "An efficient Reconfigurable Multiplier Architecture for Galois Field $GF(2^m)$", Microelectronics Journal, volume 34, pp.975-980, 2003.
8. Laubenbacher and B. Stigler, "Dynamic networks," Adv. in Appl. Math Vol.26, pp. 237-251, 2001.
9. R. Laubenbacher, J. Shah, and B. Stigler, "A Computational Algebra Approach to the Identification of Gene Regulatory Networks", Proc. Third International Congress on Systems Biology, Stockholm, 2002.
10. Lee, T.I., Rinaldi, N.J., Robert, F., Odom, D.T., Bar-Joseph, Z., Gerber, G.K., Hannett, N.M., Harbison, C.R., Thompson, C.M., Simon I., Zeitlinger J., Jennings, E.G., Murray, H.L., Gordon, D.B., Ren, B., Wyrick, J.J., Tagne, J., Volkert T.L., Fraenkel, E., Gifford D.K., and Young, R.A. Transcriptional Regulatory Networks in Saccharomyces cerevisiae. Science 298: 799-804 (2002).
11. E.D. Mastrovito, "VLSI Architectures for Computation in Galois Fields", PhD thesis, Dept. of Electrical Eng., Linkvping Univ., Linkvping, Sweden, 1991.
12. O. Moreno, D. Bollman, and M. Aviñó, "Finite dynamical systems, linear automata, and finite fields", 2002 WSEAS Int. Conf. on System Science, Applied Mathematics & Computer Science and Power Engineering Systems, pp 1481-1483.
13. H. Ortiz, "Analysis of Gene Regulatory Networks Using Finite-Field models", PhD. Thesis Proposal, University of Puerto Rico, 2005.
14. B. Sunar and Ç. K. Koç "Mastrovito Multiplier for All Trinomials", IEEE Transactions on Computers", volume 48, number 5, pp. 522-527, May 1999.

Two Challenges in Genomics That Can Benefit from Petascale Platforms

Catherine Putonti[1,2], Meizhuo Zhang[1], Lennart Johnsson[1],
and Yuriy Fofanov[1,2]

[1] University of Houston, Department of Computer Science, 218 Philip G. Hoffman
Hall, Houston, Texas 77204-3058 USA
[2] University of Houston, Department of Biology and Biochemistry, Houston, Texas
77204-5001 USA
putonti@bioinfo.uh.edu, mzhang@bioinfo.uh.edu, johnsson@cs.uh.edu,
yfofanov@uh.edu

Abstract. Supercomputing and new sequencing techniques have dramatically increased the number of genomic sequences now publicly available. The rate in which new data is becoming available, however, far exceeds the rate in which one can perform analysis. Examining the wealth of information contained within genomic sequences presents numerous additional computational challenges necessitating high-performance machines. While there are many challenges in genomics that can greatly benefit from the development of more expedient machines, herein we will focus on just two projects which have direct clinical applications.

1 Introduction

Recent advances in sequencing techniques have lead to an explosion in the amount of biological data available. The number of sequences made publicly available is increasing exponentially. The whole genome sequencing strategy fragments the genomic sequence into many overlapping sequences, sequences these smaller segments, and then assembles these shotguns into contiguous sequences. Assembling the shotgun sequences produced necessitates high-performance machines. It was estimated that the assembly of the human genome took approximately 20,000 hours of CPU time [1]. In addition to the human genome, one must mention the ongoing sequencing efforts of several other organisms such as chimpanzee, chicken, rat, mouse, cow, and dog, just to name a few. With the advent of the recently developed 454 sequencing technique by 454 Life Sciences (Branford, CT), one may expect the availability of sequence data to proceed even more rapidly. Sequencing of the corn genome, considered the most complex sequencing project attempted to date, is possible thanks to IBMs Blue Gene/L supercomputer, capable of a peak performance of 5.7 teraflops (http://www.iastate.edu/%7enscentral/news/2006/jan/supercomputer.shtml).

In addition to sequencing projects, identification of single nucleotide polymorphisms (SNPs), which appear throughout the genomic sequence, is still underway. SNPs are responsible for the variations among individuals and have

W. Lehner et al. (Eds.): Euro-Par 2006 Workshops, LNCS 4375, pp. 313–322, 2007.

been shown to be directly correlated with an individuals susceptibility to disease, response to vaccines and medications, as well as the success of blood, tissue, and organ transplantations. To date 11,961,761 SNPs, 5,646,244 of which have been validated, have been identified in the human genome and are publicly available from the National Center for Biotechnology Information (NCBI). The International HapMap Consortium was formed specifically to catalog genetic these variations in order to determine the similarities and differences in human beings [2]. All of the data from the HapMap project is publicly available from http://www.hapmap.org and includes the genotypes available from the Affymetrix (Santa Clara, CA) GeneChip® Human Mapping 500K Array Set which is comprised of two arrays, each of which is capable of genotyping 250,000 SNPs.

These genomic sequences and identified SNPs contain a wealth of information including indicators for an individuals immunity and susceptibility to disease. Mining such data is not a trivial task. Translating the information found to real applications in the medical and clinical arena is of great importance. In recent years, various diagnostic assays have been developed using nucleic acid-based technologies including the polymerase chain reaction (PCR), microarrays of cDNA and oligonucleotides, and nucleic acid sequence-based amplification (NASBA) assays, amongst others. Nucleic acid-based methods are founded on the principles of hybridization and include primer and/or probe sequences which are complementary to a region of the genomic material of the target region, e.g. a particular gene, mRNA, etc., such that in its presence, the primer/probe will hybridize to the targets DNA/RNA. The results of such assays also contribute to the overwhelming amount of data presently available.

The high-throughput nucleic acid-based microarray format was originally utilized exclusively for the monitoring of gene expression. As is evidenced by the Affymetrix GeneChip®, this is no longer the case. Microarrays, commercially produced as well as those produced in-house, are now being employed for gene expression profiling, sequencing and resequencing efforts, genotyping, DNA-protein interactions, pathogen–host interactions, diagnostics, etc. Countless publications have been dedicated to discussing the applications of microarrays, e.g. [3,4,5,6]. Commercially available arrays can contain thousands to hundreds of thousands of probe sequences, each of which contains information specific to the experimental design. Through gene expression experiments, one can gather information about an organisms cellular functions, regulatory mechanisms as well as biochemical pathways. Microarrays used in such experiments, however, produce only an image representing the hybridization of the microarray probe sequences and the target mRNA. The task of translating this hybridization image into the actual gene regulatory and proteomic networks is far from simple and computationally intensive. Numerous approaches and applications have been developed for inferring these networks. The power of the computational resources available is of primary concern for all such techniques. As the cost of microarrays continues to decrease and the number of probes which can be accommodated increases, the number of assays produced and thus information generated will most certainly continue to increase.

While there are many challenges in genomics that can greatly benefit from the development of more expedient machines, herein we will focus on just two projects having direct clinical applications. Both are based on the analysis of genomic sequences for the design of probe sequences for the identification of single somatic base mutations and SNPs.

2 Genomic Signatures for Monitoring the Rate of Accumulation of Somatic Mutations

In the living cell, DNA undergoes frequent chemical changes, most of which are quickly repaired. Those that are not result in a mutation. Evolution absolutely depends on mutations because this is the only way that new alleles can be created. At the same time, however, most of the mutations observed are harmful or, at best, neutral. Mutations are relatively rare events. Humans inherit about 3×10^9 base pairs of DNA from each parent. Just considering single-base substitutions, this means that each cell has approximately 6×10^9 different base pairs that can be the target of a substitution. Single-base substitutions are most likely to occur when DNA is being copied. It has been estimated that in humans and other mammals, uncorrected errors occur at the rate of 1 in every 5×10^7 nucleotides [7].

The ability to evaluate the rate at which these mutations are accumulating is advantageous for many facets of research. Firstly, it would be possible to estimate a tissues specific biological age as well as predict the risk of somatic mutation-related disorders such as those that cause certain types of cancer. For instance, retinoblastoma, cancer of the retina, typically affects children. The development of a tumor occurs as a result of somatic mutations in both copies of the RB1 gene or through the inheritance of a mutation in a single copy of the RB1 gene and a somatic mutation in the other. Thus, by monitoring the mutation rate within the RB locus of children having a hereditary predisposition could provide a means of early detection. Another important application for the estimation of the mutation accumulation rate is for monitoring the deviation of cell lines from their ancestors. Because at the present time cell lines are less expensive than laboratory animals, human (and other organisms, e.g., mouse, rat, etc.) cell lines are commonly used in cancer research, drug design, and drug screening. Because mammalian cells usually cannot perform more than 50–60 divisions, many cell lines are artificially immortalized. Technically, human cells growing in a culture are different from normal human cells in their natural environment, and this difference becomes more and more significant in time because of the mutation accumulation process. Therefore, research as well as drug development and testing would greatly benefit from early detection of the genetic drift of cell lines. The current method to estimate the mutation rate is based on analysis of a particular part, usually within the coding region, of a genome. Such an approach requires PCR amplification of the genomic region of interest followed by sequencing, which can be both time consuming and expensive.

Recently, utilizing our computational abilities we have performed analysis of subsequences of length n (n-mers) located in the area of "one mismatch distance for several microbial genomes. The one mismatch distance corresponds to sequences that are not present in the genome but given any one base change or mismatch are present in the genomic sequence. Such sequences have a higher probability of appearing as a result of a single point mutation than, for example, sequences 2, 3, etc. mismatches away. For each n-mer located in the one mismatch distance area, we were able to compute the exact number ancestor sequences in the genome which can "mutate to this n-mer. As a result of these calculations, we observed a large variation in the number of ancestors for different sequences: from 1 or 2 (expected) to thousands for microbes and presumably dozens of thousands for humans. This new observation leads us to the idea of using a set of sequences having a high number of ancestors as an indicator of the accumulations of mutations. This approach is based on the assumption that sequences with a higher number of ancestors have a much higher probability to appear as the result of random mutations.

For microbial genomes, identifying the n-mers within the areas of 1, 2, and 3+ mismatch(s) is computationally feasible due to the fact that these genomic

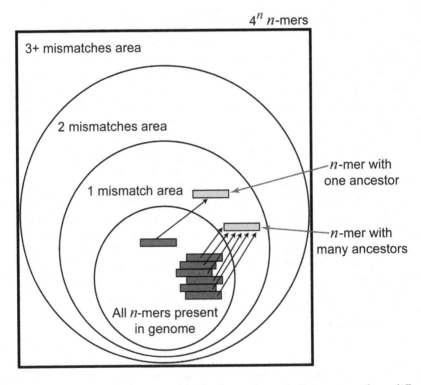

Fig. 1. Those n-mers that are located in the "one mismatch distance area have different probabilities of appearing as result of random mutation

sequences are relatively short. For instance, the *Escherichia coli* K12 genome (NCBI: NC_000913) is 4,639,675 bp. Considering both strands of this genome, all n-mers are expected to appear only once at most for $n \leq 12$. Thus, for 12-mers one may expect that some sequences are absent from the genomic sequence which may have ancestors present. This prediction is made under the assumption that the appearance of n-mers is independent which is known not to be completely true; it does, however, provide a reasonable estimation. There are in fact several 10- and 11-mers which are absent from the *E. coli* K12 genome. For a single n-mer S of length n, there are $3n$ possible n-mers which can be created by changing only one base, $3n * 3(n-1)$ possible n-mers by changing any two bases, etc. For each of the n-mers absent from the genome, there are

$$\prod_0^c 3(n-c) \tag{1}$$

possible ancestors which can be present given c base changes. For the human genome (≈ 3 Gbp), much longer n-mers must be examined. As a result, many more possible combinations of mutations must be considered. It is important to mention that rather than looking exclusively at the human genome sequence, SNP information must also be included as the appearance of such a variation would be the result of the individuals particular genotype rather than the occurrence of a mutation.

Using such sequences in an assay, e.g. as microarray probes, would provide a means to rapidly and conclusively examine the rate of the accumulation of mutations in microbial and eukaryotic organisms, including microbial cultures, strains of model organisms, and human cell lines. We expect that the same approach will also be useful in distinguishing individual humans, as well as other eukaryotes (including organisms of economic importance, e.g., pigs, cows, chickens, etc.) based on small DNA samples. Furthermore, we anticipate that the sensitivity of this technique will be sufficient to monitor the rate of somatic mutations accumulating in different tissues during the lifetime of an organism. Such an opportunity is likely to be of great benefit to cancer related research. In contrast to existing methods of mutation rate monitoring, our approach takes into account entire genomes, including noncoding sequences which in the case of human cover 97% of genome. A calculation of such sequences and their ancestors becomes exponentially more difficult as the n-mer size being considered increases. Due to the large number of calculations that are required, petascale computing provides a solution in which such computations can be conducted in an acceptable amount of time.

3 Optimal Combinations of Genomic Signatures for Human HLA Typing

The HLA (human leukocyte antigen) system, the group of genes in the human MHC (major histocompatibility complex) located on chromosome VI, encodes the cell-surface antigen-presenting proteins. HLA antigens are the major determinants used by the body's immune system for recognition and differentiation of self from nonself (foreign) substances. This system consists of numerous SNPs encoding 2435 known alleles according to the latest statistics available from the IMGT/HLA Database as of August 3, 2006 [8]. Since the previous release (2.13) a month earlier, 125 new alleles were added [8]. The allelic composition in the HLA loci, or the HLA type, varies significantly within the population. A direct correlation has been observed between the variation present in the HLA system and ones genetic susceptibility to diseases [9,10,11,12], response to infection [13], response to drugs and vaccines [14,15], as well as the success of blood and tissue transplants [16]. Significant research has been devoted to the development of HLA typing techniques for determining the combination of alleles responsible for particular responses. Numerous nucleic acid-based approaches have been employed for genotyping the HLA loci by designing primer/probe sequences complementary to the SNPs present in particular alleles. The design of an assay which can type all of the known alleles as well as be used for the discovery of new HLA alleles would dramatically improve current typing methods. The high-throughput microarray format is ideal, offering the convenience of miniaturization and the ability to perform thousands of hybridizations in a single experiment. Previously documented microarray-based typing methods were intended to provide a low resolution typing and are therefore able to identify only 6% to 33% of the allelic variations in the loci of interest [17,18,19,20,21]. In order to achieve higher resolution typing, more alleles must be targeted. At the same time, however, one most consider the complexity of the assay design as it may impact the expediency in which a diagnosis can be made as well as the cost and the level of expertise required for technicians to correctly interpret the results. Thus, a simplistic design is preferable. Minimizing the number of probes in the assay, while maintaining or improving the assays resolution and reliability, further complicates the task of designing typing tests thus necessitating rigorous computations.

Alleles	P_1	P_2	P_3	P_4
A_1	✓		✓	
A_2		✓		✓
A_3	✓	✓	✓	✓
A_4		✓	✓	✓
A_5	✓	✓		

Fig. 2. Determining the optimal probe set. Each allele is expected to hybridize with a subset of the probes $\{P_1, P_2, P_3, P_4\}$. Not all probes, however, may be necessary in order to maintain the same resolution. Thus, the optimal set $\{P_1, P_3, P_4\}$ provides the same resolution while reducing the complexity and cost of the assay.

Different probe sets for HLA typing can be created providing variable resolutions. Given a set of alleles, all of the sequences containing the polymorphisms can be selected as candidate probes. While all of the candidate probes can be included in the design of the assay, it is likely that some of the probes do not contribute to the informativity or increase the resolution of the typing. To illustrate such an instance, Figure 2 shows four probes (P_1, P_2, P_3, and P_4) which can be used to distinguish between five different alleles (A_1, A_2, A_3, A_4, and A_5). In order to be able to distinguish one allele from another, the set of probes which hybridizes to each allele must be unique resulting in a unique hybridization pattern on the array. The inclusion of P_2, however, is not necessary as all five alleles will still be expected to hybridize with a unique set of probes. By removing any of the other probes, it will no longer be possible to distinguish between the alleles and thus the resolution of typing is decreased. Therefore, an "optimal set containing only three probes can be used to distinguish between the alleles. Although a rather simple example, the optimal assay provides the same resolution at a reduced cost and complexity. As one can imagine, identifying the minimum number of probes necessary for distinguishing between 2000+ HLA alleles is a significantly more complex problem.

To optimize an assay, one can either search for the maximum coverage of the targets using: (1) a predetermined number of probes or (2) the minimum number of probes. If there are only a few candidate probes, it is feasible that one can iterate through each possible combination in order to identify the combination of probes with the maximum coverage. However, as the number of candidate probes increases so too does the number of possible combinations. For instance, to identify the set of probes having the maximum coverage for an assay of the predetermined size of 60 probes from 100 candidate probes, 1.37×10^{28} different combinations must be examined. Approximately 1.27×10^{30} combinations exist for the same set of 100 candidate probes for all possible probe set sizes. If analysis of each combination requires 1 millisecond, iterating through all of the combinations to find the optimal set using the minimum number of probes will take 4.02×10^{19} years!

Computing the minimum number of probes in a realistic time is nontrivial (an instance of the minimum set cover problem which is NP-complete). Optimization of the probe set design problem was first discussed in 2000 by Herwig et al. [22] in which a greedy heuristic was introduced based on clustering and entropy. Formulation of the problem was further refined by Borneman et al. [23] to the Minimum Cost Probe Set (MCPS) and Maximum Distinguishing Probe Set (MDPS); MCPS searches for the minimum number of probes necessary to distinguish all target sequences while MDPS maximizes the number of distinguished pairs of target sequences for a set of k probes. Here Borneman *et al.* [23] developed a Lagrangian relaxation algorithm to approximate the MCPS problem and a simulated annealing algorithm for the MDPS problem. While successful in designing a smaller probe set, certain sacrifices were made for efficiency by considering only one length of probes ($n = 8$) and predetermining the set size. Two approaches were also developed based on the Integer Linear Program (ILP) formulation. The method of Rash et al. [24] uses suffix trees to solve the minimization problem. Their solution is based upon the concept of a unique barcode. This barcode is a binary vector consisting of 0s and 1s where 0 means that the probe sequence will not hybridize with the sequence of interest and 1 means that the probe sequence will hybridize with the sequence of interest. In this ILP implementation, each sequence (genome) being considered must be uniquely identified by at least one probe under the assumption that only one target sequence is present in the sample [24]. The second approach of Klau et al. [25] consists of three steps: (1) computing the target–probe incidence matrix, (2) computing a design matrix, and (3) decoding the result for identification of the sequence(s) present in the sample [26]. The design is computed using a branch-and-cut algorithm (http://www.inf.fu-berlin.de/inst/ag-bio). This algorithm proves more robust than that proposed by Rash et al. [24] by taking into consideration during design the set size, the probability of hybridization errors, and the case in which multiple targets are simultaneously present in the sample (*d-separability*) [25]. All of these approaches [22,23,24,25,26], can only approximate the best solution within the space and time allotted to the probe set design process. Therefore, the optimal set identified by any such approach may not in fact be the true optimal set having the minimum number of probes for the maximum coverage. With respect to designing the optimal probe set for high resolution HLA typing, certain alleles may occur with very low probability. If it is possible to identify all alleles except this rare allele with a higher level of resolution and a smaller probe set, such a solution may be preferred, thus adding another dimension to the optimization problem.

4   Conclusions

The rate in which new data is becoming available far exceeds the rate in which one can perform analysis. Sequence data as well as the results of microarray experiments of gene expression profiling, genotyping, and diagnostics further contribute to amount of data to be examined. Analysis of large, complex genomic sequences such as the human genome necessitates high-performance computing resources. The projects discussed here are just two of many that are currently underway in research laboratories throughout the world. Due to the limitations of current systems, it has only been possible to analyze a fraction of the vast amount of biological data currently available. The development of cutting edge computational resources, both in terms of the memory available and the precision and speed in which calculations can be performed, is likely to dramatically impact biotechnology, human health as well as our general understanding of mechanisms of disease development, vaccine development, aging, and evolution.

References

1. Walgate, R.: Weapons lab to develop Celeras new supercomputer. Genome Biol. (2001)
2. The International HapMap Consortium: A haplotype map of the human genome. Nature 437 (2005) 1299–1320

3. Jares, P.: DNA microarray applications in functional genomics. Ultrastruct. Pathol. 30 (2006) 209–219
4. Lockhart, D.J., Winzeler, E.A.: Genomics, gene expression and DNA arrays. Nature 405 (2000) 827–836
5. Peeters, J.K., Van der Spek, P.J.: Growing applications and advancements in microarray technology and analysis tools. Cell Biochem. Biophys. 43 (2005) 149–166
6. Geschwind, D.H.: DNA microarrays: translation of the genome from laboratory to clinic. Lancet Neurol. 2 (2003) 275–282
7. Kimball, J.W.: Biology. 6th edn. Wm. C. Brown, Iowa (1994)
8. Robinson, J., Waller, M.J., Parham, P., de Groot, N., Bontrop, R., Kennedy, L.J., Stoehr, P., Marsh, S.G.E.: IMGT/HLA and IMGT/MHC: sequence databases for the study of the major histocompatibility complex. Nucleic Acids Res. 31 (2003) 311–314
9. Diepstra, A., Niens, M., Te Meerman, G.J., Poppema, S., van den Berg, A.: Genetic susceptibility to Hodgkins lymphoma associated with the human leukocyte antigen region. Eur. J. Haematol. Suppl. 75 (2005) 34–41
10. Saftlas, A.F., Beydoun, H., Triche, E.: Immunogenetic determinants of preeclampsia and related pregnancy disorders: A systematic review. Obstet. Gynecol. 106 (2005) 162–167
11. Ahmedov, G., Ahmedova, L., Sedlakova, P., Cinek, O.: Genetic association of type 1 diabetes in an Azerbaijanian population: the HLA-DQ, -DRB1*04, the insulin gene, and CTLA4. Pediatr. Diabetes 7 (2006) 88–93
12. Listi, F., Candore, G., Balistreri, C.R., Grimaldi, M.P., Orlando, V., Vasto, S., Colonna-Romano, G., Lio, D., Licastro, F., Franceschi, C., Caruso, C.: Association between the HLA-A2 allele and Alzheimer disease. Rejuvenation Res. 9 (2006) 99–101
13. Keet, I.P., Tang, J., Klein, M.R., LeBlanc, S., Enger, C., Rivers, C., Apple, R.J., Mann, D., Goedert, J.J., Miedema, F., Kaslow, R.A.: Consistent associations of HLA class I and II and transporter gene products with progression of human immunodeficiency virus type 1 infection in homosexual men. J. Infect. Dis. 180 (1999) 299–309
14. Ovsyannikova, I.G., Vierkant, R.A., Poland, G.A.: Importance of HLA-DQ and HLA-DP polymorphisms in cytokine responses to naturally processed HLA-DR-derived measles virus peptides. Vaccine 24 (2006) 5381–5389
15. Ovsyannikova, I.G., Pankratz, V.S., Vierkant, R.A., Jacobson, R.M., Poland, G.A.: Human leukocyte antigen haplotypes in the genetic control of immune response to measles–mumps–rubella vaccine. J. Infect. Dis. 193 (2006) 655–663
16. Morishima, Y., Sasazuki, T., Inoko, H., Juji, T., Akaza, T., Yamanoto, K., Ishikawa, Y., Kato, S., Sao, H., Sakamaki, H., Kawa, K., Hamajima, N., Asano, S., Kodera, Y.: The clinical significance of human leukocyte antigen allele compatibility in patients receiving a marrow transplant from serologically HLA-A, HLA-B, and HLA-DR matched unrelated donors. Blood 99 (2002) 4200–4206
17. Haddock, S.H., Quartararo, C., Cooley, P., Dao, D.D.: Low-resolution typing of HLA-DQA1 using DNA microarray. Methods Mol. Biol. 170 (2001) 201–210
18. Consolandi, C., Frosini, A., Pera, C., Ferrara, G.B., Bordoni, R., Castiglioni, B., Rizzi, E., Mezzelani, A., Bernardi, L.R., De Bellis, G., Battaglia, C.: Polymorphism analysis within the HLA-A locus by universal oligonucleotide array. Hum. Mutat. 24 (2004) 428–434
19. Palmisano, G.L., Delfino, L., Fiore, M., Longo, A., Ferrara, G.B.: Single nucleotide polymorphisms detection based on DNA microarray technology: HLA as a model. Autoimmun. Rev. 4 (2005) 510–514

20. Bang-Ce, Y., Xiaohe, C., Ye, F., Songyang, L., Bincheng, Y., Peng, Z.: Simultaneous genotyping of DRB1/3/4/5 loci by oligonucleotide microarray. J. Mol. Diagn. 7 (2005) 592–599

21. Wells, D.: Advances in preimplantation genetic diagnosis. Eur. J. Obstet. Gynecol. Reprod. Biol. 115 (2004) S97–S101

22. Herwig, R., Schmidt, A., Steinfath, M., OBrian, J., Seidel, H., Meier-Ewert, S., Lehrach, H., Radelof, U.: Information theoretical probe selection for hybridization experiments. Bioinformatics 16 (2000) 890–898

23. Borneman, J., Chrobak, M., Vedova, C.D., Figueroa, A., Jiang, T.: Probe selection algorithms with applications in the analysis of microbial communities. Bioinformatics 17 (2001) S39–S48

24. Rash, S., Gusfield, D.: String barcoding: uncovering optimal virus signatures. In: Myers, G., Hannenballi, S., Istrail, S., Perzner, P., Waterman, M. (eds): RECOMB '02: Proceedings of the Sixth Annual International Conference on Computational Biology. ACM Press, New York (2002) 254–261

25. Klau, G.W., Rahmann, S., Schliep, A., Vingron, M., Reinert, K.: Optimal robust nonunique probe selection using integer linear programming. Bioinformatics 20 (2004) I186–I193

26. Schliep, A., Torney, D.C., Rahmann, S.: Group testing with DNA chips: generating designs and decoding experiments. Proc. IEEE Comput. Soc. Bioinform. Conf. 2 (2003) 84–91

High Throughput Image Analysis on PetaFLOPS Systems

Robert Henschel, Matthias Müller[1], and Yannis Kalaidzidis[2]

[1] Technische Universität Dresden,
Zentrum für Informationsdienste und Hochleistungsrechnen (ZIH),
01062 Dresden, Germany
Robert.Henschel@zih.tu-dresden.de
http://www.tu-dresden.de/zih
[2] Max Planck Institute of Molecular Cell Biology and Genetics,
Pfotenhauerstr. 108,
01307 Dresden, Germany

Abstract. Today's state of the art high-throughput screening facilities can produce tens of thousands of images of cells per day. Analyzing images from high-throughput screening experiments is very time consuming and computationally demanding. Researchers are currently limited not by the availability of experimental data, but by the computing resources for the image analysis. The Max Planck Institute of Molecular Cell Biology and Genetics Dresden, Germany, (MPI-CBG) and the Center for Information Services and High Performance Computing at the Technische Universität Dresden (ZIH) are working together to integrate high performance computing systems into the workflow of biologists. The MPI-CBG has developed software that biologists use for their image analysis work. The software can utilize local workstations and remote HPC systems for image analysis. Currently the software is used successfully on small clusters and PC-Farms. Most parts of the image analysis workflow of screening experiments can be performed in parallel and is ideal for distribution on large systems. With a few modifications and a new approach to data management, the software should be able to scale to PetaFLOPS systems.

1 High Throughput Image Analysis

Determining the DNA sequence and the genes of an organism is an automated process. The DNA of a number of organisms has been successfully sequenced and is now available for further research. In contrast, assigning a function to a gene is far more complicated and involves a lot of human interaction.

High-Throughput screening experiments are used to search for gene functions. These experiments do also include a large number of control experiments to cope with off-target effects and other side-effects that occur in complex screening experiments. Because of the large number of images that high-throughput screening experiments produce, automated image analysis software is used to analyze the images and detect phenotypes.

W. Lehner et al. (Eds.): Euro-Par 2006 Workshops, LNCS 4375, pp. 323–329, 2007.

The research group of Professor Dr. Marino Zerial at the MPI-CBG develops its own image analysis software called Motion Tracking. The project was started because there was no standard software package available on the market that would suit the needs of the biologists, in terms of ease of use, feature set and scalability. The application that is developed at the MPI-CBG is used for the whole image analysis process and despite its name, is not limited to just motion tracking. It is capable of performing image analysis on a large number of images in batch mode. It can also visualize the input and the result data as well as perform statistical calculations. To decrease the overall runtime of image analysis tasks, the application can distribute the calculation on small clusters of workstations running Microsoft Windows or on IA32/AMD64 HPC systems running Linux. Motion Tracking is successfully used in the analysis of current experiments such as research in endocytosis and endosomes [4].

The general image analysis workflow for screening experiments at the MPI-CBG is show in fig. 1. It consists of six distinct phases. At first, the biological experiment is prepared in plates with, for example, 96 or 384 wells. Each of those wells contains a slightly different experiment, for example different knocked down genes or a control experiment. Depending on the objective of the experiment, the number of required wells can vary quite a lot. In a small screening experiment that is only concerned with 120 genes, 12 96-well plates are required. For a genome wide screening experiment, more than 200 384-well plates would be required. In the next step, confocal images of the cell cultures or single cells are taken with an automated microscope. The microscope acquires images at different positions in each well and stores them in image packages.

In the preprocessing phase the images are extracted from the image packages, separate images for the red, green and blue color channel are combined into one image, out of focus images are removed and images that show no or only very few cells are filtered out. This phase is done in parallel for all images on a small cluster of eight dual XEON workstations at the MPI-CBG. The small cluster is used because the actual compute phase is very small and a lot of I/O operations are performed. Since the small cluster can be accessed faster than the HPC system at the ZIH and a more efficient communication protocol is used for

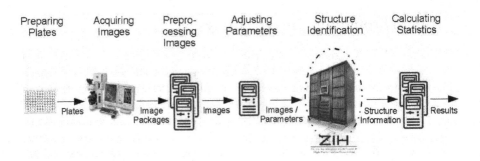

Fig. 1. Image Analysis Workflow at the MPI-CBG

communication to the small cluster, this phase is performed fastest on the small cluster.

In the next phase, the parameters for the structure identification algorithm are adjusted. Since the optimal parameter values vary between individual image sets, for example depending on the characteristics of the microscope that was used for the screening, this step requires human interaction.

With the parameter file created in the previous phase, all images are searched. Since most analysis tasks have no dependency on images besides the one that is analyzed, this process can be performed in parallel. The image files and the parameter file are transferred to the HPC system and analyzed. This phase is the most computational intensive. The analysis time for a single image can range from a few minutes to one or two hours depending on the analysis that is performed. The structure information is transferred back to the main workstation after the analysis.

The statistics calculation represents the final analysis step that provides answers to the underlying biological question of the screening experiment. It is done on the MPI-CBG cluster again since the calculation time is very small compared to the overhead that is incurred by transferring input files and results to the HPC system at the ZIH and back.

2   Performance and Limitations

Motion Tracking is written using a combination of the programing languages C++ and PLUK. The performance critical parts of the application as well as the PLUK interpreter are written in C++ while large parts of the application, the graphical user interface and utility functions are written in PLUK. PLUK is an interpreted programing language that was developed in the mid to late 1990s at the Lomonosov Moscow State University [3]. It was created as a fourth generation programing language, designed to make prototyping of algorithms easy and aid rapid application development. PLUK is not meant as a high performance computing tool, but is targeted at users that need to test algorithms and develop small to medium applications for example to process and analyze data of experiments.

Motion Tracking was developed as a tool for biologists to work with images from screening experiments. When the experiments got larger, it was enhanced to distribute calculation tasks among workstation to shorten the overall analysis time. At that time, the front-end of the application and the underlying PLUK library were only available for Microsoft Windows 2000 and later versions. This meant that the calculation tasks could only be distributed to workstations running a compatible version of Microsoft Windows. For the communication between the calculation workstations and the master workstation, a custom protocol was developed that is implemented using the Winsock library. One communication channel per workstation is used to dispatch tasks and to send and receive data. This channel is kept open for the entire runtime of a task. The solution works nicely for a limited number of workstations. The MPI-CBG uses this on a local cluster of 8 dual XEON workstations.

As the experiments got larger, the need for more compute resources arose. The ZIH provided access to its HPC systems, a PC-Farm with 64 dual Opteron nodes and an SGI Altix 3700 system with 192 Itanium 2 CPUs. For Motion Tracking to utilize those systems, the PLUK interpreter and associated libraries were ported to Linux. The graphical user interface that controls the calculation remained on the Microsoft Windows platform. The communication channel was replaced with an implementation that uses the OpenSSH binaries on the Windows platform to perform file transfer and remote job submission. The current version works satisfactory on the PC-Farm with Opteron CPUs. To use the Altix HPC system, the PLUK interpreter must be adapted to the memory alignment requirements of the Itanium 2 CPU which is still an open issue.

The PC-Farm at the ZIH was added to the image analysis workflow with as little modifications to existing software as possible. This means that the existing communication pattern was adapted to work over SSH connections but not really changed to reflect the properties of the new connection. The communication speed to the PC-Farm at the ZIH is slower than to the local workstations at the MPI-CBG. The latency and the startup time for a file transfer are a lot higher as well. To start a job on the PC-Farm, all input data has to be moved to the PC-Farm and a job has to be submitted to the batch system. At the end, the results of the calculation have to be moved back to the MPI-CBG for visualization in Motion Tracking. The input data and the results for each job are in the order of a few megabytes. All this contributes to a very high startup time for a calculation that can only be compensated by running large calculations.

Equation 1 can be used to calculate the number of processors (n) that can be used in parallel on the PC-Farm. If not enough jobs can be submitted from the master workstation at the MPI-CBG to the HPC system at the ZIH, not all available processors can be utilized. The maximum number of processors that can be utilized depends on the average runtime (T_R) of an image analysis job, the number of communication channels (c) that are available to send data and the transfer time per image (T_I) using such a channel. Every communication channel is used on the master workstation by a thread. Every thread prepares the input data, sends it to the HPC system and submits a job to the batch system. Currently, the number of threads is set to 7. It is limited by the number of concurrent open SSH connection to the HPC system and the load that those connections induce on the master workstation.

The transfer time per image varies between 4,5 and 5 seconds. It can be calculated using (2). The preparation time (T_P) is required to transfer the image to the master workstations memory and repackage it for sending it to the HPC system. The actual transfer time is comprised of the time to open the connection (T_O) and the time to transfer the image, which can be calculated by dividing the image size (I_S) by the transfer rate (R) of the channel. The time to submit a task to the batch system is denoted with T_S. Values for the individual times can be found in table 1.

Table 1. Key parameters of image Analysis jobs

Item	Value
Average runtime of image analysis jobs	≈1800 seconds
Number of channels	7
Transfer time of one image	≈4.75 seconds
Preparation time	≈2 seconds
Time to open a connection	≈0.5 seconds
Image size	2 MiB
Transfer rate	≈1.6 MiB/seconds
Submission Time	≈0.5 seconds

$$n = T_R * \frac{c}{T_I} \tag{1}$$

$$T_I = T_P + (T_O + \frac{I_S}{R}) + T_S \tag{2}$$

To utilize the full PC-Farm with 128 CPUs, the average runtime of an image analysis job must be at around 90 seconds. If the runtime is shorter the entire PC-Farm will not be filled with jobs before the first job has finished and makes the CPU available again. If the runtime is longer, jobs will pile up in the batch queue and are ready to be dispatched as soon as a CPU becomes available. Fig. 2 shows that the workflow software is able to utilize basically all available CPUs of the 128 CPU cluster.

Image analysis jobs are always longer than 90 seconds, so the overhead of the network transfer is no problem. The runtime of jobs in the preprocessing phase is largely determined by I/O operations of reading image packages and writing single images. This phase is not done on the PC-Farm. The runtime of jobs in the statistical calculation phase depends on how many statistical calculations are performed per image and how complex they are. The more calculations are performed, the better is the chance that such a job can be run efficiently on the PC-Farm.

Overall, the current solution works nicely for the current size of the PC-Farm. However, when the PC-Farm is upgraded to 2500 Opteron cores the submission rate has to be increased to be able to utilize the whole PC-Farm. If the submission rate stays at 1.5 jobs per second, the average job length would have to be more than 1300 seconds to utilize the whole PC-Farm.

3   Suggested Changes to Scale Up to PetaFLOPS Systems

None of the limitations outlined above are due to the underlying image analysis workflow, they exist because of the current implementation. The current version of Motion Tracking has grown over time and was successfully adapted to new requirements. With the growing number of images that screening experiments produce, not only Motion Tracking must be adapted, but also the underlying storage concept must be changed to better fit the workflow.

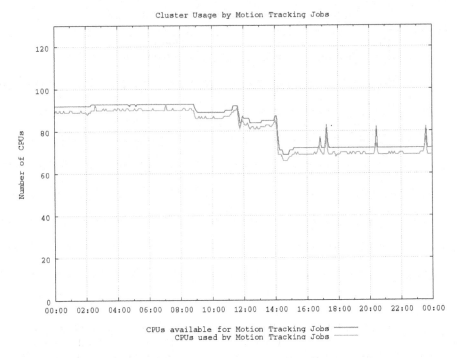

Fig. 2. Available and used CPUs during one day

In the structure identification phase, all images could be analyzed in parallel as there are no dependencies between images. The degree of parallelism is only limited by number of available CPUs and the ability to move the required input data to those CPUs. The same is true for the statistical calculation phase only that the computing time is generally shorter in that phase. To achieve such a high degree of parallelism, a number of changes would have to be implemented.

3.1   Introduce Distributed Storage of Data

In the current version, all data is stored at a central location and is only transferred to the HPC system for the period of the analysis. When the structure identification phase needs to be repeated, because different parameters are selected, the raw images have to be transfered again from the MPI-CBG to the ZIH. This can be avoided by introducing a resource broker or a global database that can be queried for the location of particular image files.

Also, the results that are created in the structure identification phase are the input data for the statistical calculation phase. Currently the results are transferred back to the MPI-CBG because they are needed for further analysis that can either be performed in parallel on a cluster or by the master workstation alone. Also, for exporting results they have to be available locally at the MPI-CBG.

Raw images and the result files could be replicated between both locations. That would guarantee fast access and availability on both locations. For this approach to be successful, a layer that takes care of data management, such as tracking and replicating images, must be added.

3.2 Improve Communication and Job Control

Currently, the communication with the HPC systems is built around the OpenSSH binaries for transferring files and executing remote commands. This implementation was choosen to support as many Linux based HPC systems as possible. The solution is very flexible and can be adapted quickly to different environments.

The disadvantages are limited transfer rates and only basic job control and status information about running jobs on HPC systems. When using Windows workstations as compute resources, Motion Tracking uses a custom communication protocol that allows for example to query jobs status information. This communication protocol is not yet ported to Linux.

To increase the transfer rates ssh/scp can be replaced with mechanism like GridFTP[1], that provide higher utilization of the available network bandwidth. To improve job submission and control on large HPC systems, the use of grid middle-ware such as UNICORE[5] or Globus[2] as alternative to the proprietary protocol of Motion Tracking will be examined.

With the above mentioned changes in place, Motion Tracking should be prepared to benefit from large HPC systems.

References

1. W. Allcock, J. Bresnahan, R. Kettimuthu, and M. Link. The globus striped gridftp framework and server. In *SC '05: Proceedings of the 2005 ACM/IEEE conference on Supercomputing*, 2005.
2. Ian Foster. Globus toolkit version 4: Software for service-oriented systems. *J. Comput. Sci & Technol.*, 21(4):513–520, July 2006.
3. Y.L. Kalaidzidis, A.V. Gavrilov, P.V. Zaitsev, A.L. Kalaidzidis, and E.V. Korolev. PLUK - an environment for software development. *Programming and Computer Software*, 23(4):206–212, 1997.
4. J. Rink, E. Ghigo, Y. Kalaidzidis, and M. Zerial. Rab conversion as a mechanism of progression from early to late endosomes. *Cell*, 122(5):735–749, 2005.
5. A. Streit, D. Erwin, Th Lippert, D. Mallmann, R. Menday, M. Rambadt, M. Riedel, M. Romberg, B. Schuller, and Ph Wieder. Unicore - from project results to production grids, 2005.

Author Index

Lecture Notes in Computer Science

For information about Vols. 1–4374

please contact your bookseller or Springer

Vol. 4424: O. Grumberg, M. Huth (Eds.), Tools and Algorithms for the Construction and Analysis of Systems. XX, 738 pages. 2007.

Vol. 4423: H. Seidl (Ed.), Foundations of Software Science and Computational Structures. XVI, 379 pages. 2007.

Vol. 4422: M.B. Dwyer, A. Lopes (Eds.), Fundamental Approaches to Software Engineering. XV, 440 pages. 2007.

Vol. 4421: R. De Nicola (Ed.), Programming Languages and Systems. XVII, 538 pages. 2007.

Vol. 4420: S. Krishnamurthi, M. Odersky (Eds.), Compiler Construction. XIV, 233 pages. 2007.

Vol. 4419: P.C. Diniz, E. Marques, K. Bertels, M.M. Fernandes, J.M.P. Cardoso (Eds.), Reconfigurable Computing: Architectures, Tools and Applications. XIV, 391 pages. 2007.

Vol. 4418: A. Gagalowicz, W. Philips (Eds.), Computer Vision/Computer Graphics Collaboration Techniques. XV, 620 pages. 2007.

Vol. 4416: A. Bemporad, A. Bicchi, G. Buttazzo (Eds.), Hybrid Systems: Computation and Control. XVII, 797 pages. 2007.

Vol. 4415: P. Lukowicz, L. Thiele, G. Tröster (Eds.), Architecture of Computing Systems - ARCS 2007. X, 297 pages. 2007.

Vol. 4414: S. Hochreiter, R. Wagner (Eds.), Bioinformatics Research and Development. XVI, 482 pages. 2007. (Sublibrary LNBI).

Vol. 4412: F. Stajano, H.J. Kim, J.-S. Chae, S.-D. Kim (Eds.), Ubiquitous Convergence Technology. XI, 302 pages. 2007.

Vol. 4411: R.H. Bordini, M. Dastani, J. Dix, A.E.F. Seghrouchni (Eds.), Programming Multi-Agent Systems. XIV, 249 pages. 2007. (Sublibrary LNAI).

Vol. 4410: A. Branco (Ed.), Anaphora: Analysis, Algorithms and Applications. X, 191 pages. 2007. (Sublibrary LNAI).

Vol. 4409: J.L. Fiadeiro, P.-Y. Schobbens (Eds.), Recent Trends in Algebraic Development Techniques. VII, 171 pages. 2007.

Vol. 4407: G. Puebla (Ed.), Logic-Based Program Synthesis and Transformation. VIII, 237 pages. 2007.

Vol. 4406: W. De Meuter (Ed.), Advances in Smalltalk. VII, 157 pages. 2007.

Vol. 4405: L. Padgham, F. Zambonelli (Eds.), Agent-Oriented Software Engineering VII. XII, 225 pages. 2007.

Vol. 4403: S. Obayashi, K. Deb, C. Poloni, T. Hiroyasu, T. Murata (Eds.), Evolutionary Multi-Criterion Optimization. XIX, 954 pages. 2007.

Vol. 4401: N. Guelfi, D. Buchs (Eds.), Rapid Integration of Software Engineering Techniques. IX, 177 pages. 2007.

Vol. 4400: J.F. Peters, A. Skowron, V.W. Marek, E. Orłowska, R. Słowiński, W. Ziarko (Eds.), Transactions on Rough Sets VII, Part II. X, 381 pages. 2007.

Vol. 4399: T. Kovacs, X. Llorà, K. Takadama, P.L. Lanzi, W. Stolzmann, S.W. Wilson (Eds.), Learning Classifier Systems. XII, 345 pages. 2007. (Sublibrary LNAI).

Vol. 4398: S. Marchand-Maillet, E. Bruno, A. Nürnberger, M. Detyniecki (Eds.), Adaptive Multimedia Retrieval: User, Context, and Feedback. XI, 269 pages. 2007.

Vol. 4397: C. Stephanidis, M. Pieper (Eds.), Universal Access in Ambient Intelligence Environments. XV, 467 pages. 2007.

Vol. 4396: J. García-Vidal, L. Cerdà-Alabern (Eds.), Wireless Systems and Mobility in Next Generation Internet. IX, 271 pages. 2007.

Vol. 4395: M. Daydé, J.M.L.M. Palma, Á.L.G.A. Coutinho, E. Pacitti, J.C. Lopes (Eds.), High Performance Computing for Computational Science - VECPAR 2006. XXIV, 721 pages. 2007.

Vol. 4394: A. Gelbukh (Ed.), Computational Linguistics and Intelligent Text Processing. XVI, 648 pages. 2007.

Vol. 4393: W. Thomas, P. Weil (Eds.), STACS 2007. XVIII, 708 pages. 2007.

Vol. 4392: S.P. Vadhan (Ed.), Theory of Cryptography. XI, 595 pages. 2007.

Vol. 4391: Y. Stylianou, M. Faundez-Zanuy, A. Esposito (Eds.), Progress in Nonlinear Speech Processing. XII, 269 pages. 2007.

Vol. 4390: S.O. Kuznetsov, S. Schmidt (Eds.), Formal Concept Analysis. X, 329 pages. 2007. (Sublibrary LNAI).

Vol. 4389: D. Weyns, H.V.D. Parunak, F. Michel (Eds.), Environments for Multi-Agent Systems III. X, 273 pages. 2007. (Sublibrary LNAI).

Vol. 4385: K. Coninx, K. Luyten, K.A. Schneider (Eds.), Task Models and Diagrams for Users Interface Design. XI, 355 pages. 2007.

Vol. 4384: T. Washio, K. Satoh, H. Takeda, A. Inokuchi (Eds.), New Frontiers in Artificial Intelligence. IX, 401 pages. 2007. (Sublibrary LNAI).

Vol. 4383: E. Bin, A. Ziv, S. Ur (Eds.), Hardware and Software, Verification and Testing. XII, 235 pages. 2007.

Vol. 4381: J. Akiyama, W.Y.C. Chen, M. Kano, X. Li, Q. Yu (Eds.), Discrete Geometry, Combinatorics and Graph Theory. XI, 289 pages. 2007.

Vol. 4380: S. Spaccapietra, P. Atzeni, F. Fages, M.-S. Hacid, M. Kifer, J. Mylopoulos, B. Pernici, P. Shvaiko, J. Trujillo, I. Zaihrayeu (Eds.), Journal on Data Semantics VIII. XV, 219 pages. 2007.

Vol. 4379: M. Südholt, C. Consel (Eds.), Object-Oriented Technology. VIII, 157 pages. 2007.

Vol. 4378: I. Virbitskaite, A. Voronkov (Eds.), Perspectives of Systems Informatics. XIV, 496 pages. 2007.

Vol. 4377: M. Abe (Ed.), Topics in Cryptology – CT-RSA 2007. XI, 403 pages. 2006.

Vol. 4376: E. Frachtenberg, U. Schwiegelshohn (Eds.), Job Scheduling Strategies for Parallel Processing. VII, 257 pages. 2007.

Vol. 4375: W. Lehner, N. Meyer, A. Streit, C. Stewart (Eds.), Euro-Par 2006: Parallel Processing. XV, 332 pages. 2007.